perils of
progress

perils of
progress

THE HEALTH AND ENVIRONMENT HAZARDS
OF MODERN TECHNOLOGY
AND WHAT YOU CAN DO ABOUT THEM

John Ashton & Ron Laura

ZED BOOKS LTD
London & New York

FERNWOOD PUBLISHING
Canada

UCT PRESS
Cape Town

The Perils of Progress: The Health and Environment Hazards of Modern Technology and What You Can Do About Them was first published in the UK by Zed Books Ltd, 7 Cynthia Street, London N1 9JF and Room 400, 175 Fifth Avenue, New York, NY 10010, USA, in 1999.

Distributed in the USA exclusively by St Martin's Press, Inc., 175 Fifth Avenue, New York 10010, USA.

Published in Canada by Fernwood Publishing Ltd, P.O. Box 9409, Station A, Halifax, Nova Scotia, Canada, B3K 5S3.

Published in South Africa by University of Cape Town Press (Pty) Ltd, Private Bag, Rondebosch 7701, South Africa.

Published in Thailand by White Lotus Company Ltd, GPO Box 1141, Bangkok, 10501.

Not available in Australasia.

Copyright © J.F. Ashton and R. S. Laura 1998

Cover designed by Andrew Corbett

Printed and bound in Malaysia

Canadian Cataloguing in Publication Data

Ashton, John, 1947-

   Perils of progress

   Includes bibliographical references and index.

   ISBN 1 55266 012 5

1. Environmental health -- Popular works. I. Laura, R. S. (Ronald S.) II Title.

RA565.A75 1999   613   C99-950075-9

A catalogue record for this book is available from the British Library.

ISBN: 1 85649 696 1 hb
ISBN: 1 85649 697 X pb

Canadian ISBN: 1 55266 012 5 pb
South African ISBN: 1 919713 40 9 pb

contents

foreword

This is a book whose time has come. Indeed it is overdue. Who has not read the day's newspaper only to find that many of the changes we have made to our lifestyles, which were are told are healthier, are in fact hazardous to health.

Where then can we find the truth about a healthy lifestyle? This book attempts to give us the answers. It is a guide to a healthy lifestyle, based on the best information available today. Of course, some of the recommendations may change as more data becomes available. Most importantly, we must always be on our guard. We can do that by turning to the guardians of our health, such as the authors of this book, who have brought together in one volume a wealth of information to help us on our way.

This book covers widely the effects technology has had on our environment, from the food we eat to the radiation we are exposed to. Those of us who live in cities will find out just how much of our environment is artificial. This book proposes that we return as far as it is possible to a natural environment by taking account of modern hazards and trying to avoid them. We can think back to the water that was pure, the soil that was sustainable to healthy plant growth and the drugs we did not need to take to keep us healthy. We can then ask how can we best return to the advantages of a natural lifestyle and at the same time take advantage of the new technologies that surround us. It is both a personal challenge and a challenge to society and its controlling bodies such as councils and governments.

I believe that everyone who reads this book will be challenged to a newer, healthier lifestyle presented here. *Perils of Progress* provides the data, but is up to each one of us to weigh the advantages and disadvantages of the options it presents, and act accordingly.

Charles Birch
Professor Emeritus of the University of Sydney and Challis professor of biology for 25 years, he was awarded the Templeton Prize for progress in religion in 1990 and is the author of Confronting the Future, On Purpose, Regaining Compassion *and* Feelings.

preface

Much research is now being devoted to improving our understanding of the role of lifestyle factors such as diet, exercise, rest and recreation in maintaining our health. With the proliferation of health reports of all kinds, the problem is how to determine the best course of action when there is apparently conflicting research. Even doctors and dietitians find themselves modifying the advice they were giving patients a few years ago and in some cases, only a few months ago. The butter versus margarine controversy serves to illustrate this point. For the last couple of decades we have been urged to use margarines to help lower our blood cholesterol levels and prevent heart disease. In recent years, however, the position is no longer so clear, and there are many researchers who would now urge us to limit our consumption of margarines because of the harmful effects of the trans-fatty acids which they contain. According to some reports it seems that trans-fatty acids may even be associated with an increased risk of heart disease and cancer. As a result, some well-known nutritionists are now advocating a return to the use of butter in moderation. In some instances it is argued that butter is a more 'natural' product compared with margarine, which is produced by quite involved chemical processes.

Some of the greatest challenges confronting those who seek to be healthy is to be able to sort through the mass of sometimes contradictory recommendations deemed to be requisite for good health. The aim of this book is to empower readers to a whole new understanding of health, an understanding which supplies the insight needed to know when the advice of experts is worth heeding, and when such advice requires further reflection than it is usually given. For some people the illumination is simple enough, deriving from their deepest intuition that the highly technological world in which they find themselves represents a threat of monumental proportions to human health. Others only dimly perceive that the radically synthesised world modern society has created for itself, is not a world which we are very well adapted to live in—either physiologically or psychologically. The truth enshrined by these intuitions presents itself starkly: in general, the more synthesised the environments to which we expose ourselves, the less efficient we become at staying healthy within them.

In the chapters that follow we shall try to show that the predilection for the 'things of nature' may well be an instinctive clue of enormous

importance in the quest for human survival. The burden of this book is
to help provide the reader with a few basic principles that can be used
to sort through the conflicting advice on how to live a healthy life.
It is difficult to live well if we live unsure about whether the advice
we follow today is bound to mislead us tomorrow. If you would like
to know whether the things you do now will make you healthier or
sicker in five years time, you will need a guide for healthy living that
explains in simple and convincing terms why living differently is
important to living well. It is hoped that this 'Handbook of Life
Enhancement' achieves this goal.

John Ashton
Ronald Laura

introduction

The technological society in which we live offers us many great benefits. With the flick of a switch we can now light up our houses, indeed even light up our cities; receive television programs from around the globe; have hot water, household heating and airconditioning, all upon command. We can microwave our meals in minutes, if not seconds, slip into pre-heated waterbeds or warm ourselves in the comfort of electric blankets. If we suffer aches and pains, we turn either to our medicinemen, or our medicine cabinets, to gain instant relief. If we are bored with our surroundings, the technology is available to whisk us away by air to the other side of the world and exotic places in less than a day. We now enjoy living-out the dreams of the kings and queens of ages past without even knowing it. The plain truth is that the average-income person of today is probably surrounded by more comforts, conveniences and luxuries than most of the kings and queens of previous centuries. But just as there was a price to be paid by the monarchs of the past for their indulgences, so we too pay a price for the conveniences which flood our lives. The price paid for some of these conveniences is not only high, but also a well-kept secret.

There are those who would prefer that you did not know how much you are paying, particularly in terms of your health, for basic requirements such as food and water, which in our culture have themselves come to function almost as symbols of health. To challenge their integrity is, in a sense, to challenge the integrity of the technological society itself. This is one reason why there are public health secrets which have been kept well-hidden, and it is one reason why, when the truths are occasionally revealed, they are so fervently denied. Let us now mention a few of these well-kept secrets which in the chapters to follow will be explored in depth.

One would expect that we all have a right to healthy drinking water. Because we have polluted the once pristine water supplies, largely through technological development in western countries, we have adopted the practice of adding chemicals such as chlorine to our water in the hope of restoring it to its original purity. Surely chlorine is a good and necessary thing to add to water, after all, doesn't it kill the germs that might threaten to kill us, or at least make us sick? While it is true that chlorine does serve to purify drinking water, it is also a well-kept secret that evidence exists to suggest a correlation between the extent to which municipal water supplies are chlorinated and an increased incidence of bladder cancer in those places where people drink chlorinated water. Not only may chlorinated drinking water be contributing to your increased risk of cancer and possibly even heart disease, but it also turns out that the chlorine in the water you drink may be destroying some of the vitamins which would otherwise be derived from the food you have eaten. If you still need to be convinced, consider the evidence that the chlorine in our municipal water supplies, including our pools, may be responsible for causing premature aging of the skin, if not also increase the photosensitivity of the skin, thereby contributing to a greater risk of skin cancer and melanoma.

Well, even if chlorination of water is problematic, surely fluoridation has proved to be a good thing? What secrets could possibly be hidden here? Did you know that fluorides have traditionally been used as insecticides for killing cockroaches and also as a main constituent in rodenticides for the extermination of rats? Fluoride is also the active ingredient in the widely used poison known by the numbers 10–80. You might also be interested to find that the fluoride wastes derived from aluminium production have in some countries provided the fluoride source used in the manufacture of toothpaste. Appreciation of the fact that fluoride has long been used as a poison will make it easier to understand why the evidence is mounting to suggest a likely association between the intake of fluorides and an increased risk of hip fracture and even bone cancer.

The next time you drink a glass of water, you might think differently about the price we pay for technological progress. If you read on, you will become privy to well-kept secrets about the chemicals we put into foods, not to mention the highly synthesised food substances we substitute for natural foods. You will doubtless be surprised to hear that a number of the foods you eat are either wholly or partially fabricated in laboratories or in laboratory-like factories. Needless to say, the process of refining foods almost always considerably reduces their nutritional content. This is why chemical nutrients need to be added to such foods as a way of compensating for what has been lost by refining them. Chemical colours and chemical flavours are added so that the refined product will neither look nor taste unfamiliar to you. Chlorine is even used to bleach some naturally coloured flours to a uniform shade of virginal white. How do such chemical substances, either ingested directly with our foods or as a substitute for them, affect our bodies and the delicate hormonal and other physiological processes that keep them in balance? To survive technology, we need to know how technology may be threatening us. The problem is that those who stand to benefit most from selling technology to the public seem to be the most reluctant to admit that there could be any problem, especially to human health, with what they sell. The explorations of this book will redress this imbalance and provide a clearer picture of the problems.

Would it surprise you to hear that margarine may contain photo-sensitising agents which make your skin more vulnerable to the harmful rays of the sun and thus more prone to skin cancer? What should you know about the chemical sweeteners which are added to refined foods? While some readers may have heard about the health problems of cyclamates, the health risks associated with aspartame, marketed under the brand name of 'Nutri-Sweet' remain a better-kept secret. But if the evidence that aspartame has the potential to alter brain chemistry can be believed, the question of whether brain development in the growing foetus may be threatened by the mother's consumption of aspartame deserves far more attention than it has yet been given.

As the health problems associated with the use of insecticides have become increasingly publicised, the move to irradiate food as a means of eliminating insect and bacterial damage has received a favourable response in many countries. Once again, however, there are secrets which are closely guarded whose revelation may significantly change our attitude to the process of irradiating food with atomic energy. Did you know, for example, that irradiation has been implicated in mutations of certain bacterial components of food, some of which are believed to be harmful to the human body? Even in countries where irradiated foods are not available for retail sale, such as Australia, New Zealand, Canada, the United Kingdom and North America, some food packaging may be irradiated. Certain imported foods such as spices are commonly irradiated to kill bacteria before being sold on the international market, as are some medicines. It is in your interest to know about such things and to know what you can do to survive the technological onslaught of our times. While we add chemicals to our food on the one hand, we bombard it with nuclear energy on the other, with the overall result that the things we eat and drink are very different from the things of nature which we are well-adapted to consume.

One of the most controversial areas of study arising out of technological development relates to the adverse effects of electromagnetic fields upon humans. In recent years, the level of electromagnetic field radiation has grown exponentially. Ever-new construction of television and radio towers, radar installations, electric power stations, transformers and the accompanying grid of high and lower voltage power lines, makes it clear that those of us who live in or around cities are surrounded by an invisible electromagnetic smog. A substantial research literature has now accumulated, though it is seldom made public, to show that our increasing exposure to environmental electromagnetic fields presents a new risk of disease and especially of cancer. The recent introduction of UHF (ultra high frequency) television and particularly mobile phones has contributed a new and potentially dangerous level of electromagnetic pollution to the current microwave environment. Did you know that the microwaves characteristic of mobile phone communication systems produce electromagnetic radiation fields for which the human body can effectively act as an antenna? This means that those who rely upon certain kinds of mobile phones to conduct their daily business may be exposing themselves to inordinately high radiation levels. Recent research studies show that the brain, eyes and other organs of the body are especially prone to absorb the radiation in the spectral region deployed by mobile phones and UHF television. Some evidence exists to suggest a link between the increased incidence of brain tumours, and certain other forms of cancer, and high levels of exposure to low frequency magnetic fields such as those associated with telecommunications. If you use a mobile phone or are involved in telecommunications, it is imperative that you familiarise yourself with the potential risks involved. This book is written to assist you in making the informed judgement required for this task.

Other potential electromagnetic field risks to be considered in the chapters that follow include: computer visual display units (VDUs) and television, electric blankets and waterbeds, and even microwave ovens. The potential hazards of airconditioning and obtrusive noise levels, will also be discussed. Finally, serious attention will be given to the environmental factors involved in the increased incidence of melanoma. Evidence will be marshalled to show, for example, that the connection between artificial light exposures and melanoma may be as important in understanding the aetiology of melanoma as is the conventional theory of over-exposure to natural light.

The point of this wide-ranging deliberation on the hidden hazards of our times is to heighten public awareness of the adverse impact upon health which results from our obsession to synthesise and technologise our world. The technologisation of society does not come without a price, but unless we are informed of the true cost of our indulgences, we can never know whether the price we are indeed paying is too high. In the end, the real question is not how technology can help us survive, but rather how we can survive technology?

PART one

Technology and the quest for Utopia

chapter 1

The paradox of progress: Is the price we pay for technology too high?

> *Man has lost the capacity to foresee and to fore-stall. He will end by destroying the earth.*
>
> **Albert Schweitzer**

Despite the fact that modern technology has undoubtedly improved the standard of living for many people in the Western world, the cost both to the natural environment and to human health has been prohibitive. The cost of our technological indiscretions now constitutes a massive debt to nature which, if we are to survive on this planet, will have to be paid with interest.

In this chapter we argue that much of what has passed for progress in our society is illusory. Motivated by an obsession with unlimited economic growth, the science of technology has been corrupted and shaped into a tool for the rape of the earth. Having confused the difference between an improved standard of living and an improved quality of life, we have unwittingly sacrificed the latter for the former. While the health and environmental risks associated with technological progress are slowly becoming recognised, the belief remains that new and improved technology will provide the way to resolve the massive ecological problems which technology has in the first place created. Against this conventional view, it will be argued here that the commitment to ever-more intensive forms of technology is stultifying and inherently disruptive of the established harmony and rhythm of nature. In this chapter, our aim is to show by reference to a number of concrete examples that the driving force behind technology is the lust for control over the environment. By hoping to satisfy our insatiable appetite for dominance of the world in which we live, we have separated ourselves from nature and thus made it easier to exploit and degrade it without conscience. We have failed to see that the idea of rectifying the disastrous side-effects of scientific technology by developing more sophisticated forms of it is rather like trying to put out a fire by pouring kerosene on it.

The problem is that preoccupied by the desire for dominance, technological intervention will continue to be fundamentally alienating and destructive. The commitment to a technology of power may turn us into giants, but we will inevitably finish by being no more than blinded giants. The paradox of progress is that our best efforts to make ourselves infinitely strong have in the end made us infinitely weak. Let us now see why.

Lust for power: From steam engine to nuclear bomb

Before the development of the steam engine, all living things depended upon nature for their daily allocations of available energy. The rate at which solar energy was converted via plants into fuel energy for humankind and beasts of burden was determined predominantly by nature. Simple machines such as the water wheel, windmill and sail, eventually enabled our ancestors to harness the energy stored in water and wind, but again, it was nature which largely determined when, and how much of these resources were at any one time available. Innovation has always been the mother of invention however, and it was not long before our predecessors found new ways to loot and store the hidden treasures of nature for their own convenience.

The introduction of the mechanical steam engine in 1769 by James Watt, afforded the means by which solar energy from ages past, stored as coal, could be used to provide seemingly limitless power at the beck and call of anyone in possession of the required technology. Man, not nature, would now determine how the resources of nature were to be used, and a new dimension to the goal of unlimited economic growth was uncovered.

The power of the steam engine ensured that a single machine could do the work of virtually a hundred people. Created to exercise the power over nature we would ourselves have wished to exercise directly, the machine was created in the image of man on the one hand and in the image of a god on the other. In respect of the production of labour, the birth of the engine was in essence the birth of 'superman'. Bedazzled by our ingenuity, we were quickly deluded into thinking that the power we sought was finally in our own hands; that we, rather than the machines we had created, were the true 'supermen'—but we were wrong. The technology of power is a power unto itself. In our desperate efforts to avoid being controlled by nature, the human species was being progressively controlled by machines. To get machines to do what we wanted them to do, we began to reorganise society in ways that made it more suited to what our machines could do rather than to people.

Factories were built to house these energy machines so that they could join systematically to transform the crude resources of nature into material goods at an unprecedented rate. The speed at which machines could do this in turn required that the resources of nature be harvested in increasing quantities, and then concentrated to provide the raw material needs of these energy-based industries. New machines would have to be invented to perform these tasks, and the tasks of humans would increasingly become focused upon interaction with machines, rather than with nature. Machines which could run all day and night, required people to work them all day and all night. Because of our devotion to the machine, we would come to have less and less contact with nature. To gain the power over nature desired, we unwittingly enslaved ourselves to the machines of our own creation. The needs and goals of technology had covertly influenced the very structure of society, along with our relationship to nature, and this influence would continue to grow.

The size of cotton and wheat fields, for example, soon came to reflect the capacity of the steam mill to process what was harvested. The more machines to fuel and the bigger their fuel needs, the deeper and more extensive became the coal mine. Similarly, forests were sacrificed to meet the capacity requirements of the steam-powered sawmill, and new technologies were required to cut the forests as fast as mills could process what was cut.

The concentration of energy in a particular location required a concentrating of resources, employment and housing in that location, and resulted in the growth of industrial cities and slums. Geographical concentration required transportation systems which could move increasing volumes of material from their place of origin to and from factories. Again the steam engine was harnessed and railways opened up vast tracts of land whose resources were soon transformed into more grist for the industrial mill.

Our obsession with controlling energy continued unabated. During the nineteenth century the efforts of S Carnot, AB de Rochas, NA Otto, R Diesel and others, enabled petroleum to be harnessed as a fuel, using the internal combustion engine which now dominates our sea, land and air transport systems. The power of the steam engine was thus supplanted by the power of the internal combustion engine. Society and nature would be similarly reorganised to meet its needs and capacity for devouring and transforming the earth's resources.

The nineteenth century also saw Michael Faraday discover the laws of magnetism and electricity, and in 1879 Thomas Edison successfully developed the electric light bulb. The development of electric power reached a transmission point when on 16 May 1888 Nikola Tesla read his famous lecture 'A New System of Alternate Current Motors and Transformer' before the American Institute of Electrical Engineers.[1] His discovery of the use of a rotating magnetic field as a means of generating alternating current electricity made possible the practical, economic production and transmission of electrical power and is the basis of practically all the commercial power systems used in the world today.

Within half a century of Tesla's discovery, the energy initially provided by the steam engine became available in almost every factory and home of the developed world in the form of electrical energy along power lines.

The availability of concentrated energy in our homes reinforced the assumption that technology had bequeathed a new inheritance of vast control over nature. With only the flick of a switch the darkness of night could instantly be transformed, almost magically, into the light of day. By simply pressing a button the cold of winter could be subdued and transformed into the warmth typical of a fine summer day. Or conversely, as the seasons would have it, the heat of summer could just as easily be displaced by cooling breezes from our airconditioners. With the advent of electricity, a whole range of energy-consuming gadgets proliferated from electric razors and hair dryers to electric can-openers and heated waterbeds.

Just five decades after Tesla's historic lecture, Leo Szilard filed a patent with the British Admiralty: 'a description of the release of energy in a chain reaction'. The culmination of the lust for power gave birth to the nuclear age.[2] Eight years later, on 2 December 1942 in the United States, the first sustained atomic reaction was unleashed. The goal of unlimited energy supplies to effect the total control of nature now seemed palpably within the reach of mankind, and scientific knowledge was elevated to a new position of eminence among human values. President Dwight Eisenhower, speaking on the Atomic Energy

Commission's tests of hydrogen bombs in 1956, stated: 'The continuance of the present rate of H-bomb testing, by the most *sober and responsible scientific judgement* [our italics] ... does not imperil the health of humanity.'[3]

Today, nuclear power stations provide energy in many developed countries, though the promise of a safe and healthy energy source has never been realised, as Chernobyl and even the Three Mile Island disasters make plain.

Chemical control of nature: To dope or not to dope

The industrial growth of the nineteenth century was spurred on, not only by the utilisation of steam power, but also by the growth of chemical knowledge.

In 1774, Antoine Lavoisier demonstrated the difference between physical and chemical change and articulated the notion of a chemical reaction. Following this work, John Dalton put forward the theory that the elements of matter are made up of small separate particles called atoms, and that all atoms of any one element are the same and differ from the atoms of all other elements. He suggested that chemical compounds are formed by the combination of atoms of different elements in simple ratios.

This atomic theory gave fresh impetus to the known chemistry of that time and the following one-and-a-half centuries saw thousands of researchers discover, analyse and synthesise millions of chemical reactions and compounds. One important example of chemical research was the work of Justus von Liebig. In 1835, Liebig discovered that ethylene dichloride ('oil of the Dutch chemists'), combined with alcoholic potash produced a chlorine-containing gas, subsequently named vinyl chloride. This compound later became the basis of the now extensively used plastic, polyvinyl chloride or PVC.

In 1840, Liebig published his classical monograph on agricultural chemistry titled 'Chemistry in its Applications to Agriculture and Physiology'. Liebig had found that phosphates mixed into the soil would promote the growth of plants. He also found that the ashes of plants gave useful information as to the requirements of crops and that a watery extract of humus gave little or no residue on evaporation. From this work the importance of nitrogen (N), phosphorus (P) and potash (K) in the soil was perceived, and the modern NPK mentality of chemical agriculture was born.[4]

By the mid-twentieth century, the science of chemistry had enabled metals like aluminium and uranium to be won from the earth and utilised. From oil, via chemical processes, a vast array of plastics has been developed which could be moulded into an endless variety of 'things', or extruded into synthetic yarns. It was chemical technology that has enabled forests to be converted to paper and cardboard. Pesticides, herbicides and fungicides have been produced in the vain and futile hope of changing the balance of nature to fight insects and check plant disease. Chemical knowledge has now become so specialised that drugs, chemicals and materials can be designed and synthesised to meet specific applications in medicine, industry, agriculture and the home.

Our technological power over nature has increased to such an extent that the best landscapes of nature can be cleared bare within hours, dug up and removed or diverted within days. To meet the insatiable demands of technological growth we have disrupted and decimated the very ecosystems upon which all economic growth ultimately depends. We have displaced nature with our cities, airports, freeways, farms and mines, and poisoned the air, streams, rivers and oceans either deliberately with pesticides or unthinkingly with our wastes.

In our technological efforts to gain ultimate control of nature we have transformed our society and ourselves along with nature. We have unwittingly become the servants of the technology we have created, and we have fashioned a throw-away society to sustain the goal of unlimited growth. Ultimate control and domination of nature is represented economically by our willingness not only to 'use nature up', but to throw it away. The commercial viability of a product on this model is not that it lasts, but that *it is disposable*. If you can convince people that the products of technology can with impunity be thrown away, then you have already convinced them that the products of technology can with impunity be bought again. The power comes in feeling that technology can produce all you need to buy, and dispose of all you need to get rid of. The truth is that it can do neither. Nature is neither an infinite well nor an infinite drain. Let us see why.

Air pollution: The energy–health connection

By the end of the eighteenth century, the proliferation of steam power and associated industrial development, particularly in Great Britain, was such that it had already begun noticeably to reduce the quality of human health. As Rene Dubos observed:

> England had known first and on the largest scale the prosperity but also the terrific destruction of human values that accompanied the first phase of the Industrial Revolution. Laymen as well as physicians could not fail to recognise that disease and physical frailty were most

common among the poor classes. In *Conditions of the Working Man in England* Engels spoke of the 'pale, lank, narrow-chested, hollowed-eyed ghosts' riddled with scrofula and rickets, which haunted the streets of Manchester and other manufacturing towns. If ever men lived under conditions completely removed from the state of nature dreamed of by the philosophers of the Enlightenment, it was the English proletariat of the 1830s.[5]

These changes in public health had become so noticeable that in 1819 the British parliament appointed a Select Committee to enquire:

how far it may be practicable to compel Persons using Steam Engines and Furnaces in their different works to erect them in a Manner less prejudicial to public Health and public Comfort.[6]

Despite an active anti-smoke campaign from the 1840s to the 1890s, the obsession with steam power and industrial development prevailed, and the Industrial Revolution gripped Europe and America in a cloud of pollution. The burning of coal not only put smoke into the air but also released gases such as carbon dioxide, water vapour and sulfur dioxide. In London, in particular, industrial emissions of smoke and gases were so high that harmful smogs became intolerable and extremely dangerous under certain weather conditions.

In December 1873, a week-long London smog is believed to have claimed at least 500 lives through respiratory failure alone. Scarcely six years later a three-week-long smog claimed about 2000 lives. Despite pronouncements that the problem was under control, the death rate from respiratory failure caused by polluted air continued to climb. As recently as December 1952, a smog engulfed London which, according to a subsequent Royal Commission, claimed about 4000 thousand lives.[7]

The growth in air pollution produced a parallel increase in the incidence of rickets, particularly in children of the industrial cities. It turns out that smog absorbs the natural ultraviolet solar radiation which is essential for human vitamin D synthesis and calcium metabolism.[8]

The advent of coal-fired steam electric power generation, while gradually replacing individual steam plants, enabled concentrated power in the form of electricity to be readily accessible to the community. The more readily available electricity became, the greater the electricity usage which in turn resulted in a steady increase in carbon dioxide and sulfur dioxide gas emissions. In 1860, about 2 million tonnes of sulfur dioxide were released into the atmosphere from the burning of coal and the smelting of ores. One hundred and twenty-five years later, global sulfur dioxide emissions from burning coal and oil for energy and copper smelting had reached 90 million tonnes.[9]

Within a few days of being in the atmosphere, sulfur dioxide gas is converted to sulfuric acid which is ultimately precipitated as acid rain. This rain has acidified extensive regions of North America and Europe, causing fish populations in freshwater lakes to dwindle, and toxic metals to be leached from the soil. Around the world, efforts have been made to reduce the effects of acid rain by liming lakes. However, it is now realised that liming does untold damage to freshwater biosystems and is not the simple panacea it was once thought to be.[10] This is just one of many examples which demonstrate that further technological intervention is not the way to solve the problems created by technological intervention.

Instead, placing limits on the release of pollutants into our environment is becoming increasingly recognised as the only way we can save our environment. For example, many countries in Europe and North America have now signed a treaty to curb emissions of gases which cause acid rain as part of a Convention on Longrange Transboundary Air Pollution. It is planned that over the next 13 years up to 2010, acid rain-producing pollution from factories and power stations in Europe and North America should drop by up to 87 percent from 1980 levels.

However, in 1996 Jan Thompson, chairman of the Convention's executive body, pointed out that the contribution of shipping to worldwide sulfur emissions could more than double by the year 2010 as ships burn cheap dirty oils with a high sulfur content. Shipping is already responsible for 7 percent of global sulfur emissions.[11] While the average sulfur content of ship's fuel is about 2.8 percent, marine fuel oil may contain in excess of 5 percent sulfur. In this case each tonne of oil burnt, in combination with atmospheric moisture produces an equivalent of more than one tonne of 5-normal sulfuric acid (the concentration of acid commonly used in laboratories). Friends of the Earth International claims that the sulfur content of ship's fuel should be limited to 1 percent or less, if environmentally sensitive areas are to be protected from acid rain.[12]

Air pollution from modern power stations is usually reduced by devices called electrostatic precipitators which can remove up to around 99.97 percent of the particulate matter called fly ash. However these may not always be maintained to peak efficiency and can breakdown releasing large quantities of pollutants into the air.

One such incident occurred in October 1996, when two of the fire precipitator units at the 324 megawatt Indra Prastha power station in the heart of the city of Delhi in India broke down. The power station's chimneys were spewing an estimated 1500 tonnes of fly ash into the atmosphere of the city every day.[13] A recent World Bank study shows that air pollution levels in 36 Indian cities is responsible for 40 000 deaths and 20 million hospital admission each year.[14] In addition to the contribution to Delhi's air pollution from power stations, about 70 percent of the smog comes from vehicles powered by internal combustion engines. The development of the internal combustion engine as a source of concentrated power for transport added an extra dimension to air pollution. For the past eight decades, vehicles on land, sea and in the air, have pumped ever-increasing quantities of combustion gases into the atmosphere. These gases include water vapour, carbon dioxide, carbon monoxide, unburnt hydrocarbons and oxides of nitrogen. In the presence of sunlight, some of these gases, namely hydrocarbons and nitric oxide are photo-oxidised in the air to form ozone, peracylnitrates (PANs), nitric acid and aerosol. This resulting mixture of chemicals is known as photochemical smog.[15]

This form of air pollution is now believed to be one of the main environmental factors causing dieback throughout the forests of Europe which has now reached epidemic proportions.[16] In 1989, West Germany reported that 53 percent of its forests showed signs of dieback damage[17] (see Figure 1.1).

Figure 1.1 Forest damage from pollution in Europe

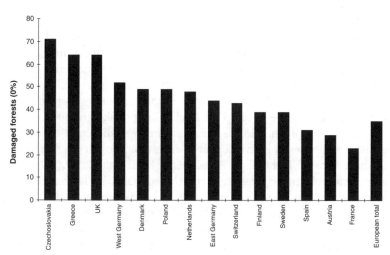

LR Brown, *State of the World*, Allen and Unwin, Sydney, 1990

As we have seen, trees aren't the only life affected by smog. In 1994, French researchers reported that more people die from heart and lung diseases on smoggy days. Increases of 100 micrograms per cubic metre in the black smoke (from diesel engines) content of city air can lead to a 6 percent increase in hospital admissions for heart attack and a 30 percent increase in asthma attacks. An equivalent increase in nitric oxide levels was found to lead to a 63 percent rise in asthma attacks. A 100 micrograms per cubic metre increase in sulfur dioxide levels was linked to a 10 percent rise in heart attacks. When a similar increase in the level of ozone in city air occurred, the admissions of elderly people with chronic breathing problems rose by 20 percent and lower respiratory tract infections in children rose by 24 percent.[18] The French Public Health Society have estimated that between 1987 and 1990 as many as 350 people in Paris died prematurely each year from heart problems triggered by air pollution.[19]

The internal combustion engine and ozone

Ozone is needed in the upper atmosphere to protect the earth's environment from harmful ultraviolet radiation, but in the air layer it is a poison to both plants and animals. It is formed in the lower atmosphere when unburnt hydrocarbons from vehicle and industrial emissions react with nitrogen oxides. While it has been known that ozone impairs breathing and irritates the eyes, nose and throat, recent studies throw new light on its toxicity. Human subjects exposed to ozone levels below the United States Environment Protection Agency (EPA) recommended maximum exposure level of 120 parts per billion (ppb) were found to have increased levels of chemicals in the lungs, which produce inflammation. Macrophages in the lungs were found to be less capable of consuming and destroying bacteria than normal. Furthermore, the ozone exposure increased the levels of the hormone-like substance prostaglandin E2 in the lung, producing a suppression of the immune response in this organ.[20] In the meantime, other studies have shown that ozone is a pulmonary carcinogen even at relatively low levels.[21] Findings of the Office of Technology Assessment reported in 1989, show that more than half the population of the United States are now exposed to levels of ozone higher than the concentrations recommended by the EPA.[22] In 1995 a study for the United States Department of Health's National Toxicology Program found that mice exposed to 2000 micrograms per cubic metre of ozone developed lung tumours at twice the rate of unexposed mice, and some scientists are now recommending that ozone be reclassified as a suspected carcinogen.[23]

A poisonous gas produced directly by internal combustion engines is carbon monoxide. This gas interferes with the blood's capacity to carry and circulate oxygen. Studies in Washington DC in 1983, showed that almost 20 percent of city commuters who travelled 6–10 hours per week were exposed to levels of carbon monoxide exceeding the National Air Quality Standard.[24] In the 1920s, Thomas Midgley, a chemist working for General Motors, discovered that tetraethyl lead could raise the octane rating of fuel and prevent engine 'knock'. Lead has been added to petrol since that time. However, in the 1940s, more efficient high compression petrol engines were developed. This necessitated the addition of higher levels of 'anti-knock' agent to the fuel to prevent preignition. For over three decades, increasing quantities of lead were finely dispersed as dust into the air around the world. Analysis of a Greenland glacier at dated levels showed an exponential increase in lead concentration after 1950.[25] In 1980, approximately 1000 tonnes of poisonous lead dust from vehicle exhausts were deposited annually over cities such as Sydney, Australia,[26] with about 250 000 tonnes of lead dust being released into the air of the cities of North America each year.[27]

This anthropogenic or human-made atmospheric lead has now found its way into water, food and human lungs. Lead causes injury to bone marrow where red blood cells are produced, to the kidneys and to the nervous system including the brain. Intelligence quotient (IQ) deficits in children have been associated with blood lead levels as low as 100–150 micrograms per litre,[28] and there appears to be no evidence of a threshhold for the effect.[29] Hearing is also adversely affected by increased lead exposure, again with no evidence for a threshhold.[30] Elevated blood lead levels are also associated with human developmental abnormalities including foetal neurologic damage, reduced birth weight, reduced stature and slower attainment of developmental milestones.[31] Several epidemiological investigations have noted an association between elevated blood lead levels and blood pressure.[32]

In 1996, the National Institute of Environmental Health Sciences near Washington DC released details of a study which suggests that women who once lived in an environment polluted with lead may pass on the poison to their unborn offspring. It appears that lead stored in a women's bones may be released into the bloodstream during pregnancy.[33] Children are particularly susceptible to lead poisoning,[34] and Jerome Nriagu at the University of Michigan points out that when childhood lead poisoning was widespread in the United States in the early 1980s, unleaded petrol was originally introduced 'not to protect public health, but to protect catalytic converters'.[35] Between the late 1970s and the end of the 1980s, the average blood lead levels in American children aged between one and five years declined by 77 percent. However, even with stringent limits on industrial and vehicle emissions, 20 000 tonnes of lead are still exhausted into the air over North America annually.[36]

In 1996, an estimated 70 000 tonnes of lead was still being added to petrol worldwide, with Russia, Nigeria, South Africa and Saudi Arabia among the leading users.[37] Lead is now banned from petrol in the United States, Japan, Canada, Austria, Sweden, Brazil, Colombia and South Korea. In Australia and most European Union countries roughly half the cars run on unleaded petrol. However in more than 50 countries including 20 in Africa, petrol containing 0.8 grams of lead per litre (more than five times the European Union limit) is still sold. Children in many of the cities of these countries will continue to suffer permanent mental retardation given that most research studies show an average IQ deficit of two to four points for every 100–150 micrograms per litre increase in blood lead levels[38] (see Figure 1.2).

Figure 1.2 Blood lead levels in children in cities where leaded petrol is still used

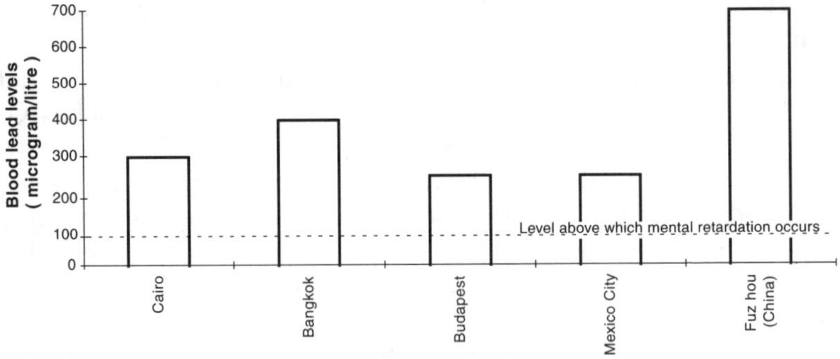

F Pearce, *New Scientist*, 27 July 1996

In Sydney, Australia, where leaded petrol is still sold alongside unleaded, children living within 10 kilometres of the Central Business District between 1992 and 1994 still had an average blood lead level of 81 micrograms per litre with 7 percent of children having blood lead levels above 150 micrograms per litre.[39] In 1996 in the United States, 15 years after unleaded petrol was introduced, the mean blood level in children had dropped to 29 micrograms per litre.[40]

Another side to the petrol saga is now emerging as leaded petrol is being phased out. Unleaded petrol can contain up to 2 percent of the chemical benzene which is both carcinogenic and neurotoxic.[41] The change to unleaded fuels is clearly helping to reduce the lead problem, however, it should also be mandatory that refineries fractionate out benzene and use non-aromatic additives to raise octane levels such as methyl tertiary butyl ether (MTBE).

A new side to the internal combustion engine saga has emerged since the early 1990s. Fine particulates less than 10 micrometres (µm) across are emitted by the exhausts of cars, trucks and buses. In 1991, Joel Schwartz at the United States government's Environment Protection Agency (EPA) studied the concentration of these particles (referred to as PM10 levels) in the air of cities, and death rates, and found there was a strong correlation. After further data was collected, Schwartz estimated in 1994, that each year 60 000 people die in the United States and 10 000 people die in England and Wales as a result of these vehicle emissions. The American studies suggest that there are no safe levels of PM10 and that when the concentration in the air of a city increases by 10 micrograms per cubic metre, the death rate rises by 1 percent.[42] In the United States, the safety level for PM10 has been set at 150 micrograms per cubic metre, but in 1995 the United Kingdom government's Expert Panel on Air Quality Standards recommended a maximum standard of 50 micrograms per cubic metre averaged over 24 hours.[43] In London between 1992 and 1994, this limit was exceeded on 139 days. In December 1991, London experienced a particularly bad episode of air pollution and the levels of PM10, the most worrying of the cocktail of pollutants emitted by road traffic, were estimated to be as high as 800–1000 micrograms per cubic metre.[44]

Diesel engines emit up to 80 percent more particulate matter than petrol engines. Some of these particles produced by incomplete burning of fuel in diesel engines are less than 2.5 micrometres across and can penetrate deep into the lungs. These particulates can also carry carcinogens such as benzopyrene, naphthalene and phenanthrene absorbed on the surface, and the International Agency for Research on Cancer is now warning that the exhaust from diesel engines probably causes cancer.[45]

Not surprisingly, a 1996 Danish study found that diesel bus drivers in Copenhagen had a 60 percent higher incidence of lung cancer than average. Copenhagen bus drivers operating in the city centre were also found to have significantly more chromosomal aberrations and more DNA damage than bus drivers in the suburbs. Approximately 60 percent of the vehicles in Copenhagen are powered by diesel engines.[46]

The enormous problems of air pollution in the cities around the world and the serious health consequences are becoming increasingly apparent. Reducing the number of vehicles in our cities and the distance that they travel is a fundamental requirement of saving our environment and reducing the toll of human disease. The trend of increased commuting by car which took place in the 1980s (see Figure 1.3) must be reversed by imposing limits on our use of internal combustion engine powered vehicles in city and urban areas.[47]

Figure 1.3 Increase in car usage in major cities, 1980 and 1990

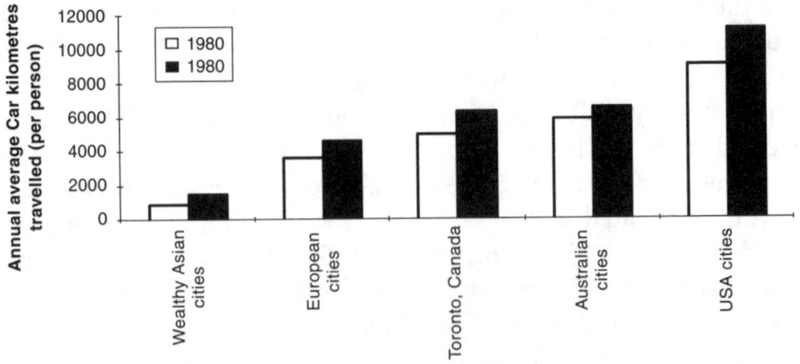

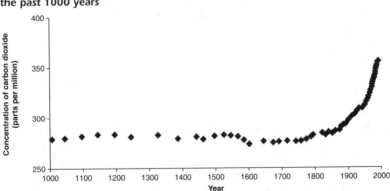

P Newman, Murdoch University, Western Australia, 1997

The greatest pollution produced by our power stations and vehicles is the emission of carbon dioxide into the atmosphere. It is interesting to note that this by-product of fossil fuel combustion was once considered innocuous. In 1860, about 100 million tonnes of carbon dioxide were released into the atmosphere from fossil fuel combustion. Since that time, our use of fossil fuels to produce power has increased rapidly. In the 1990s, only 130-or-so years later, the quantity of carbon dioxide being discharged into the atmosphere from vehicles and power generation has now reached a rate of around 20 billion tonnes per year.[48] An estimation of the variation in carbon dioxide content of the atmosphere over the past 1000 years calculated from the ice core data is shown in Figure 1.4.

Figure 1.4 Variation in carbon dioxide levels in the Earth's atmosphere over the past 1000 years

DM Etheridge et al., *Journal of Geophysical Research*, vol. 101, no. D2, 1996, pp. 4115-4128 (copyright by the American Geophysical Union)

The greenhouse effect

Much of the carbon dioxide accumulated in the atmosphere, eventually alters its heat insulation properties and produces what is now commonly known as the greenhouse effect. As the concentration of carbon dioxide and other greenhouse-type gases in the atmosphere increases, less heat from the sun is reradiated back into space, with the result that the earth's temperature slowly rises. The variation in the global mean temperature over the past 130-or-so years is shown in Figure 1.5, page 17. Since the early 1900s, the earth's temperature has shown an alarming heating trend.

In 1994, the atmospheric concentration of carbon dioxide (CO_2) was calculated to have reached 358 parts per million (ppm)(corresponding to around 760 billion tonnes of carbon dioxide), compared with carbon dioxide concentrations of around 280 ppm (equivalent to approximately 590 billion tonnes of carbon dioxide) before industrial activity.[49] This change has increased the total solar energy absorbed by the Earth by 1 percent.[50]

Recognition of the greenhouse effect by bodies such as the World Meteorological Organisation and the United Nations Environment Programme has led to the formation of the Intergovernmental Panel on Climate Change (IPCC). The IPCCs research led to the Climate Change Convention which was signed by a number of nations at the Earth Summit in Rio in 1992 and later ratified in 1993. The Convention's aim is to stabilise greenhouse gases—which are now known to include chloroflurocarbons (CFCs), methane and nitrous oxide—in the atmosphere at 1990 levels by the year 2000.[51]

One of the direct effects of this temperature rise has been the expansion of the oceans and the melting of polar ice. Since the beginning of this century, sea level has risen by about 15 centimetres.[52] Over the next century, the Intergovernmental Panel on Climate Change (IPCC) estimates that sea level could rise by a further 31–110 centimetres.[53] Rises of this magnitude could have drastic consequences around the world, especially in areas such as the Pacific Islands, the Netherlands and Bangladesh. There is also evidence that atmospheric warming has contributed to the relatively large 2.5 degrees Celsius increase in the average temperature of the Antarctic region over the past 50 years.[54] Recent satellite data has also shown that the rate of melting of Arctic Sea ice has increased from 2.5–4.3 percent per decade.[55]

Global warming is likely to have both direct and indirect effects on human health. The First International Conference on Human Health and Global Climate Change was held in Washington DC in September 1995. Dr Anthony McMichael from the London School of Hygiene and Tropical Medicine told the Conference that severe heat waves could cause an increase in mortality similar to the 500 or so deaths seen during a heat wave in Chicago earlier that year.[56] He also pointed out that because the rate of development of parasites accelerates as temperature rises, diseases such as dengue fever are likely to invade the United States. The range of malaria, which now affects 45 percent of the world's population is likely to increase its potential range to affect 60 percent of the population by the middle of the next century.[57]

Dr Paul R Epstein, a Professor of Tropical Medicine at Harvard University, reported to the Conference that many climate models indicate global warming also makes the cyclical warming of the tropical Pacific Ocean and associated weather disturbances around the world (known as the El Niño phenomenon) both more intense and more frequent. Definite connections between cholera and malaria outbreaks have been established and this could mean serious consequences for human health if El Niño becomes an almost constant feature of global warming.[58]

Insect attack on crops may also be increased by global warming. For example, the peach potato aphid, *Myzus persicae* is a major agricultural pest in the United Kingdom, attacking potatoes, oilseed rape and sugar beet as well as spreading plant viruses. Aphid clones have now emerged which are resistant to major pesticides, though interestingly, these clones suffered far greater losses than non-resistant aphids during winter. It appears that the biggest factor in the aphid's demise was the amount of time they spent below 2 degrees Celsius. As global temperatures rise, the number of frosty nights per year will decrease which, in turn, reduces this natural insect control.[59]

While the greenhouse effect has been recognised by many governments, political promises and actions can, however, be quite different, especially under the continual economic pressure to increase economic growth. For example, between 1990 and 1994 carbon emissions from fossil fuels rose 5 percent in North America and in 1995, the United States government's Climate Action Plan was not on track to meet the year 2000 goal.[60] In July 1996, F Pearce reported that stonewalling by the United States, Australian and other governments could sink negotiations aimed at halting global warming.[61] Australia and New Zealand have been the only Organisation for Economic Cooperation and Development (OECD) countries not to support calls for legally binding targets and timetables to reduce greenhouse gas emissions at the United Nations Convention on Climate Change.[62] The greenhouse effect is a clear example of where we must limit our use of technology in the name of conserving human health and the environment of our planet. John Burton has suggested that a steep carbon tax would effectively lower carbon dioxide emissions.[63] Such a measure would also be a step towards counting the true economic cost of our modern lifestyles. Such actions may not be 'too soon'. In November 1996, a new climate model proposed by J Sarmiento and C Le Quere of Princeton University, suggests that global warming may be happening much faster than climate researchers had feared. They found that over the next 350 years the amount of carbon dioxide absorbed by the oceans could be 28–49 percent less than assumed in current models, because rising temperatures could reduce the ocean's ability to absorb the gas. As a result of this 'positive feedback' effect, global warming would increase much faster than the IPCC predictions.[64]

The bold assumption that we can continue to dump our technological waste into the atmosphere and the oceans without adverse environmental changes needs urgently to be challenged, as can be observed from the graph below (see Figure 1.5).

Figure 1.5 Observed anomalies in Australian mean temperature from 1910 to 1994

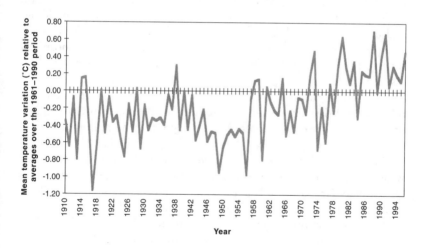

Note: Sea level has risen about 15 cm since 1900 as the result of global warming.
National Climate Centre, Bureau of Meteorology, Melbourne, Australia, 1997

The nuclear energy problem

Since the 1950s a steadily increasing proportion of the world's electricity demand has been produced by nuclear power stations. In 1986, the year of the Chernobyl Nuclear Power Station disaster, the United States had 98 operable nuclear reactors generating about 16 percent of total electricity production, with another 32 plants in some stage of planning or construction.[65] Worldwide in 1985, some 528 nuclear power generating plants were either in existence or under construction.[66]

Originally proposed as a 'clean' and safe form of power generation, the major health risks associated with nuclear energy can no longer be denied or covered up. Over the four decades of nuclear power up to 1986, there have been numerous incidents where radioactive material was released into the environment.[67] On 26 April 1986, the world's most disastrous nuclear accident occurred at Chernobyl in the Ukraine. An experiment in using residual kinetic energy to generate emergency power at the Chernobyl Nuclear Power Station went tragically wrong. The resulting explosion and fire released a cloud of radioactive material which was carried by prevailing winds to contaminate much of northern Europe.

The blasts destroyed one of the four reactors in the power station and extensively damaged the building that housed it. Pieces of hot, radioactive debris rained down on the roofs of neighbouring buildings and fires broke out. The graphite core of the reactor became a raging inferno and spewed radioactive gas and particles into the atmosphere for ten days. During the aftermath, 31 people, most of them firefighters, died of acute radiation sickness.[68]

The World Health Organisation (WHO) estimates that Chernobyl released 200 times as much radioactivity as the uranium-based bomb dropped over Hiroshima on 6 August 1945 and the plutonium bomb dropped over Nagasaki on 9 August 1945, put together.[69]

Air, food and milk around Chernobyl were heavily contaminated with isotopes of iodine for weeks after the accident, and countries as far away as Greece were heavily contaminated by radioactive fallout. Years after the incident, the frightening extent of the harm to human health caused by this accident is still being revealed.

At an International Seminar on Humanitarian Medicine and Human Rights held in April 1990, for example, researchers reported that up to 250 000 people may be living in areas still too polluted from radioactive fallout for normal life and are therefore in need of urgent evacuation. In Byelorussia, which received 70 percent of the radiation fallout, common illnesses have suddenly become 70 percent more prevalent amongst the population. Research efforts to explain the radical increase in common illnesses have come to the conclusion that the immune systems of the inhabitants were badly weakened by radiation exposure. It is estimated that one-fifth of the children in this region have received radiation exposure of 10 gray (Gy; or 1000 rad) or more since the accident. This level of exposure is ten times the figure associated with radiation illness and about 100 000 times the level from normal background radiation. In one region, Palessie, 50–70 percent of the children now have health problems, with about 8 percent having thyroid complications in forms that had not been observed before Chernobyl. In another region, Minsk, the incidence of tuberculosis has risen by 14 percent since the accident.[70] Nor is the problem confined to political boundaries. In West Germany, investigations have revealed that the number of infants dying within seven days of birth in the south of that country was significantly higher in the months following the disaster.[71]

More health problems are emerging. At a WHO conference on Chernobyl held in Geneva in November 1995, the world's leading experts on thyroid cancer reached a consensus that the epidemic of thyroid cancer among children in the three states closest to Chernobyl (Belarus, Ukraine and Russia) has been caused by radiation from Chernobyl. At Gomel in southern Belarus, more than 200 kilometres north of the Chernobyl reactor, rain is believed to have deposited heavy iodine contamination. Here the number of children who have contracted the disease every year has increased from less than one per million before 1986 to more than 200 per million in 1994. At the time of the conference, 680 thyroid cancers had been diagnosed in children, ten of whom had died.[72] Six months later in April 1996, the number of confirmed thyroid cancer cases in children had reached 800.[73]

D Williams, Professor of Histopathology at the University of Cambridge, believes that as many as 40 percent of the children exposed to the highest levels of radiation fallout from Chernobyl when they were infants, could develop thyroid cancer as adults.[74] Given these stark warnings, the French health ministry announced in April 1996 on the eve of the tenth anniversary of the Chernobyl disaster, that it planned as a preventative measure, to distribute iodine tablets to thousands of homes within a five kilometre radius of the country's 19 nuclear reactor sites. However such a plan has limited merit, as fallout from Chernobyl reached homes more than 2000 kilometres away.[75] There are also cancers other than thyroid cancer which are associated with exposure to radioactive fallout, such as breast cancer, all of which have been reported to have risen in the regions around Chernobyl since the explosion.[76] The first evidence that radioactivity from the Chernobyl accident may have caused leukemia in young children as far away as Greece, has been reported recently by Professor D Trichopoulos of the Harvard Centre for Cancer Prevention in Boston. Babies exposed to radiation in the womb after the 1986 explosion were on average 2.6 times more likely to develop leukemia than children who were not exposed to radiation from the accident.[77] Children exposed to radiation in the womb may also suffer brain damage. Preliminary findings from a WHO-sponsored study comparing 2189 children born in contaminated areas of Belarus, Ukraine and Russia, with 2021 children from nearby uncontaminated areas were released in November 1995. The findings suggest that children from the contaminated areas were suffering more mental retardation, more emotional disorders and more behavioural disorders, with approximately 50 percent of children having experienced some kind of mental disorder.[78] These findings reinforce the results of a study of 1100 Hiroshima children who were exposed to radiation in the womb and also suffered a higher than average rate of mental retardation.[79] The joint United States–Japanese Radiation Effects Research Foundation in Hiroshima has also found evidence which links radiation exposure to strokes, heart disease and cirrhosis of the liver.[80]

A joint British–Russian–Belarussian research team has recently found evidence that exposure to radiation may damage sperm and ova and cause permanent genetic damage. The researchers found that children whose parents were exposed to radioactive fallout in Belarus after the Chernobyl disaster have twice as many mutations in their DNA as British children.[81]

The environment has also been extensively contaminated by the Chernobyl nuclear accident. Pine forests in Belarus contain high levels of radioactivity which could take centuries to decay naturally to acceptable levels. Plans are now underway to burn the pine trees in specially designed power stations. It is hoped that 99.9 percent of the radionuclides will be trapped in the ash and that in this way the contaminated forests may be removed within 40 years.[82]

Radioactivity from the Chernobyl area is contaminating large areas of Europe as a result of rain runoff and floods. Radioactive waste dumped around the reactor after it exploded over a decade ago has been washed down the Pripyat River to Europe's third-longest river the Dnieper, which flows 800 kilometres to the Black Sea. More than 30 million people are now considered at risk. Reservoirs downstream from the damaged reactor provide drinking water for 9 million people in Ukraine, as well as irrigation and fish for another 23 million people. There is also concern that radioactivity from these dumps could reach into groundwater and further contaminate waterways in 10–15 years time.[83]

One report has estimated that the Chernobyl explosion deposited 380 terabecquerels (TBq) of strontium and plutonium on the flood plain around the reactor. Given that a terabecquerel represents 1012 radioactive disintegrations each second, an enormous amount of radioactivity was deposited into the environment. In Lake Kojanovskoe, 250 kilometres away, the level of radioactivity is 60 times higher than the European Union (EU) safety limit, and perch and pike from the lake contained radioactivity as high as 40 000 becquerels per kilogram of caesium-137 in 1994. (The EU safety limit is 600 becquerels per kilogram).[84]

Workers in nuclear power stations and their children are also exposed to increased risks of cancer, even when these stations are functioning properly. In 1995, the maximum radiation dose generally considered safe for workers was still 50 millisieverts (mSv) in any one year.[85] However, research by M Gardner and co-workers, has revealed that where British workers received doses of radiation as low as 10 millisieverts in the six months prior to conception, their children faced a six- to eight-fold increase in the risk of developing leukemia.[86]

Much of the technology used in nuclear power generation has been derived from nuclear weapons research and testing. This too has taken its toll on human health. The Chukot Peninsula in the former USSR is located in the vicinity where atmospheric tests were carried out in the 1950s and 1960s. It is not accidental that the population of this peninsula has the highest death rate from oesophageal cancer in the world. An investigation in 1989 by a commission of Soviet deputies and scientists found that more than 90 percent of the population had chronic lung disease. In one village visited by the investigating team, every adult suffered from cancer and members of the commission calculated the life expectancy in the region to be around 45 years. The infant mortality was estimated to be 70–100 per 1000 live births, which was four times the official government figure.[87] Workers at nuclear weapons sites in the United States have received similar radiation exposures. In Nevada, of 218 workers present at early bomb tests, 200 have since died of cancer.[88] As early as the mid-1960s the hidden cost of progress in nuclear technology was being revealed. Less than eight years after President Eisenhower's assurance of the safety of atomic testing, President Lyndon B Johnson, speaking of the new Nuclear Test Ban Treaty on 12 October 1964, announced:

This treaty has halted the steady, menacing increase of radioactive fall-out. The deadly products of atomic explosions were poisoning our soil and our food and the milk our children drank and the air we all breathe. Radioactive deposits were being formed in increasing quantity in the teeth and bones of young Americans. Radioactive poisons were beginning to threaten the safety of people throughout the world. They were a growing menace to the health of every unborn child.[89]

Poisoned air, water and soil: A legacy of chemical progress

Emissions into the air from our industries of chemical technology over the past 150 years have now reached frightening proportions. M Lippmann and R Schlesinger list over 50 major air contaminants produced by industry during the extraction of raw materials and the manufacture of modern commodities.[90] Industrial hygienists are now concerned over the increasing levels of more than 1000 hazardous compounds which are released daily into the air of the workplace and nearby environment.[91] A hundred years ago, the poisoned vegetation around copper smelters and related industries aroused little cause for concern. In recent times, environmental tragedies have forcefully borne witness to the price paid in terms of human health from the production of the chemical wares of progress. One such incident occurred at Bhopal in India. About midnight on 2 December 1984, poisonous methyl isocyanate leaked from a nearby giant chemical plant and engulfed the town. Over 2000 people died and an estimated 250 000 people suffered health effects.[92]

Much of our modern technology is dependent on the production of chemicals which now return a massive financial income to the global economy. In 1994, total chemical sales by the world's top-50 chemical companies exceeded US$330 billion dollars.[93] However, the cost to the environment of this massive profit is also high. For example, in the United States, manufacturing industries generated 27 million tonnes of waste. Only about 12 million tonnes of this waste was recycled, treated or burned for energy recovery on site.[94] The top-15 chemicals released into the United States environment are listed in Figure 1.6. The majority are highly volatile and toxic. In 1993, United States industries released nearly 1.3 million tonnes of these chemicals into air, land or water.[95]

Figure 1.6 Top 15 chemicals released into the US environment in 1993

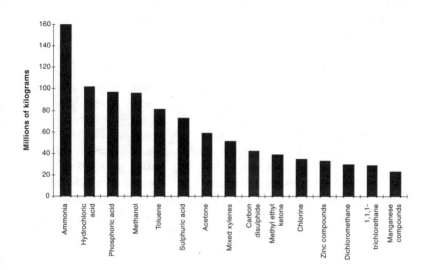

D Hansen, *Chemical and Engineering News*, 3 April 1995

While we have come to accept chemical pollution as an inevitable consequence of cities, we may also tend to reassure ourselves that there are still many parts of the world which are largely unaffected by toxic wastes. One such example might be New Zealand, which has relatively few smokestack industries. The apparently pristine forests and lakes give the country its reputation as a part of the world still relatively untouched by the ravages of industrial pollution and attracts tourists from all over the world. However in 1993, the New Zealand Ministry for the Environment released two reports which showed that more than 10 000 sites, including New Zealand's tourist jewel, Lake Rotorua, are contaminated with a legacy of toxic wastes from agriculture, forestry and industry.[96]

Lake Rotorua was found to contain levels of the highly toxic organochlorine chemical dioxin, which were 50 percent higher than the levels in the lower Rhine in Europe, and twice as high as the levels in the heavily polluted Lake Ontario. Sewage, industrial waste and polluted water were found being discharged into Wellington Harbour, and pentachlorophenol, other organochlorines and industrial effluent were being washed into Manukau Harbour near Auckland. At the southernmost part of the beautiful South Island there were accumulations of hazardous waste and air pollution from an aluminium smelter. One of the main toxic chemicals found to be contaminating more than 420 timber-producing sites was the organochlorine chemical pentachlorophenol which is used extensively as a wood preservative.[97]

Chlorine compounds

Chlorine is widely used in industry to make paper, to treat water and in the manufacture of organochlorine chemicals. These have been used as solvents and as raw materials in a wide variety of industrial processes and products, including the manufacture of plastics. Examples include chloroform, carbon tetrachloride, dichloromethane, chlorofluorocarbons, polychlorinated byphenols, pentachlorophenol and vinyl chloride, just to name a few. In times past, chloroform was used as an anaesthetic while carbon tetrachloride was extensively used in fire extinguishers. What is little realised is that large quantities of these compounds, many of which are now suspected or known carcinogens, are evaporated into the atmosphere. Dichloromethane, also known as methylene chloride, is a case in point.

Methylene chloride is widely used in industrial processes, food preparation and agriculture. It is used as a solvent in paint removers, degreasing agents and aerosol propellants; as a blowing agent in flexible urethane foams; and as a process solvent in the manufacture of steroids, antibiotics, vitamins and tablet coatings. Methylene chloride is used as an extraction solvent for spice oleoresins, hops, and caffeine from coffee. Historically, it has been used as an inhalation anaesthetic and as a fumigant for grain and strawberries. In 1980, the world production of this chemical was reported as 517 000 tonnes and it has been estimated that 80 percent of this production was released into the atmosphere.[98] At levels as low as 500 parts per million, this commonly used solvent has been found to produce mammary gland tumours in rats [99] while higher exposure levels produced various cancers in mice.[100]

The organochlorine compounds of industry are produced from the raw material chlorine which, in turn, is manufactured by the electrolysis of sea salt. An estimated 41 million tonnes of chlorine are produced worldwide each year, with North America producing 12 million tonnes, Western Europe producing 9 million tonnes and Australia 140 000 tonnes each year. Australia also imports each year 80 000 tonnes of chlorine in the form of vinyl chloride monomer for the manufacture of PVC plastics. About 60 percent of the chemical manufacturing industry in Australia is dependent on chlorine and chlorine products.[101] Figure 1.7 shows a breakdown of the global use of chlorine.[102]

Figure 1.7 Global chlorine use (total production, 41 million tonnes per annum)

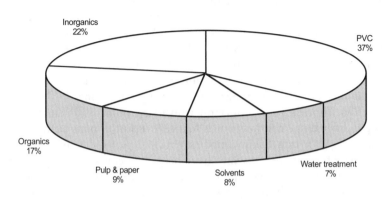

Chlorine in Perspective, CSIRO Division of Chemicals and Polymers, 11 June 1996

There is growing evidence that the use of chlorine by industry can harm vertebrates, including humans. In February 1994, the United States–Canada Joint International Commission on the Great Lakes published evidence which showed that environmental exposure to chlorinated compounds harms humans. The Commission found that women living around the lakes who ate large amounts of lake fish, had acquired high levels of dioxins and related chlorinated compounds. The children of these women, had low birth weights and were less intelligent and less healthy than the children of women who had not been exposed to such high levels of chlorinated compounds.[103] These findings support the call by the Commission, two years earlier in 1992, for a ban on the use of chlorine around the Great Lakes because of the high concentrations of chlorinated compounds in lake water.[104]

Chlorinated compounds containing two-ring structures, such as dioxins and polychlorinated biphenyls, can bind to a receptor on the nucleus of vertebrate cells, where they mimic or block oestrogen and other steroid hormones. These findings together with those of the Commission have prompted proposals to ban or limit chlorine.[105] The industry federation has responded by arguing that many organochlorine compounds occur naturally in the environment.[106] GW Gribble from Dartmouth College in Hanover, New Hampshire, has also published a detailed study of the production of chlorinated compounds in nature.[107] Gribble suggests, for example, that dioxins are produced naturally by forest fires and that other chlorinated compounds are formed by the enzymatic reactions between chloride ion and phenolic humic acids.[108] However, B Commoner from the City University of New York, points out that Gribble's study actually supports the view that these compounds are dangerous because none of the natural organochlorine compounds occur in vertebrates. Many are intermediates in chemical reactions within cells, and are secreted specifically as toxins by moulds, plants, algae and marine invertebrates. Vertebrates cannot have natural polycylic chlorinated compounds because they would interfere with the animal's use of steroid hormones.[109] TG Spiro and VM Thomas from Princeton University, point out that modern human liver tissue contains much higher levels of dioxins than that of 400-year-old frozen Eskimo bodies or 2800-year-old Chilean mummies. The former are known to have been high consumers of fish and animal fat which concentrate organochlorine compounds, and the latter are known to have been frequently exposed to wood smoke (a reported natural source of dioxins) from interior fires.[110]

Sediment cores from lakes also provide a historical record of dioxin deposition. For example, in Siskiwit Lake, dioxin levels were low before 1940, increased rapidly during the industrial expansion of the 1950s and 1960s, and declined since the 1970s in parallel with the decreased production of chlorinated phenols. Spiro and Thomas suggest that:

> It is hard to see how these trends can be reconciled with predominantly natural sources.[111]

In May 1995, the United States Environmental Protection Agency (EPA) circulated a draft report which found that low doses of dioxin, similar to the background levels now found in most United States citizens, can cause cancer, infertility and interfere with foetal development.[112]

Chlorofluorocarbons

The magnitude of the impact of anthropogenic chemicals is illustrated by the production of chlorofluorocarbons (CFCs). These chemicals were originally developed by chemists to replace ammonia in refrigeration plants. CFCs subsequently found extensive use as propellants in aerosol cans, and as the bubbling agent in the manufacture of certain plastic foams used to make such items as furniture, and so-called 'disposable' polystyrene cups and food trays. Originally, these compounds were believed to be inert but, as E Longstaff points out, CFCs are not biologically inert. Some CFC compounds have been shown to be bacterial mutagens, cell-transforming agents and rodent carcinogens.[113]

The emissions of CFCs into the atmosphere worldwide in recent decades have been enormous, with possibly 1 million tonnes being released annually during peak production in the mid-1980s. Like carbon dioxide, these gases are greenhouse gases, changing the heat insulating properties of the atmosphere. However, *one molecule of CFC produces the same greenhouse warming effect as 10 000 molecules of carbon dioxide.*[114] We also know that CFCs may remain in the atmosphere for long periods of time. The most abundantly produced classes of these compounds, known as CFC-11 and CFC-12, are estimated to remain in the atmosphere for approximately 75 and 110 years respectively. CFC pollution is a long-term phenomenon ensuring that we will be reminded of the folly of our throw-away cups for decades to come.

There is also another side to the CFC story. In 1974, F Rowland and M Molina propounded their well-known hypothesis of ozone depletion by CFCs.[115] Further elucidation of the mechanisms of the destruction of the ozone layer by chlorine atoms released by CFCs in the upper atmosphere have since been presented.[116] It is well known that the ozone layer shelters all life on this planet from harmful ultraviolet solar radiation. Diminishing this protective mantle threatens the health balance of all living plants and animals exposed to sunlight.[117] Indirect and synergistic consequences may also impact on biological and physical processes such as the role of the oceans as a carbon dioxide sink or the formation of cloud condensation nuclei.

Phytoplankton in the oceans, for example, fix more than half the carbon dioxide produced globally each year. However, these marine organisms are highly sensitive to increased levels of ultraviolet radiation because they lack the protective outer layers enjoyed by higher plant and animal species. If the populations of phytoplankton decrease as a result of increased levels of ultraviolet radiation, the decreased carbon dioxide uptake by the ocean will result in much higher levels of carbon dioxide remaining in the atmosphere. This, in turn, will accelerate the greenhouse effect and result in rises of sea level.[118]

Similarly, the increase in ultraviolet (UV) radiation and resulting reduction in phytoplankton populations may, in turn, significantly alter the balance of solar radiation reaching the earth. The synergistic effect involved here is complicated, but it can be simplified as follows. Phytoplankton have been identified as the source of dimethyl sulfide (DMS) which they release into the atmosphere. The flux of DMS from the ocean to the atmosphere is climatically important because the particles of sulfur released become the nuclei or droplets upon which clouds form. By altering the phytoplankton population, we are thus inadvertently modifying the density of clouds. This is important because as marine clouds become thinner, they become less effective at regulating the amount of solar radiation which hits the earth. The more ultraviolet radiation which gets through to the ocean's surface, the greater the degree of destruction to the phytoplankton—thus reinforcing a negative feedback loop with disastrous consequences for the ecosystem. It is worth noting also that marine cloud formations figure as an integral factor in the generation of ocean turbulence associated with sea storms. Turbulence is itself crucial to the survival of phytoplankton, inasmuch as storms churn up the ocean, drawing food for phytoplankton from the cooler lower layers of ocean to the warmer surface areas where phytoplankton feed. Thus, the decrease in cloud formation—which, in turn, leads to a reduction in sea storms and ocean turbulence—reduces the available food for phytoplankton and diminishes the phytoplankton population in consequence, further reinforcing the vicious cycle referred to above.[119] Appreciation of this point about the synergistic connections among ecosystems makes it easier to see that the destruction of the ozone layer by CFCs is, in turn, disrupting one of the most important systems of climatic regulation on the Earth.

The growing scientific consensus of the dangers of CFCs and also halons (bromofluorocarbons) in the environment, led to an agreement between 49 countries, meeting in Vienna in 1985, to protect the ozone layer. In 1987, 24 nations signed an agreement to reduce the production and use of CFCs, which has become known as the Montreal Protocol on Substances that Deplete the Ozone Layer. As scientific understanding of the processes and chemicals that destroy the ozone layer has grown, the number of countries signing the Protocol had increased to 96 by 1996.[120] In the meantime, several amendments have been made. The so-called London Amendments were agreed to in 1990. These required more stringent controls on the production and use of other ozone-depleting chemicals, such as carbon tetrachloride and methyl chloroform. The 1992 Copenhagen Amendments were added following the National Aeronautics and Space Administration's (NASA) finding that ozone depletion was occurring at a faster rate than was previously estimated.[121] These amendments included restrictions on the use of hydrochlorofluorocarbons (HCFCs) which are now being substituted for CFCs, hydrobromofluorocarbons, and the pesticide methyl bromide. The CFCs phase-out has been affected in most countries however, world production of CFCs in 1995 was estimated to still be about 360 000 tonnes.[122] Meanwhile the ozone layer is now believed to be thinning twice as fast as originally predicted,[123] and satellite data have confirmed that manufactured chlorofluorocarbons are the predominant source of the chlorine that is eroding the Earth's protective ozone layer.[124]

The unhealthy food chain

In January 1940, Sir Albert Howard wrote in the preface of his seminal work, *An Agricultural Testament*:

> Since the Industrial Revolution the processes of growth have been speeded up to produce the food and raw materials needed by the population and the factory. Nothing effective has been done to replace the loss of fertility involved in this vast increase in crop and animal production. The consequences have been disastrous. Agriculture has become unbalanced: the land is in revolt; diseases of all kinds are on the increase; in many parts of the world Nature is removing the worn out soil by means of erosion.[125]

Howard goes on to argue in his book that the use of inorganic fertilisers, together with excessive irrigation and over-cultivation, is destroying the fertility of the soil, and that as a consequence there is increased disease in plants, animals and man.

The over-cultivation and over-irrigation of the soil gradually occurred as agricultural progress came to be defined in terms of increased yield per hectare. The development of machines to convert concentrated energy to mechanical power gave a whole new sense to the concept of efficiency. Efficiency, the watchword of the mechanical paradigm and the Industrial Revolution, was carried into the realm of agriculture. Machines became increasingly employed to divert and pump water in vast quantities and to cultivate the soil on a scale never before possible.

Liebig's observations of the increased growth of plants which occurred when phosphates were mixed with the soil, led to a proliferation of patent chemicals whose specific purpose was to increase the soil crop yield per hectare. One such example was the use of sulfuric acid to increase the solubility of phosphates, patented by JB Lawes in 1842.[126] This form of phosphate later became known as superphosphate. Within a few years, increasing quantities of potassium salts from deposits at Stassfurt in Germany and nitrate salts from Chile were being transported to various parts of the world.

Given this worldwide dissemination of superphosphates the composition of soil was rapidly altered on a worldwide basis. Fertilisers containing a mixture of nitrogen, phosphorus and potassium compounds became known as 'complete' fertilisers,[127] reinforcing and entrenching in the minds of farmers the misleading NPK mentality discussed earlier.

As research on the chemical control of plants continued, agricultural scientists discovered that individual types of crops were found to respond to particular ratios of NPK fertiliser mixes. For example, the ratio for potatoes was 4:8:10, for wheat 2:12:6, clover 0:12:15 and strawberries 5:8:7. These differing fertiliser requirements, together with the development of particular cultivation configurations of farm machinery for different crops, led to the practice of cultivating ever-larger areas of land for a single type of crop. Monoculturing was thus incorporated as part of the all-pervasive Western perspective devoted to growth and efficiency. F Graham comments:

> By the end of World War II, farming had evolved from a way of life to an industry. The farmer acquired the approach and the philosophy of the manufacturer. The small family farm, which once had produced a variety of fruits, vegetables, and dairy foods, now merged with a dozen of its fellows to form a single vast 'factory farm' producing a single crop. In the interests of economy, every move was directed, every machine was adapted, to plant, nourish, and harvest that crop— whether it was celery, apples, corn, or soy beans. A new word—mono-culture—found a place in the dictionary.[128]

Biodiversity is a life-supporting characteristic of nature.[129] Since the introduction of modern farming the genetic diversity of our agricultural plants has been rapidly vanishing. This century, the United States has lost more than 90 percent of its 20 000 varieties of agricultural plants (see Figure 1.8). Since 1930, 80 percent of the varieties of maize in Mexico have been lost and China now only grows 1000 strains of wheat, compared with the 10 000-or-so strains it grew before 1949. This decline in agricultural biodiversity is leaving our crops prone to pest and plague.[130]

Figure 1.8 Lost diversity of vegetable varieties in the USA

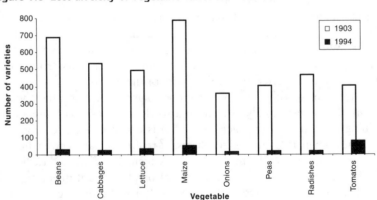

R Edwards, *New Scientist*, 17 August 1996, p. 14

Monoculture is a distortion of nature which, in a sense, produces a 'plague' of plants. What was missed in the equation was that this extreme ecology inevitably results in a plague of insects. Given the commitment to technology as the panacea of all ills, chemical controls or pesticides were quickly enlisted in the battle against the armies of invading insects. It is revealing that the rapid growth of monoculture after World War II, which was paralleled by an inordinate expansion in farm machinery and fertiliser sales, was also accompanied by an expanding use of pesticides.[131] What is even more apocalyptic is the evidence which establishes that concomitant with the growth of technological farming, the percentage of crops lost to pests and disease increased accordingly.

Giving evidence in 1951 before the Delaney Committee inquiry into pesticide use in the United States, RL Webster described how, in 1909, only single sprayings of arsenate were used each year. By 1926, pests had become so much more troublesome, that five or six additional sprayings were required during the season. He went on to say:

> Worm control in the Pacific North-west was becoming more difficult, year by year. We now know that the codling moth was acquiring a degree of resistance ... More and more lead arsenate was used every year and at one time Washington State was using 25 percent of the entire production in the United States. The peak came in 1943, when more than 17 000 000 pounds (lbs.) were sold in the state.[132]

Webster's observations contrast revealingly with those of Howard. Howard had reported that crops grown in soils rich in humus around 500 000 villages in India, suffered remarkably little from pests.[133] He also had argued that the replacement of animal manure and compost by artificial NPK fertilisers was directly linked to the increased susceptibility of plants to disease and insect attack.

> With the spread of artificials and the exhaustion of the original supplies of humus, carried by every fertile soil, there has been a corresponding increase in the diseases of crops and of the animals which feed on them. If the spread of foot-and-mouth disease in Europe and its comparative insignificance among well-fed animals in the East are compared, or if the comparison is made between certain areas in Europe, the conclusion is inevitable that there must be an intimate connection between faulty methods of agriculture and animal disease. In crops like potatoes and fruit, the use of the poison spray has closely followed the reduction in the supplies of farmyard manure and the diminution of fertility.[134]

As nature revolted against the mechanical and chemical transformation of the environment, an extensive arsenal of poisons was developed to subdue the revolt. In 1939, P Muller discovered the insecticide properties of DDT, a chlorinated organic compound first synthesised 65 years earlier. Since that time there has been an exponential increase in the number of synthetic insecticides, herbicides and fungicides available.[135] In 1986, some 1500 active pesticide ingredients were registered in the United States,[136] with about 360 000 000 kilograms of these chemicals being applied annually in the agricultural sector alone.[137]

The link between monoculture, fertiliser requirements and pesticide use is well illustrated by the farming of corn in the United States. This single crop consumed 44 percent of the total United States fertiliser use, with the application rate being as high as 340 kilograms per hectare.[138] At the same time, to control insect attack, a massive 43 percent of the total pesticides used in the United States were applied to this same crop.[139] Despite this huge application of poisonous pesticides in the name of technological progress, R Nader points out that the percentage of pre-harvest crops lost to insect damage has increased from 9 percent in 1945 to 12 percent in 1989.[140] This is due, in part, to the increasing resistance of many pests to pesticides[141] and also possibly to a decreased immunity of plants to pests, which Howard suggests is a consequence of the use of chemical fertilisers.

There is a mordant irony in the fact that in India, where in the 1930s crops grown on fertile soils suffered remarkably little from pests, the emphasis on economic growth and chemical technology has led to a tenfold increase in pesticide consumption during the past two decades. Motivated by the technological imperative, India in 1990–91, consumed an estimated one million tonnes of pesticides.[142]

Howard also noted that plants grown for food on fertile soil, free of chemical fertilisers but rich in humus, confer upon the animals and humans which eat them the health benefit of increased immunity to disease.[143] Among the many examples of communities whose immune functions have been improved, Howard cites the Hunzas in Pakistan. The health, beauty, fitness and longevity of these people has provided discussion topics in numerous articles in the popular media.[144] It was not until 1964, however, that medical research began to confirm certain of the claims which figured in these popular reports.[145]

Recent findings have added further support to Howard's claims for compost-grown foods. D Pitt and co-workers at Exeter University have found that garden compost can suppress economically significant plant diseases by between 60 and 80 percent, and can sometimes eliminate them altogether. It was found, for instance, that an 80 percent reduction in the disease 'take all' in wheat, could be achieved using compost, while red core in strawberries was reduced by 90 percent. Club root in cabbages was eliminated completely. It appears that organisms in the compost suppress the growth of disease-causing bacteria and fungi.[146]

The adverse effects of the worldwide use of chemical fertilisers for the health of the world's inhabitants are still only being recognised. For instance, the high-yield rice, wheat and maize varieties of the 'Green Revolution' which depend on the liberal use of inorganic fertilisers, are usually low in minerals and vitamins. Because these new 'technology' crops have displaced local foods, the diet of many people in the developing world is now dangerously low in iron, zinc and vitamin A, and other micronutrients.[147]

The extent to which artificial fertilisers undermine the microbial fungal ecosystems of the soil is still not adequately appreciated, though research in the area has established a connection between chemical fertilisers and soil erosion.[148] In 1950, J Russel Smith, Emeritus Professor of Economic Geography at Columbia University, warned that modern crop-based agriculture, such as that used in the production of corn, cotton and tobacco, was not only destroying soil fertility but was, in addition, contributing substantially to soil erosion. He argued urgently for the need to develop a sustainable 'permanent agriculture' to reduce these problems. He made some startling comparisons. The average annual yield of corn in the Appalachian Mountains was 500 kilograms per acre. The yield of chestnut orchards in the mountains of Europe is approximately the same. The chestnuts, however, yield year in and year out for centuries. Corn 'burns out' the soil rapidly. Yet the food value and palatability of the chestnut to livestock are comparable to that of corn.[149] Hundreds of other tree crops could be employed by American farmers for producing food for livestock and humans. Unfortunately, little attention has been paid to Smith's suggestions. In the meantime, the rapid depletion of natural soil conditioners such as humus, fungi, bacteria, earthworms and insects—all of which have in recent years been displaced largely by chemical fertilisers and poisonous sprays—has, according to the technological view, necessitated additional mechanical tilling to achieve soil fertility. However, this additional technological intervention has, in turn, modified the physical structure of the soil, as well as its surface profile and has resulted in increased erosion by wind and rain.[150] Soil erosion is now a major threat to crop production around the world, with a loss of some 80 billion tonnes of topsoil every year. This is the total amount of topsoil which underlies the whole of Australia's wheatlands.[151]

According to the CSIRO report *State of the Environment Australia 1996*, which reviewed current farming methods, many current practices are not sustainable, and biodiversity-based industries such as agriculture, pastoralism, forestry, fisheries and tourism often erode the very resources upon which they depend.[152]

Farm chemicals have also become a major source of environmental pollution.[153] The application of nitrate fertilisers is a case in point. In the early 1900s, F Haber and C Bosch developed a process by which nitrogen from the atmosphere could be synthesised with hydrogen (produced by reacting steam with red hot coke) to form ammonia. Since ammonia could be easily oxidised to nitric acid for fertiliser production, the Haber operation conveniently enabled the commercial industry to tap the huge reservoir of nitrogen in the atmosphere. Clever as this sounds, the problem is that the process bypasses nature's own bacteria-moderated nitrogen-fixation processes in the soil itself, which protect the soil from nitrogen saturation.

Huge quantities of nitrogen fertilisers are now applied to the soil around the world each year to promote plant growth. In the United States in 1985, for example, 11 million tonnes of nitrogen were applied to the soil in the form of fertilisers.[154] Once again, this massive manipulation of the natural environment has not been without adverse effects. Nitrate pollution of groundwater and drinking water, for instance, is now a major problem in many countries.[155]

Excessive nitrate levels in vegetables and plants have also become a problem. Attempts have been made to develop varieties of plants with low nitrate accumulation rates in the hope of reducing the health risk of nitrogen poisoning.[156] Nitrates are converted by bacteria in the body to nitrites, which then bind to haemoglobin, impairing the capacity of blood to transport oxygen. Excessive nitrogen intake can thus result in a condition of the blood known as methaemoglobinaemia.[157]

Another possible health aspect of the application of nitrate fertilisers to our soils involves an indirect contribution to the greenhouse effect. PA Steudler and co-workers have reported that the application of nitrogen fertilisers may reduce the capacity of the soil to act as a sink or absorption table for methane gas.[158] This being so, nitrate fertilisers may actually inhibit the subtle mechanisms in the soil which assist in regulating the atmospheric levels of this naturally produced greenhouse gas.

Since highly soluble chemical fertilisers such as nitrates leach from the soil, they enter the waterways and seas producing a new and inharmonious ecological mix. In some areas, particularly in the North Sea, this new mix has encouraged the phenomenal growth of algae. When the algae die, the biomass is often so large that the bacteria breaking down the dead algae consume all the oxygen in the water, thereby causing fish and other species to suffocate.

Other research has shown that algae give off a sulfur-containing chemical, dimethyl sulfide, which oxidises in air to form sulfur dioxide, the principal cause of acid rain. It has been estimated that in summer, 25 percent of the acid rain falling over Europe now comes from huge algae masses in the North Sea.[159] This observation illustrates the difficulty in anticipating the extent of environmental disruption and degradation caused by technological intervention. Given the interconnectivity and delicate interrelationships which we now know characterise nature, the more intensive the technology, the more it is bound to impact synergistically across the network of these relationships.

Disease: The price of progress

Over the past two centuries, we have used our increasing scientific knowledge to unlock the concentrated stores of energy in the earth's crust. This energy has been tapped and distributed for conversion into power, which has, in turn, enabled us to exploit to the fullest the resources of nature. With this empowerment we have altered in subtle but radical ways, the structure of our environment. We have changed not only the constitution of our air, water, soil, food and even our electromagnetic environment, but we have also affected the existence and distribution patterns of plant, animal and microbial species on this planet.

The desire for mastery over the environment has led to the evolution of a union between knowledge and technology which seeks to maximise our expropriation of the earth's resources, while minimising the time and effort devoted to the task. Rather than seeking to dominate and control the environment, we need to seek to connect and empathise with it.

Although we shall in subsequent chapters address more thoroughly the whole range of health issues raised by technological development, there is a philosophical principle implicit in this chapter which should be made explicit. Characteristic of the human race and its patterns of 'progress' has been what might be called an attitude of 'species arrogance'. We see ourselves as separate from nature and superior to the extraordinary and vast array of creatures which it sustains. Because we see ourselves as a group apart, we have construed our survival as a species on the metaphor of a battle against nature. The security of our species, we have come to believe, is established through the domination and control of nature.

The truth of the matter is quite different. In general, we have as a species tried to control nature, either by transforming it to suit our needs (for example, the addition of chemical fertilisers to enrich soil), or by insulating ourselves from it, as in the construction of homes and offices which stand as fortresses against nature. The attitude of species arrogance notwithstanding, the problem is that we are in truth a part of nature, and thus contact with nature is as crucial to our health as is our contact with each other. Trying to make life easier, we have attempted to make nature adjust to us so that we would not have to adjust to it. The world we have therefore created for ourselves, either by way of our chemical interventions in nature or by the creation of antiseptic internal environments away from it, is, in essence, an *artificial world*. What we did not foresee is that while we may have succeeded in minimising the necessity of adjusting to the natural world, we have unwittingly maximised the necessity of having to adjust to the artificial world we have created for ourselves. The clean, crisp air which once refreshed the places in which we lived and worked has now been replaced by an atmosphere of chemical pollutants. Although we are physiologically adapted to meet most of the demands of our interface with nature, we have little provision to adapt to the chemical environments within which we are now held capture. The problem is that we are not made for the world we have made for ourselves.

To free ourselves, we need first to acknowledge the errors of our ways. The domination of nature and the economic goal of unlimited growth are two sides of the same coin. The time has come to institute a new form of currency and to reconceptualise our relationship with nature. The changes required doubtless challenge the many vested interests which depend upon pursuing the course of environmental exploitation already begun. In the chapters that follow, however, we shall see that what is at stake in the argument is not just the vested interests of a chemical company here or an electricity company there. What is at stake are the vested interests of the general public in their own personal health and in the health of their children. If our commitment to technological progress is systematically undermining the health of the community, then we need to re-think our commitment to the symbols of progress which define the world in which we presently live.

What you can do

It is of fundamental importance that we limit the wastes of our technology which are released into the environment. This can be done in a number of ways.

1 You can reduce your demand for goods that produce waste.
For example, you can refuse to use disposable cups, plates and other throw-away items. You can buy well-made items that will last and take care to maintain existing ones.

2 You can reduce personal sources of pollution such as the motor car and electricity usage.
When a litre of petrol or gasoline is burned, approximately 2.4 kilograms of oxygen is used up and about 2.1 kilograms of carbon dioxide is produced. A small tree with a foliage volume of 0.5 cubic metres produces about 0.053 kilograms of oxygen per week.[160] In comparative terms this means that 1 litre of petrol uses up the oxygen produced by about 45 small trees during a week. If you run a motor car, keep a record of the number of litres of gasoline you use in a week. Multiply that number by 45 and arrange to have that number of trees planted and maintained. Planting fruit trees would have an additional advantage of contributing to our food supply. If you run the average car, which runs 10 kilometres on one litre of petrol and you travel 10 000 kilometres per year, you will need to plant and maintain 865 trees to balance out the oxygen you are using and to reabsorb the carbon dioxide you are producing.

Economic incentives to discourage car use can also be introduced such as:
a) Having a carbon tax on all fuel used. This tax could be used to plant and maintain new areas of forest;
b) Offering a tax rebate for people who use public transport to commute to work;
c) Reducing toll charges on freeways for vehicles carrying three or more passengers;
d) Allowing only specially licenced vehicles to enter central city areas.

When you get your electricity bill, calculate the number of kilowatt hours of electricity you use each week on average. About 0.41 tonnes of coal are burned to produce and distribute one million watt hours of electricity. This requires that one tonne of oxygen also be consumed.

Thus, for every kilowatt hour of electricity you consume per week you should plant and maintain 19 small trees. For a home that consumed 100 kilowatt hours of electricity per week you should plant and maintain 1900 small trees.

By planting trees you can help reverse the greenhouse effect. As the trees grow larger they will help use up some of the carbon dioxide you generated before you were supporting sufficient trees.

3   *You can campaign to have the mindless clearing of rainforests stopped.*

A large rainforest tree with a crown span of 14 metres produces about 1.7 kilograms of oxygen per hour.[161] For every hectare of rainforest we allow to be cleared, at least 630 cars should be taken out of circulation, or we should cut back our annual electricity production by about 1000 megawatt hours.

4   *You can encourage recycling of wastes with economic incentives.*

All recyclable goods should carry a 5–10 percent deposit which is redeemable at special waste stations. This process places value on waste and offers opportunity for people to capitalise on keeping our environment clean as a form of income. Newspapers, paper and cardboard, glass, aluminium cans, car bodies and tyres are just some of the items which could be recycled far more efficiently than they presently are.

Consider, for example, that by recycling one stack of newspapers only 2 metres high, you could personally save two trees. Recycling four glass bottles could save energy equivalent to running a television for several hours. It takes about 15 megawatts of electricity to produce 1 tonne of aluminium. Therefore a 15 gram aluminium can requires 225 watt hours of electricity to produce. This amount of electricity uses up the oxygen produced by four small trees over the period of a week. It requires 28 small trees to support the habit of discarding an aluminium can each day. Recycling your can, can save the oxygen produced by these trees.

5   *We **must stop** the dumping of **all** wastes into waterways such as rivers, lakes and seas.*

Fresh water is one of the most precious of the life supporting components of our environment, and will become increasingly difficult to find anywhere on earth, if we continue with our rape of nature. The purity of our waterways must be restored and protected with the utmost urgency and priority.

No chemical or nuclear wastes should ever be allowed to reach our waterways. We must pay the cost of rendering them harmless or contain them on land. No sewage must ever reach a waterway. We must pay the cost of treating any waste water to the point that it is fit for human consumption before it is released into our environment. Such expenditure must be taken into account when all our industrial and urban developments are costed. Only in this way can the real cost of our progress be evaluated.

6 *We must limit community energy use by placing a carbon or nuclear tax on all electricity generated by these means.* The income from this tax should be used to research and build alternative energy sources such as using solar, wind and tidal energy.

It should be a fundamental human right that no company, organisation or government has the right to increase the level of pollution to which individuals are exposed without their informed consent. This means that individuals should have the basic human right to veto any development that can be shown to affect adversely their quality of life. Such developments include even the building of an airport, a freeway, a microwave transmission tower, as well as the building of any power station, industry or sewage plant which would release pollution into their environment.

While some may argue that it is impossible to halt progress, our reply is that given the current state of environmental degradation, it is imperative that the value of progress, and thus what we do in the name of progress, be defined in accord with its environmental sustainability.

It is only by placing these limits on ourselves that we will provide the incentives to research and develop alternative processes which are in harmony with nature and thereby reinvent the future of our planet in ways that make living on it worthwhile.

PART two

Electrical technology

chapter 2

The new electrical environment: The shocking truth

In London on 12 January 1882, Thomas Edison commissioned the first public electricity generation system. Later that same year commercial electric power generation commenced in New York and Appleton in the United States. The electrical age had been born. These original systems involved 110 volts of direct current (DC) distributed over a relatively short distance by underground cables.

Four years later the first commercial single phase alternating current (AC) distribution system was introduced by G Westinghouse. The following year, Westinghouse patented an improved transformer for stepping up and down AC voltages. Tesla's polyphase AC system was introduced commercially in 1893 and DC was soon replaced by AC which could be easily transformed to the high voltages required for economic distribution, on pole-supported wires.

The biological effects of electricity depend on the magnetic field, electric field and frequency effects produced in the electrical environment. Magnetic fields are produced by a magnetic dipole such as the earth, or by an electrical current passing through a conductor. The larger the electrical current, the larger the conductor or cable required, and the larger the associated magnetic field. If the current in the wire fluctuates with time, such as in AC systems, then the magnetic field also fluctuates. The number of fluctuations per second is called the frequency and is measured in cycles per second or hertz (Hz). The frequency of the magnetic fields in our homes produced by domestic wiring, or in the vicinity of powerlines, corresponds to the power generation frequency which is usually 50 or 60 hertz. A fluctuating magnetic field can induce a current in a nearby conductor, whereas a non-fluctuating or constant magnetic field cannot. This is highly significant when considering the possible health impact of magnetic fields, as different frequencies may produce different biological effects. Magnetic fields are commonly measured in milligauss (mG) or microteslas (μT). Ten milligauss equals 1 microtesla. The magnetic field produced by an electrical current has an associated electric field which is measured in volts per metre (V/m). The value of this electric field depends on the voltage applied to the conductor. Very high voltages (110–765 000 volts) are used to transmit electrical energy, because power can be carried in association with a smaller electrical current, and therefore a smaller and lighter cable may be used. Generally speaking, the higher the voltage, the greater the amount of electrical energy which can be distributed by wires.

Electricity was seen as a clean, efficient, convenient form of power. The demand for electrical energy to carry out mechanical, chemical and physical transformations of the environment, as well as provide power for technology, grew rapidly. The proliferation of electrical gadgets, heaters and lighting appliances in our homes, along with the continued expansion of electricity based technologies followed.

Paralleling the growth of electricity production has been the increase in the networks of wires and cables distributing this energy. In 1989 in the United States, there were about 560 000 kilometres of 69–765 kilovolt (kV) transmission line and about 3.2 million kilometres of 5–35 kilovolt distribution line taking electrical energy to homes and industry.[1] In our homes, these voltages are stepped down to 240 or 110 volts to run domestic appliances and lighting. Similar electricity networks in every country enable this concentrated form of power to be widely distributed over the inhabited regions of this planet.

These networks of metallic cables and their supporting structures not only visibly change our environment but also produce in their vicinity an array of invisible electric and magnetic fields which pulsate at 50 or 60 hertz, creating a new electrical environment thousands to hundreds of millions of times stronger than the corresponding pulsating fields that occur in nature.

Both electric and magnetic fields exist naturally at the earth's surface. In fair weather there is an essentially steady, or almost constant, atmospheric electric field of about 150 volts per metre which may increase up to 10 kilovolts per metre (kV/m) in thunderstorms.[2]

Pulsating electric fields with frequencies of 50 or 60 hertz do exist naturally at field strengths of about 0.1 millivolt per metre (mV/m). In contrast, the fields produced in the close vicinity of high voltage power lines are approximately 100 million times stronger, and the pulsating fields introduced into homes by wiring or appliances are about a thousand to a million times stronger than natural background.[3] Even 90 metres away from a 500 kilovolt transmission line, the pulsating electric field can be as high as 100 volts per metre,[4] which would be around one million times background.

Thus in nature, any strong electric fields are virtually constant and any rapidly fluctuating electric fields are relatively weak. Modern electrical technology produces rapidly fluctuating electric fields which are relatively strong and therefore create a new and different electrical environment from that found in nature. The same situation is also produced with respect to magnetic fields.

It is generally accepted that the planet on which we live is bathed in a relatively strong but virtually constant magnetic field. In most Western countries this field happens to have a strength ranging from 400 milligauss (mG) to 600 milligauss depending on geographical location. Field strengths as low as 250 milligauss occur off the east coast of South America, while fields up to about 700 milligauss occur near the North Magnetic Pole. During the day, the earth's field undergoes a slight variation of about 0.4 milligauss. The moon also causes a daily magnetic variation in the order of 0.03 milligauss. Research from the relatively recent field of chronobiology reveals that these weak time varying components of the earth's magnetic field interact with the brain and in particular the pineal gland, thereby contributing to the entrainment of circadian rhythms. Studies of men in underground bunkers for 3- to 8-week periods, showed different body rhythms depending on whether the bunker was shielded from natural electromagnetic fields or not. Subjects living in the shielded bunker showed a tendency to have desynchronised temperature and activity periods, with significantly longer circadian periods than those who continued to be exposed to natural fields.[5]

Very weak fluctuating magnetic fields known as magnetotelluric fields are produced naturally by lightning and by interactions between solar radiation and the earth's magnetosphere. The 60 hertz component of these fields is in the order of 0.2 microgauss (μG).[6] In comparison, the 60 hertz magnetic fields produced in homes from distribution lines and household wiring is in the range of 0.1–10 milligauss.[7] That is, the lowest levels of pulsating magnetic fields found in electrically wired homes are about 500 times background levels, and can range up to 50 000 times background levels.[8]

In the close proximity of electrical appliances, the fields can be much stronger still. Fifteen centimetres from common household electrical appliances such as can-openers, electrical shavers and hair dryers, the 50–60 hertz magnetic field can be as high as 400 milligauss, or

2 million times natural background levels.[9] Measurements of the magnetic fields found near common household appliances in a typical Australian home were measured for the authors by engineers from TransGrid the organisation responsible for maintaining the high-voltage distribution lines in New South Wales. The values are shown in Figure 2.1. Of the 12 appliances tested, nine gave readings above 2 milligauss, or more than 10 000 times natural background levels.

Figure 2.1 Magnetic fields measured near some common household appliances at normal operating distances

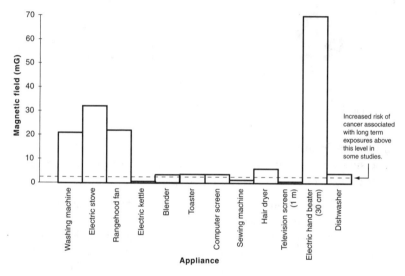

Note: Readings were taken in the author's home by Transgrid, Sydney, Australia, in 1995. JF Ashton, University of Newcastle, NSW, 1995

The magnetic fields produced by domestic appliances and household wiring are similar to the magnetic fields produced near high voltage transmission lines. RJ Caola and co-workers, after making detailed measurements of electric and magnetic fields in and around urban homes near a 500 kilovolt transmission line in New Jersey, reported the surprising finding that 'Inside homes, the magnitudes of electric fields produced by house wiring were in the same range as those produced by the transmission line'.[10] Typical magnetic field levels near transmission lines[11] are shown in Figure 2.2.

In 1989, what appears to be the first study of the levels of 60 hertz magnetic field levels on suburban footpaths and street corners where children run and play was reported. This study, by LJ Dlugosz and

Figure 2.2 Typical magnetic fields near high voltage transmission lines

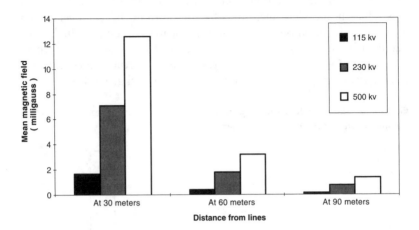

US National Institute of Environmental and Health Sciences, 1995

co-workers in urban Buffalo, New York, indicated that during child-hood substantial magnetic field exposure may occur outside the home. Their measurements revealed, for example:

> a child who spends four hours per day on the sidewalk (one hour com-muting to and from school and three hours playing) in an area with average ambient magnetic flux density of 9 milligauss, six hours in a school at 1 milligauss, and 14 hours at home at 2 milligauss will receive the majority of daily exposure to 60 hertz magnetic fields while on the sidewalk.[12]

The authors concluded:

> Research is also needed to investigate the importance of contributions to exposure by service wire configurations and appliances such as elec-tric blankets and heated waterbeds.[13]

Electromagnetic fields: The cancer connection

Evidence that the electric and magnetic fields produced in association with the distribution and utilisation of electric power may have adverse health effects began to be reported in the mid-1960s. At that time, sev-eral Russian researchers had noted vague medical symptoms in electri-cal switchyard workers. Subsequent studies in Spain, Canada and Sweden failed to detect any significant health effects.[14] Other research was, however, beginning to reveal the extent of the interactions of non-ionising electromagnetic fields with biological tissue. In the late 1970s, WR Adey and co-workers published data which showed that very weak oscillating electromagnetic fields could affect the efflux of calcium ion from the brain tissue of baby chickens.[15] Further research showed that weak fields interacted with tissue to produce chemical, physiological and behavioural changes only within what are called windows or distinct ranges in frequency and incident energy.[16]

Other studies have since confirmed this fascinating frequency and intensity window effect.[17] These findings were particularly important because they indicated that contrary to intuition, in some situations weak electromagnetic fields may produce more significant interactions than a stronger field. From 1981–83, JMR Delgado and co-workers reported the highly significant findings that weak pulsed electromagnetic fields produced inhibitory effects on chicken embryogenesis.[18] The foetal abnormalities (teratogenic effects) were produced by field intensities as low as 10 milligauss, and window effects were observed.

Studies have also shown that the pineal gland is sensitive to electromagnetic field exposure. Human and avian circadian rhythms or daily physiological cycles are lengthened in shielded environments that exclude natural and artificial electric fields.[19] Electric fields of 60 hertz have been observed to depress pineal melatonin levels in animals,[20] and pineal melatonin depression has been associated with cancer growth.[21] Other hormone effects manipulated by electric fields have also been observed [22] and experiments have shown that at certain intensities, 60 hertz electric fields may suppress T-lymphocyte cytotoxicity which is part of the body's natural defence against cancer.[23]

While these important research findings offer sufficient grounds for concern, there are also increasing numbers of epidemiological studies associating electromagnetic field exposure and increased rates of disease, particularly cancer.

In 1979, N Wertheimer and E Leeper reported the following observation after studying the incidence of childhood cancers:

> In 1976–77 we did a field study in the greater Denver area which suggested that, in fact, the homes of children who developed cancer were found unduly often near electric lines carrying high currents.[24]

They also noted:

> The finding was strongest for children who had spent their entire lives at the same address, and it appeared to be dose-related. It did not seem to be an artifact of neighborhood, street congestion, social class, or family structure.[25]

Wertheimer and Leeper noticed that homes close to transformer points where distribution current was stepped down to the 240 volt service lines were particularly over-represented among the observed cancer cases. (For a detailed discussion of Wertheimer and Leeper's research, see Chapter 6, page 82.) These findings have been supported by a Swedish study in the early 1980s. Children with cancer were found to live near high voltage (200 kilovolt) power lines, or at homes with high (50 hertz) magnetic fields measured at the front door, statistically more often than control children.[26]

In the meantime a number of subsequent epidemiological investigations had been made which substantiate these claims.[27] The consistency of the findings is highly significant—as we shall later see. Wertheimer and Leeper take up this point in the conclusion of the report of their second study which found that, like childhood cancer, adult cancer was found to be associated with high current electrical wiring configurations (HCCs) near the patient's residence. They argued:

> several aspects of our data suggest that the HCC–cancer association observed here may actually represent a causal link: (1) a dose-relationship was observed; (2) two different control procedures yielded essentially similar results; (3) the strength of the associations observed in different age, urbanicity and socio-economic groups was consistent with a causal interpretation; (4) a distinct pattern of latency between first exposure to the HCC and cancer diagnosis was seen which is consistent with a hypothesis of cancer promotion produced by AMF [alternating magnetic field] exposure.[28]

Numerous other studies have reported increased cancer risk associated with exposure to electromagnetic fields.

In 1991, MN Bates from the University of California, Berkeley, reviewed the epidemiological evidence available at that time and reported that the evidence is strongly suggestive though not yet conclusive that low frequency electromagnetic radiation is carcinogenic.[29] He found that the evidence was strongest for brain cancer and suggested that it may be more than a coincidence that between 1968 and 1987, mortality rates from cancers of the brain have approximately doubled in persons aged 65 and older in seven industrialised countries.[30]

In their 1995 booklet *Questions and Answers About Electric and Magnetic Fields Associated with the Use of Electric Power*, the National Institute of Environmental and Health Sciences together with the United States Department of Energy, cite 14 residential childhood cancer studies, and seven occupational electromagnetic field cancer studies, most of which found an increased risk of cancer associated with electromagnetic field exposure (see Figure 2.3 below). The 1992 Swedish cancer study of people living near transmission lines found that for magnetic fields above 3mG the risk of childhood leukemia was 3.8 times higher than expected. In the most recent major study cited, 138 000 electricity supply workers were surveyed by Savitz and Loomis from the University of North Carolina School of Public Health, who found that the results supported an association between occupational magnetic field exposure and brain cancer.[31]

A recent 1996 report by Oxford University researchers also links electromagnetic field exposure and brain cancer. In their study, they found that there was a significantly higher incidence of brain cancer and leukemia among workers in the electrical and electronic industry compared with the general population.[32] French researchers have also investigated the effects of strong electric fields as distinct from magnetic fields. In a 1996 study of electric utility workers, researchers found that those exposed to high electric fields for long periods of time (that is, greater than 387 volts per metre per year) had up to 3.7 times the risk of brain cancer compared with controls, and there was some indication of a dose response relation.[32a] Electric fields this strong occur up to about 50 metres from a 500 kilovolt transmission line and within about 30 metres from a 300 kilovolt transmission line.

Figure 2.3 Residential power-line child leukemia studies

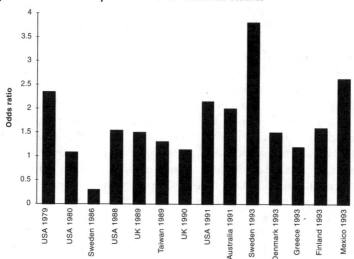

Note: An odds ratio greater than one indicates that there was an increased risk of cancer associated with high exposure to residential magnetic fields.

US National Institute of Environmental and Health Sciences, 1995

Other health effects

Many other health effects have now been linked to ELF (extra low frequency) fields including depression[33] and Alzheimer's disease.[34] (See Chapter 3, page 55, for a detailed discussion of increased risks of breast cancer associated with low frequency magnetic field exposure.)

In 1995, E Sobel and co-workers from the University of Southern California School of Medicine reported a possible link between occupational exposure to magnetic fields and Alzheimer's disease.[35] They found that people who had worked in occupations likely to result in high electromagnetic field exposure such as near electric motors, were three times more likely to have Alzheimer's disease than controls. To determine whether their findings could be replicated, they analysed a series of patients over 65 years of age from an Alzheimer's Disease Diagnosis and Treatment Center in California. In this study, patients with exposure to occupational electromagnetic fields, such as from industrial sewing machines were found to be four times more likely to have Alzheimer's disease than controls who had only low occupational exposure to electromagnetic fields.[36]

One of the interesting aspects of electromagnetic field exposure is the association with depression and suicide. Reichmanis and co-workers appear to have reported the first evidence of an association between suicide and overhead and underground high voltage power lines in 1979. They estimated the residential electric and magnetic field exposures of 598 suicide victims and found that they were significantly higher than that of controls.[37] The researchers then went back and measured the electromagnetic fields at the suicide and control addresses, confirming their earlier results.[38] One of the researchers, SF Perry together with L Pearl, went on to study the residents of high-rise apartments in England. The researchers found that significantly more people who had depressive illness than controls, had been living close to high current electric cables.[39] In another British study, people who were living close to 132 kilovolt overhead powerlines were more likely to have headaches, migraines and depression than those in the control group.[40] Other researchers reported about a twofold increase in the prevalence of depressive symptoms among residents living near powerlines.[41]

In a recent (1996) report, researchers from McGill University, Montreal and the London School of Hygiene and Tropical Medicine, studied 21 744 male electrical utility workers from Quebec, and found that there was some evidence for an association between suicide and the cumulative sum of the geometric mean magnetic field exposure value.[42]

An association between suicide and electromagnetic field exposure is made more plausible by the observations that electromagnetic fields may reduce the production of melatonin hormone by the pineal gland and disturb its circadian rhythm.[43] There is substantial evidence that disruption of pineal melatonin secretion is associated with depression [44] which, in turn, is strongly associated with suicide.[45]

The fascinating link between powerline magnetic field exposure and depression is further strengthened by the growing evidence that regular doses of certain magnetic pulses may cure depression and actually change the mood of healthy people.[46]

Policy implications

With the increasing evidence that the electromagnetic fields generated by our electricity distribution system and our electrical appliances produce harmful effects, governments around the world have been considering the situation which has enormous political, environmental and economic implications. While most Western countries, including Australia, the United States and Britain have reacted conservatively to these findings generally affirming that no conclusive evidence has been found,[47] the Swedish Government issued a public information document in May 1994 which states:

> We suspect that magnetic fields may pose certain risks to health, but we cannot be certain ... There is good reason to exercise a certain amount of caution.[48]

In July 1995, a number of pages of the United States National Council on Radiation Protection and Measurements (NCRP) draft report on the potential health effects of electromagnetic fields were leaked to the media. The report which had not been through peer review, nor received official endorsement is reported to have generally agreed upon a 2 milligauss exposure limit, especially for new day care centres, schools, playgrounds and new transmission lines near existing housing.[49] Months later in Australia, Dr R Luben, a member of the NCRP committee from the University of California at Riverside, gave evidence to a Senate Committee in Canberra, where he supported the view that we have reason to be concerned about electromagnetic field exposures. He later explained that electromagnetic fields seem to affect receptor molecules on the external surface of biological cells, and the way signals move across the cell membrane. These effects can be reproduced in laboratory experiments when cells are exposed to powerline frequencies. Fields as low as 10 milligauss produce this effect in the laboratory tests and the changes that occur are similar to the effects on cells of known carcinogens.[50]

In October 1996 however, a United States National Academy of Sciences panel released its findings that there is no conclusive evidence that electromagnetic fields are linked to cancer, reproductive and development abnormalities, or learning or behavioural problems.[51]

Conclusion

The foregoing discussion of the extent to which Western society has come to rely upon electric power makes clear that the health risks associated with exposure to certain forms of electricity and electromagnetic radiation have not yet been sufficiently appreciated. Although more detailed examination of these risks will be presented in a later chapter of this book (see Chapter 6, page 82), it is important to recognise here that the unreserved commitment of our society to electricity reflects the same lack of understanding which taints so much of technological development. The problem is that in general the more energy we extract from the natural environment (for example damming rivers to provide a source of hydroelectric power) the greater the degree of disruption to the overall system of interconnections within which any particular source of energy is just one link. It is thus salutary to remind ourselves that pollution is almost invariably the consequence of dissipated energy. The dissipated energy which results from petrochemical products is conspicuous in the form of toxic fumes. While the pollution which results from electricity is less obvious, it may in the end be just as harmful. The more massive the energy flow, the greater the disruption to the ecosystem and the greater the pollution produced.

In the case of electric power, the major form of pollution upon which we have focussed here is electromagnetic radiation. Because we cannot see it in the same way in which we can see the pollution from the smokestacks of industry, it is easier to pretend that it does not exist—but it does. For a century now we have naively exposed workers, home-makers, radio and television professionals and countless others, to the health hazards of this invisible electronic smog. We are, as a society, only now beginning to see the mammoth cost in health terms which has resulted from electromagnetic pollution.

Although this is a matter to which we shall revert in the following chapter on electromagnetic pollution, the point we are making here is that every form of energy creates its own unique form of pollution. It would be a mistake to assume, as our society traditionally has, that electricity is clean and safe. The potential health hazards associated with electromagnetic radiation can no longer be neglected.

What you can do

Many sources of high levels of electromagnetic fields such as high voltage powerlines, pole-mounted distribution line transformers, and domestic metre boxes are relatively easily located. Where possible, it may be advisable to adjust the lifestyles of our families so that we do not spend long periods of time close to these installations, where the magnetic fields may be in the order of 20–30 milligauss at ground level.

1. If our homes or schools are close to high voltage lines or transformers and especially if we have young children, it may be prudent to move to a home or school further away from these sources. If we sleep near a wall were there is a metre box or hot water heater or washing machine on the other side, we can possibly arrange the room so that we are sleeping as far away as possible from these devices.

2. If you suspect an electromagnetic field problem at home, the magnetic fields in the various rooms and yard can be measured. If the levels are above 2 milligauss in areas where you spend a considerable amount of time, you should take steps to reduce the level of the fields to the 2 milligauss level or lower.

   The 2 milligauss level has not necessarily been established as safe, however it seems to be a reasonably achievable level which is below the level believed to be associated with an increased risk of cancer. Ideally we should aim to have the magnetic fields in our homes and workplaces as low as possible.

3. Children should be discouraged from playing or spending too much time under suburban powerlines. Thick powerlines generally carry higher currents and hence produce higher magnetic fields than do thin powerlines. The magnetic fields produced by powerlines also vary during the day according to electricity usage, with the magnetic fields generally lower in the early morning and higher in the late afternoon as families begin preparing the evening meal.

4. When using electrical appliances such as stoves, washing machines, vacuum cleaners, sewing machines and power tools we should endeavour to keep our bodies as far as possible from the electric motors or heating elements, and keep their operation to a minimum.

We believe that authorities responsible for supplying electricity should be lobbied to measure magnetic field intensities under powerlines in densely populated areas and to provide the information to the public. Areas of high magnetic field intensity should be marked with warning signs, so that these areas can be avoided by the public, and in particular by children who can unknowingly play in these areas.

chapter 3

Mobile phones and UHF TV microwaves: How to minimise the risks

Natural microwave radiation reaching the earth from natural sources is in the order of 1 nanowatt per square metre, that is, one-billionth of a watt per square metre.[1] In the early 1970s in some urban areas of North America, the level of humanmade communication microwaves and high frequency radio waves in the environment was estimated to be 100–200 million times the natural radio frequency background from the sun.[2]

A decade later, a *Scientific American* review looked at the growing microwave problem resulting from earth-to-satellite television broadcast systems, long-distance telephone equipment, air traffic control systems, and radar.[3] Whether exposure to low levels of microwaves constitutes a hazard and just how strict such exposure limits should be, have remained in dispute, in part because some findings of the biological effects of microwave radiation were ambiguous.

Since that time the proliferation of UHF (ultra high frequency) TV and mobile phones has added further microwave pollution to the environment of many Western countries including Australia. What has been rarely understood is that these microwaves carry a magnetic field which is absorbed by the human body and its organs in adverse ways that are not yet fully appreciated.

What is the problem?

While much attention is currently being given to the health hazards of the magnetic fields associated with close exposure to power lines and electrical devices, there are other forms of magnetic fields radiated throughout our environment, whose adverse effects on human health are still insufficiently understood. Radiated magnetic fields from radars, microwave communication systems (including mobile phone systems), radio and TV transmissions, exhibit wavelengths in response to which the human body may act as an aerial, thus absorbing considerable quantities of radiation. The magnetic field components of these radiations have been considered as being too low to be of any health consequence. As a result, maximum levels of microwave exposure have traditionally been set for many countries around the world on the basis of the thermal heating effects of microwaves. In Australia, microwave exposure must be kept below 2 watts per square metre. This level is approximately two billion times higher than natural background microwave levels. Communication microwaves are different from the natural 'white noise' background radiation, not only because they operate at much higher power, but also because of their precise transmission frequencies, the polarisation of the waves, and the modulated digital and analog signals. These special features make them highly suspect of producing interference with biological systems.[4] In what follows, we shall argue that current levels of microwave exposure in the environment need sorely to be reconsidered and may indeed pose a serious threat to health in some locations.

The human body as an antenna

Power lines and domestic appliances produce low frequency 50 hertz (Hz) magnetic fields as a result of high electrical currents which flow within them. Low frequency 50 hertz power lines produce electromagnetic radiation of very long wavelengths, in the order of 6000 kilometres, which are believed not to be absorbed by the human body.

On the other hand, considerable evidence is accumulating to show that the microwaves used in mobile phone communication systems, together with TV transmissions and frequency modulation (FM) radio transmissions do produce electromagnetic radiation fields within the wavelength range which can be absorbed by humans. For example, a

person standing on the ground may act as a half wave antenna absorbing radio frequency magnetic fields. Maximum absorption occurs at frequencies around 70–80 megahertz in adults and from 80–150 megahertz for children depending on their height. This corresponds to the very frequency ranges used by FM radio transmitters and very high frequency (VHF) TV. Parts of the human body can also act as antennae for higher frequencies such as those used by UHF TV transmitters and mobile phones. The theory is that conductive organs or parts of the human body will act as wave antennae for electromagnetic waves of wavelengths two to five times the dimension of the body part (depending on the grounding effects of the rest of the body).

In the case of mobile phone radiation (825–845 megahertz), wavelength fields range from 355–364 millimetres. Thus, parts of the brain or organs or other conductive networks in the body whose dimensions range from one-fifth to one-half of a wavelength, that is, 70–150 millimetres, may absorb this radiation. This was confirmed in 1991 when researchers P Dimbylow and O Gandhi reported the existence of resonant peaks in the brain and eyes at the microwave frequency of 800 megahertz.[5] Given that microwaves are reflected off metal objects, it is not surprising that other researchers have found that metal framed spectacles enhanced the absorption by the eyes of the radiation emitted by a hand-held mobile phone by 9–29 percent.[6]

Melatonin production and magnetic field exposures

The finding that the brain and other organs can absorb the radiation in the spectral region used by mobile phones and UHF TV is extremely worrying. Since the brain utilises small magnetic fields as a constituent in processes associated with its own function, and given that organs within the brain such as the pineal gland are readily affected by magnetic fields, it is difficult to see how the delicate equilibrium required to preserve its harmonious balance can be maintained in the face of electromagnetic interventions. One of the roles of the pineal gland is the circadian production of melatonin, a hormone which, among other things, appears to have a powerful inhibiting effect on cancer cells.[7] There is now evidence that nocturnal melatonin production is suppressed by low frequency magnetic fields.[8] This helps to explain why melanoma has been found to be connected with high levels of exposure to low frequency magnetic fields, including may of those fields associated with telecommunications.[9] The possible effects of radio frequency or microwave fields on the pineal gland have yet to be demonstrated.

Light deprivation, melatonin and disease

Light also affects the production of melatonin. Sunlight entering the eyes passes via the optic pathway directly to the pineal gland which, in turn, regulates blood melatonin levels according to the intensity and duration of the light received.[10] The possibility that a synergy may exist between lack of sunlight exposure and exposure to high levels of magnetic fields seems likely. Two types of cancer which may provide clues to this synergy are breast cancer and melanoma.

A number of studies have reported an increased risk of breast cancer,[11] including male breast cancer,[12] associated with low frequency magnetic field exposure. The suppression of pineal melatonin due to these magnetic field exposures has been mooted as a possible mechanism.[13] Furthermore, the incidence of breast cancer also appears to vary inversely with sunlight exposure, so those populations with the highest risk of breast cancer display the lowest rates of natural light exposure. If air pollution is taken into account, the relationship between the amount of actual sunlight received at ground level and breast cancer mortality rates are even more strongly negatively correlated.[14]

The hypothesis entertained here is that sunlight exposure may be an important and confounding factor in the electromagnetic field exposure–cancer conundrum, with minimum exposure to sunlight actually exacerbating the effects of magnetic field exposure. Evidence is accumulating to suggest that telecommunication electromagnetic fields, and particularly those in the microwave region, have the potential to give rise to increased cancers in humans, partly because these frequencies are absorbed by the body but also because levels of magnetic field radiation are constantly increasing in the environment.

In 1977, JAG Holt[15] suggested that the doubling of the melanoma rate in Queensland, Australia, related to the introduction of television transmissions and the establishment of other high powered high frequency communication facilities. Holt points out that malignant tissue has a higher dielectric constant and better conductivity than normal tissue and therefore absorbs more power from electromagnetic radiation. Studies involving cancer patients have found that when cancers such as melanoma, breast cancer and lung cancer, are exposed to 434 megahertz VHF radio waves the cancers grow very much more quickly. For example, when melanoma was exposed to three doses of 10 milliwatt per square centimetre VHF electromagnetic fields, the cancer doubled in size in just a few days, whereas before exposure to VHF radiation the melanomas took around 70–120 days to double in size.[16] Data for several other cancers is shown in Figure 3.1. Further evidence comes from the survival rates of patients with chronic myeloid leukemia (CML) in Western Australia. Between 1950–59, 50 percent of patients survived 55 months. In 1960–61, three high powered television broadcast stations were commissioned in Western Australia, together with an international airport with high powered radar and radio facilities. Between 1963–67, 50 percent of CML patients survived only 21 months. The radiotherapist involved was unchanged between 1950 and 1967.[17]

Figure 3.1 Effect of microwaves on cancer growth

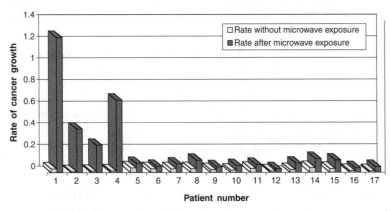

Patient details
1. Malignant melanoma
2. Malignant melanoma
3. Malignant melanoma
4. Malignant melanoma
5. Squamous carcinoma
6. Squamous carcinoma
7. Squamous carcinoma
8. Breast metastases
9. Breast metastases

10. Breast metastases
11. Breast metastases
12. Chondrosarcoma rib
13. Osteogenic sarcoma
14. Secondary adenocarcinoma: lung
15. Secondary adenocarcinoma: lung
16. Secondary adenocarcinoma: abdominal incision
17. Secondary adenocarcinoma: abdominal incision

JAG Holt, *Medical Hypotheses*, vol. 5, 1979, pp. 109-143

Broadcast towers and human health

Higher rates of cancer around broadcast towers have been reported for some years. In 1983, Dr W Morton of the Oregon Health Sciences University in Portland, found higher rates of leukemia and breast cancer near broadcast towers in Portland.[18] A few years later, Dr B Anderson and A Henderson of the Hawaii Department of Health, reported significantly higher leukemia rates in suburbs of Honolulu with broadcast towers as compared with areas without towers.[19] Clusters of leukemia around United States Navy communications installations have also been reported, one in Lualualei, Hawaii, and one in Thurson, Scotland.[20] In 1988, S Milham reported a significant mortality rate due to acute myeloid leukemia and other cancers among amateur radio operators.[21]

Recently, three other studies have found an association between exposure to radio frequency (RF) or microwave (MW) radiation. In 1996, Dr S Szmiglielski of the Center for Radiobiology and Radiation Safety in Warsaw, Poland, found that military personnel exposed to RF and MW radiation had higher rates of leukemia and lymphoma. For young soldiers the incidence of these cancers was more than eight times higher than expected and was highly significant.[22]

In the early 1990s a cluster of leukemia and lymphoma cases was reported around a four million watt TV and FM radio tower in Sutton Coldfield near Birmingham, England.[23] British Broadcasting Corporation (BBC) measurements showed that the maximum radiation at 2.5 metres above ground was 1.3 microwatts per square centimetre for TV and 5.7 microwatts per square centimetre for FM radio, and declined with distance from the transmitter. Dr H Dolk and colleagues at the London School of Hygiene and Tropical Medicine looked at leukemia rates in concentric circles around the tower and found that within half a kilometre there were nine times the expected number of leukemia cases. In the area within the next half kilometre the leukemia rate was double, with the rate declining steadily to background levels about eight kilometres from the towers.[24]

Dr Dolk and colleagues went on to study cancer rates around another 20 towers in order to confirm their findings. Within two kilometres of a tower at Crystal Place in South London, with a similar power output to the Sutton Coldfield tower, the researchers found 62 cases of adult leukemia, while a total of only 17 cases of leukemia were found within the same distance from the remaining 19 towers. However, most of these latter towers were in sparsely populated areas or did not have FM transmitters.[25]

Recently, B Hocking [26] has analysed the cancer rates for leukemia in an area within the four kilometre radius of a cluster of television and FM radio transmitters in suburban Sydney, Australia. The maximum RF/MW power level near the ground for the towers was calculated to be 8 microwatts per square centimetre and declined to 0.2 microwatts per square centimetre at a distance of 4 kilometres. A control group living 8–12 kilometres from the towers was used. Within the four kilometre zone, rates for childhood leukemia were 60 percent higher than in the control group. The mortality rate was 130 percent higher. Leukemia in adults was 12 percent higher in the region close to the towers.

UHF transmitters and the microwave environment

Television transmission towers typically transmit VHF as well as UHF stations. The latter channels operate at microwave frequencies up to 800 megahertz and can radiate up to one million watts per channel of microwave magnetic fields into the environment. One kilometre away from such a transmitter the theoretical magnetic field component of the microwave would be in the order of 0.2 milligauss. The magnetic fields in the brain are in the order of 200–1600 picogauss (pG).[27] This means that residents living close to powerful TV transmitters may be exposed to microwave magnetic fields more than 100 000 times the intensity of the magnetic field of the brain, and at levels comparable with the daily fluctuations in the earth's magnetic field due to the sun. Furthermore, these humanmade magnetic levels may interfere with the brain's reception of the solar magnetic variations of the earth's field which, as we referred to earlier, are believed to contribute to the entrainment of human circadian rhythms.[28]

In areas of very good TV reception the electric field is typically 80 decibel (dB) microvolts per metre, corresponding to a magnetic field of 0.0003 milligauss. This field is approximately 200 times greater than the normal brain magnetic field levels and readings of this magnitude may be found up to 65 kilometres from a medium-sized 300 kilowatt microwave TV transmitter. Within half a kilometre of such a transmitter the measured microwave magnetic fields near the ground may be in the order of 0.02 milligauss or 10 000 times the magnetic fields occurring in the brain. These levels will be even higher when several different channels are broadcast from the same tower or if other transmitters are nearby. UHF TV signals are radiated constantly 24 hours per day every day. Thus, not only do we work and play, but we also sleep in an environment increasingly polluted by microwave radiation, some of which is now known to be absorbed by the brain.

Mobile phone base stations and the threat to health

In August 1995, the Australian Radiation Laboratory measured the levels of microwave radiation in a kindergarten at Harbord in northern Sydney.[29] The kindergarten boundary was only 17 metres from a mobile phone base station transmitter. This transmitter was mounted approximately 10 metres above the ground on the side of a nearby building. The station provided ten 5-watt channels. With all ten channels operating, the highest microwave exposure in the play area was 0.87 microwatts per square centimetre. This value is considerably below the Australian recommended maximum exposure limit for the general public of 200 microwatts per square centimetre, and on the basis of this safety recommendation, mobile phone base stations are regarded by many health officials as posing no microwave radiation hazard to nearby communities. The 1992 American National Standards Institute (ANSI) standard allows exposures of approximately 200–600 microwatts per square centimetre at radio and TV broadcast frequencies for the general public.[30] An Australian Commonwealth Scientific and Industrial Research Organisation (CSIRO) report raises concerns over the suitability of the ANSI safety standard. The report argues that the ANSI standard is based only on well-established thermal effects and ignores the more subtle non-thermal processes that are more difficult to interpret and apply to human health.[31]

In the light of a growing body of conflicting evidence we submit that these original standards are badly in need of reassessment. For example, because the Australian Standard is based on the thermal or heating effects of microwaves, it does not take into adequate consideration the potentially harmful effects of magnetic fields. It is interesting that the phenomenon of 'microwave hearing' (that is, where certain people can hear 'clicks' in their heads when exposed to microwaves) has been reported for microwave exposure levels as low as 1 microwatt per square centimetre.[32] This is just above the level measured in the school playground mentioned above. It is accepted by many researchers that these 'clicks' are probably due to the thermoelastic expansion of brain tissue caused by the rapid but minute change in temperature produced by the microwave radiation. While these 'clicks' probably correspond to energy peaks greater than one microwatt per square centimetre, it is clear that if the brain is absorbing microwave energy for heating, it is also absorbing the associated magnetic field. In the case of the playground referred to, the magnetic field from the mobile phone base station was 0.06 milligauss. This field strength is approximately one-fiftieth of the level of 3 milligauss of low frequency magnetic fields associated with an increased risk of cancer.[33]

Microwave transmissions operating at the power limit of 200 microwatts per square centimetre would have a magnetic field component of 0.92 milligauss, which is only one-third of the level associated with increased risk of cancer.

What is not yet clearly understood is whether microwave frequency magnetic fields are stronger or weaker promoters of cancer than are low frequency fields. If microwaves are stronger promoters, which appears to be indicated by both Holt's observations and Hocking's research, then even a 0.06 milligauss exposure may be associated with a significant risk of disease. Another approach to the problem is to consider the levels of mobile phone tower magnetic fields compared with the magnetic fields which occur in the brain as part of brain function. The mobile phone base tower was beaming a microwave magnetic field into the playground and into the brains of the children playing there which was in the order of 40 000 times stronger than the levels of the magnetic fields which occur in the brain. Given the relatively high penetrating power of microwave radiation, and also that organs within the brain such as the pineal gland are especially sensitive to fluctuations in magnetic fields, exposure to artificial radiation at levels 40 000 times natural levels must surely be regarded as potentially disruptive to the established harmony of the magnetic field fluctuations associated with normal brain and hormonal functions. If the exposure persists over a long period of time, a plausible hypothesis is that such disruptions may be sufficient to increase the risk of disease.

There have been an increasing number of instances where mobile phone base stations have been erected in close proximity to kindergartens, baby health centres, day care centres and schools where children generally spend long hours exposed to higher levels of microwave radiation. Pregnant mothers dropping off or waiting to pick up children would also be exposed, though to a lesser degree. It must be remembered that the field levels considered here only reflect the magnetic field contributions from mobile phone frequencies. The children would also be exposed to magnetic fields from local television transmitters, radio stations and other microwave communication systems. The comprehensive exposure levels have yet to be determined, but the cumulative totals are likely to be staggering.

There is evidence that communication microwaves such as those used in radar and telecommunication systems can damage vegetation. In the Federal Republic of Germany 14 areas of forest damage from Federal Postal Authority transmitters and 20 instances of tree damage from military transmitters have been reported in one region alone. Trees have died *en mass* in straight lines running between transmitters.[34] Whether vegetation is more susceptible to microwave damage than humans are, is not yet fully determined, though the connection may be closer than some think. Whatever the outcome of this debate, it is clear that the evidence available constitutes strong grounds for concern regarding the level of microwave radiation damage to our environment and thus ultimately to humans.

The mobile phone connection: Should you be plugged in?

Research on mobile phones has given rise to questions concerning the safety of hand-held mobile phones which are, in effect, hand-held microwave radiation transmitters. The electric fields 5 centimetres from the antennae of a hand-held mobile phone (an average distance from the side of the head during use) can be as high as 60 volts per metre,[35] corresponding to a magnetic field of 2 milligauss or more than one million times magnetic field levels found in the brain (see Figure 3.2). Field levels of this magnitude sit precariously on the lower limit of the low frequency (LF) magnetic field exposure currently recognised as constituting an increased risk of cancer. If microwave frequency electromagnetic fields are stronger promotors of cancer than low frequency fields, then mobile phone users may be at a higher risk of developing cancer than non-users. In May 1997, strong evidence in support of this hypothesis was published in the international journal *Radiation Research*.[36] In the study, transgenic mice which had been bred to be susceptible to developing lymphoma, were exposed to microwave radiation similar to that emitted by a mobile phone when in use. The mice were exposed to this radiation for two 30-minute sessions each day for 18 months. Meanwhile, another group of the transgenic mice were used as controls and not exposed to the mobile phone radiation. At the

Figure 3.2 Microwave magnetic fields emitted from the earpiece region of a mobile phone

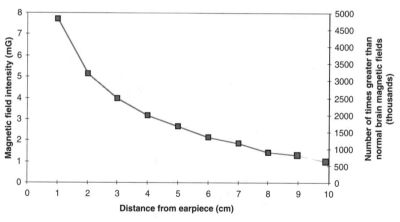

I Macfarlane, Australian Telecommunications Research, vol. 28, no. 2, 1994, p. 67

end of the 18-month period, 22 percent of the latter group had developed cancer. However, 43 percent of the mice exposed to mobile phone radiation developed cancer. These results, which add further support to JAG Holt's earlier findings (see Figure 3.1), are highly significant and indicate that small (mice-sized) biological systems absorb mobile phone-type radiation, and are affected by it. Medical practitioners in Britain, Scandinavia and Australia have reported that some regular mobile phone users are complaining of an increased incidence of headaches when using these devices.[37] A plausible hypothesis is that mobile-phone headaches represent nature's early warning of harmful magnetic field interactions with the brain.

PW French and co-workers at the Centre for Immunology at St Vincent's Hospital, Sydney, have found that when mast cells in vitro are subject to 835 megahertz radiation at power intensities similar to that of analog mobile phones, they reacted by releasing histamine. One interpretation of this phenomenon is that the microwave radiation may contribute to allergies. Indeed, the hypothesis that microwave communication magnetic fields may, in part, be the cause of the high incidence of allergies (in the order of 30 percent of the population) in Germany has been mooted.[38] Bag- and car-type mobile phones operate at five times the power of hand-held mobile phones and the associated magnetic fields produced by the antennae will be approximately twice as strong.

Given that microwaves are absorbed by the brain and that the safe levels of exposure to microwave magnetic field frequencies can be interpreted as controversial, it would seem prudent to keep the use of mobile phones to an absolute minimum.

Conclusion

In this chapter we have argued that there is now substantial evidence that the human body and its organs absorb radiofrequency electromagnetic fields. Those fields produced by UHF TV stations and mobile phones are absorbed by the brain. The levels of microwave frequency magnetic fields in the environment to which we now are exposed, range from hundreds of times natural brain levels up to one million times the natural brain magnetic field levels when using a hand-held mobile phone. Communication microwave magnetic fields inundate our environment and are on the increase. At least some initial research findings suggest that these radio waves are likely to be a causative or promoting agent for allergies, brain cancer, melanoma, breast cancer and leukemia. Unfortunately, humans cannot generally detect the presence of communication microwaves in the environment. Without testing, we have no way of knowing whether the free space in which we live and work contains high or low levels of microwave radiation, which are likely to be inimical to health.

Figure 3.3 Percentage of populations using mobile phones in1996

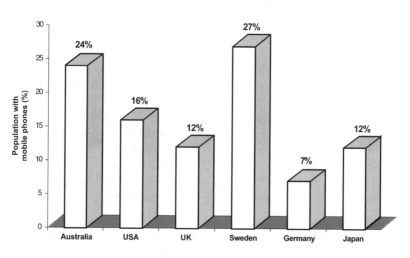

J Cadzow, *Good Weekend, Sydney Morning Herald*, 23 November 1996

In recent years, the pollution of our natural electromagnetic environment has increased dramatically with the introduction of UHF television providers and mobile phone services. In 1996 there were an estimated 4.3 million mobile phones and between 250 000 and 400 000 digital-type mobile phones in the hands of Australians[39] (see also Figure 3.3).

If the evidence cited here can be relied upon, it is clear that our commitment to this relatively new form of technological interchange will have a profoundly adverse effect on human health. To regard the electromagnetic spectrum of our environment simply or primarily as a resource to be exploited for profit is, we submit, an egregious error. The natural distribution of electromagnetic fields in our environment is part of the delicate ecological fabric of our planet necessary for the support of life. It must be protected from pollution just as the air and water must be protected, and the time is well and truly at hand for us all to recognise our responsibility in this regard.

What you can do

Mobile phone base stations may be located close to schools, day care centres, hospitals and residential areas. Find out where these microwave transmitters are in your area and if any are located close to where children learn and play.

1. Contact your local environment action group and try to have the towers relocated away from these areas.
2. Avoid using hand-held mobile phones, except in an emergency and when using the phone tilt the aerial away from you head as far as possible.
3. If you must use a mobile phone regularly choose a type which enables you to have the aerial as far away from you body as possible, such as on the outside rear mudguard of your car.
4. Find out where the nearest UHF television transmitters are located and check that your family home, schools and work places are at least 5 or 6 kilometres or more from these towers.

Reducing the microwave exposure in our environment will require a community effort. Share with others what you have learned in this chapter and encourage the reduction of the use of microwaves in your community.

chapter 4

Microwave ovens and your health: You have to know the problems to overcome them

The development of microwave ovens came as an expected spin-off from the rapidly developing post-war microwave technology, and the first commercial ovens were produced around 1962. These ovens use a device called a magnetron tube which causes an electron beam to oscillate at very high frequencies and thereby produce microwave radiation. A typical domestic unit generates a microwave frequency of 2.45 gigahertz (GHz) at an energy output of between 400–900 watts for a typical domestic unit. The oven power supply is designed to deliver 4000 volt negative pulses to the magnetron, which makes it the most dangerous power supply in any item of domestic equipment. The frequency chosen corresponds to the absorption peak for water, and thereby enables foods containing water to be heated quickly and efficiently. The microwaves are beamed from the magnetron into the oven compartment holding the food where they are contained. Microwave ovens are not permitted to leak microwaves at a power density level of more than 5 milliwatts per square centimetre at a distance of 5 centimetres from the outside of the oven. Thus small leaks below this level may occur and substantially contribute to the electronic smog in the home or office. Medical hazards reported to be associated with microwave ovens include burns, cataract formation, neurologic injury and pacemaker dysfunction.[1]

Despite the publication of these reports, microwave oven sales have continued to soar, and the majority of Australian and American homes now have these appliances. In a recent telephone survey, 1000 Americans picked at random were asked to rate the importance of certain common household technologies to their everyday lives on a scale of one to five. Microwave ovens clearly topped the poll with 63 percent of respondents giving them a full five out of five. The list of technologies covered included television, telephone answering machines, video cassette recorders and computers.[2]

The popularity of the microwave oven parallels the increasing percentage of women in the workforce and decreasing family size in many Western countries. In the words of M Doyle, president of The Consumer Network in the United States, 'There's a new kind of eater, buyer and user evolving, and there's a revolution against kitchen work and any kind of hassle anywhere. Time is so precious that "quality time" is afforded with our families not in front of the warm hearth, but around the microwave.'[3]

This presupposes, of course, the truth of the manufacturers' claims that it is safe to allow the family to gather around the microwave. We are reminded by P Brodeur, 'Not so long ago, however, it was considered perfectly safe to allow children to study the bones of their feet in the viewing windows of fluoroscope machines in shoe stores all over the country'.[4]

The microwave oven epitomises the convenience food syndrome, and hand-in-hand with increased microwave oven numbers goes already prepared prepacked frozen meals. As microwave ovens become the ideal cooking utensil of the new emancipated households, one of the last bastions of creative adult work to which children are exposed, the preparation of family meals, is being demolished. In a major review of the factors contributing to the breakdown of the family unit and the alienation between young people and adults, U Bronfenbrenner writes:

> One of the most significant effects of age segregation in our society has been the isolation of children from the world of work. Once children not only saw what their parents did for a living but also shared substantially in the task; now many children have only a vague notion of the parent's job and have had little or no opportunity to observe the parent (or for that matter any other adult) fully engaged in his or her work.[5]

Microwaves, food and health

The demand for prepacked, ready-to-heat meals places a greater emphasis on food sterilisation. In August 1989, British Government research showed that potentially fatal bacteria such as listeria can survive in microwave cooking even if instructions are followed.[6] It is obvious that food irradiation will be seen by some as a solution in part to this situation, however, this technological fix will merely add one more dimension to the health aspects of microwaves.

Microwave ovens are often used to reheat left-over food and this practice is potentially quite dangerous. This danger was made apparent to researchers who studied an outbreak of Salmonella food poisoning following a community picnic in Juneau, Alaska, in 1992. The illness was associated with eating roast pork which had been flown in from a restaurant in Seattle, Washington. Of 30 persons who took home left -over meat, all ten who used a microwave oven to reheat the pork became ill. None of the ten persons who used a conventional oven or skillet to reheat the meat became ill. Researchers who conducted a case-control study of the outbreak concluded that compared with conventional methods of reheating food, microwave ovens had no protective effect in preventing illness.[7]

Whether or not the microwave heating of food produces changes in the food which are inimical to health is a controversial issue. For example, HV Hertel and BH Blanc of the Swiss Federal Institute of Technology are reported to have demonstrated in human studies that microwaved food produced adverse changes in human blood.[8] However, D Jonker and HP Til from the TNO Nutrition and Food Research Institute in Zeist, The Netherlands, studied the effects of microwave cooked food on the blood chemistry and other health indicators of rats, and could find no adverse effects.[9]

It has been reported that the microwave heating of infant formulae can produce molecular changes in the amino acids in milk proteins causing toxicity or affecting the nutritional value of the milk formula, however the quantity of proteins changed is very small.[10] Conventional heating or pasteurisation of milk has the potential to produce greater changes in milk proteins due to the formation of hot spots.

Microwave heating human milk, such as for the thawing of frozen human milk for more rapid accessibility, causes a decrease in the level of anti-infective factors in the milk, even when low temperatures in the range 20–53 degrees Celsius are used. Microwaving at temperatures greater than 72 degrees Celsius caused a marked decrease in all the tested anti-infective factors. Researchers from the department of Pediatrics at Standford University School of Medicine, who carried out this study questioned the safety of microwaving human milk in hospitals, even at low temperatures.[11]

Microwaves, packaging and health

Another problem with microwave cooked pastry-type foods is that they are typically low in flavour and colour when compared to foods baked in a conventional oven.[12] This has encouraged technologists to develop microwavable food additives to produce artificially the flavours and colours required by consumers.

An example of one new type of flavour-producing technology designed for use in microwave ovens is susceptors. These devices are usually glued to the packaging of microwavable foods and are used to achieve local areas of high temperature. This has the effect of browning the food during microwave cooking. A subtle side-effect of some of the pre-1992 susceptor devices involved the release of small amounts of a toxic chemical bisphenol A diglycidyl ether (BADGE) into the food during microwaving. BADGE was a component of the cold cure adhesive used to fix susceptors to packaging. In one study undertaken in 1992, of 54 samples of pizzas purchased, nine samples of susceptors used in one brand contained BADGE at concentrations between 0.2–0.3 percent. This toxic chemical was found to migrate into the pizza when they were cooked in their packaging according to the instructions on the pack.[13] The use of BADGE was phased out after 1992. Since susceptors enable quite high temperatures to be generated during microwave cooking, researchers have discovered that a large number of chemicals are released from these devices which are typically packed with foods such as waffles, pizzas and French-fries intended for microwave use. In one study, 44 different volatile chemicals were identified including toxic compounds such as 1,1,1-trichloroethane, 2-(2-butoxyethoxy) ethanol and the carcinogen benzene.[14] Another toxic chemical which was found unexpectedly to migrate into food from the packaging during microwave cooking is benzophenone. This chemical is a component of the printing ink on the paper board.[15]

Dry goods such as bread and breakfast cereals are often sold in waxed bags which can be used in microwave ovens to produce 'hot buns' etc. One recent study showed that use of these waxed bags in a microwave oven as recommended on the packaging gave rise to wax contamination of the food at levels up to 0.16 percent which was equivalent to 60 percent of the wax in the packaging being transferred to the food.[16]

The PVC plastic films used to cover food during microwave cooking also have been found to release plasticisers into the food, and a 1996 study of the problem recommended that PVC should not be used in direct contact with food during microwave cooking.[17]

The bottom line of our developing microwave oven, fast food, consumer culture is reflected by the words of B Messenger, International Editor for *Food Business* magazine:

> A very creative world food industry has emerged in the 1980s. It has found new ways to preserve food, to cook it, and even to consume it. It has a high-tech underbelly nurtured by an aggressive scientific research community, which is impacting on every level of the food spectrum.[18]

One question remains: Do we continue blindly to follow the path of microwave technology? Once again our attempts to synthesise the environment have led almost imperceptibly to progressively technological ways to prepare our foods. Microwave waves today—will it be a laser beam tomorrow? As novel and wonderful as these new technological interventions may seem, they should be introduced with far greater reflection in respect of their impact upon health than has hitherto been given. Perhaps the question we should be asking is not whether such devices advance efficiency, but whether they advance health. Until this question is both asked and answered, technology is bound to serve a master other than the master of healthy invention.

What you can do

The most recently targeted market segment for microwave-ready cuisine is young children, with some microwave cooks reported to be only five or six years old. While the children's microwave meal market in the United States already constitutes in excess of $100 million in annual sales, the health cost to the population may be far greater. Exposing young, still developing children to low level microwaves from potentially leaky or misused ovens is a business too risky to be in.

Similarly, the health risks for young children associated with the long-term eating of microwaved food are unknown and may prove to be yet another disaster in the name of economic growth and blinded commercialism. Given the unknown variables in this matter, we believe it prudent for parents to discourage growing children from using microwaves and to minimise the quantity of microwaved food eaten.

There are some other things you can do to minimise the health risk to you and your children:

1. If you have a microwave oven and intend to go on using it, have it checked immediately and regularly for leaks. Microwave oven doors are especially prone to leakage.
2. Do not open the door while the oven is on.
3. Try to avoid microwave cooking of frozen foods and commercially prepared dinners whenever possible, especially if the meals are to be cooked in their packaging. There is a lower risk when microwaving fresh foods such as vegetables.
4. Where possible use non-PVC cooking containers.

chapter 5

Computer VDUs and television: Watching for the health hazards

During the latter half of this century, dramatic changes in the environment of the home and the workplace have resulted from the introduction of a vast array of technological options designed to make our lives easier and our work more efficient. We have, in a sense, become intoxicated with scientific technology and, having succumbed to its lure, we have for the most part failed to question seriously its adverse impact upon our health and the ecological integrity of this planet. Recent investigations, however, are revealing that our commitment to the gods of technology has led almost inescapably to a sacrifice of public health. Household appliances, once thought to be innocuous, are being exposed increasingly as possible menaces to the health and well-being of those who use them.

Of the range of household appliances upon which we have come to rely, television perhaps epitomises the pervasive influence of modern electronic technology on home life. In the workplace and school the badges of technological progress are now also exhibited proudly in the form of computer visual display units (VDUs). With little appreciation of the health risks involved, a rapidly growing segment of the population is spending long periods of time sitting in proximity of and looking into what is essentially a cathode ray tube (CRT).

The number of people exposed to CRT radiation is staggering. Over one billion people around the world sit for hours each day in front of a TV set. In many homes younger viewers often sit almost mesmerised very close to the screen, gazing for long periods at the ever changing artificial light patterns which portray both fantasy and real life. In the workplace the presence of the VDU screen has proliferated at an exponential rate. In 1986, there were some 15 million VDUs in use in the United States, and this swelled to about 70 million in 1990; it has been estimated to be possibly 100 million by the year 2000.[1] This growth is accompanied by a steadily increasing percentage of workforce time spent viewing these screens, as more and more office and factory functions and records are computerised. Likewise our children, even in the classroom, are spending longer periods of class time gazing at VDUs while using a range of learning packages and games. As the personal computer has become a feature in many homes, very young children are spending hours amusing themselves with computer games when not watching television. This time is spent indoors and usually under conditions of artificial lighting. In this chapter we propose to show that the progressive reliance upon TV/video entertainment, coupled with VDUs in the home and at work, is radically transforming the relationship of humans to nature in some ways which are inimical to health. By steadily increasing our exposure to human-made environmental electromagnetic radiation in the home and workplace, we have—as a species—significantly reduced our exposure to natural electromagnetic radiation, that is, sunlight, so essential to human well-being. The health and sociological issues involved in this disruption of the established patterns of interaction with nature have received far less attention than they deserve. In what follows, we shall, in the space available, be concerned to redress this imbalance. As we tease out the strands of evidence that have accumulated in the scientific and medical literature, a tapestry of conceptual connections emerge which reveal the extent to which community health depends upon maintaining the delicate equilibrium between nature and our technological transformations of it. In the words of Professor Fritz Hollwich:

> ... to define the crucial issue of today, such an important environmental factor as light must once again be correlated with the needs of man as a creature of nature. Man, in his daily routine, requires the influence of natural environmental factors to a certain degree. Among these are daylight, with its visible and invisible electromagnetic radiation, as well as the unfiltered pure air of the atmosphere and the changing temperature of the external world. Since places of work to a large extent exclude contact with nature, dining and recreation rooms on the premises should be designed to offer a certain compensation. Finally, our homes should also be planned with this purpose in mind.[2]

Development of the cathode ray tube

A century ago there was intense interest in the phenomenon produced when very high voltages were applied across two electrodes in a glass tube. Depending on the pressure and nature of the gas in the tube, a series of effects could be produced ranging from reddish streamers of light passing down the tube, to a yellowish-green fluorescence on the glass opposite the cathode or negative electrode. This greenish glow in the glass was produced by rays emanating from the cathode and were subsequently named cathode rays. They were later found to be beams of high energy electrons.

While experimenting with one of these tubes in 1895, Roentgen observed fluorescence on a nearby zinc sulfide screen. The radiation which produced this fluorescence was extremely penetrating and was found to result from the reaction of cathode rays with the metal anode producing another type of radiation which Roentgen named X-rays. These same phenomena are essential features of the modern cathode ray tube which is the module responsible for the visual images of our television sets and computer visual display units.

Prior to the 1970s, TV screens produced detectable levels of X-rays. These are a form of electromagnetic radiation with extremely high frequencies and short wavelengths which are capable of forming ions by breaking molecules or removing electrons. X-rays are therefore classified as ionising radiation. X-rays are produced within the TV or VDU when the electrons strike the front of the screen. Since the 1970s, manufacturers have improved the X-ray absorbing properties of the glass envelope to the point that from the late 1970s onwards, no detectable levels of X-rays have been found in front of TV or VDU screens by monitoring organisations.[3]

Impact of low frequency electromagnetic fields

Televisions and VDUs produce a range of pulsing electric and magnetic fields including extra low frequency (ELF) 50–60 hertz fields and radio frequency (RF) fields. Australian Radiation Laboratory measurements revealed that over 95 percent of RF emissions occurred in the frequency range 14–250 kilohertz.[4] The ELF or powerline frequency fields are similar to many other electrical appliances. For example, colour TV sets can produce magnetic fields in the range of 7–20 milligauss 30 centimetres from the screen, while at the same distance from a personal computer with a colour monitor magnetic fields are typically in the range 2–6 milligauss.[5] Consistent with our earlier discussion in Chapter 2, pages 47–48, these magnetic field strengths are close to or above those which have been associated with brain cancer and leukemia, that is, 3 milligauss. The United States has set no standards for magnetic fields from video display units (VDUs). The Swedish government has, however, issued guidelines recommending that VDUs purchased by the government produce powerline frequency magnetic fields no stronger than 2.5 milligauss at a distance of 50 centimetres from the screen. This Swedish standard has become a de facto standard for the VDU industry worldwide.[6]

High levels of radio frequency (RF) electromagnetic fields also emanate from TVs and VDUs. These fields are generated by the vertical and horizontal output transformers and the deflection yoke coils.[7] These pulsing electric and magnetic fields may have serious health implications. Many of these higher frequency sources induce more intense currents in the human body than are induced by most power (low) frequency sources. For example, the 20 kilohertz electric field at one metre from a television set, induces body currents that are comparable in magnitude to those induced by a power frequency electric field of 3000 volts per metre.[8] An electric field this high would typically be found 20 metres from a 500 000 volt high voltage transmission line.[9]

Given that several studies have reported statistically significant increased cancer risks, including leukemia, especially for children living close to high voltage transmission lines,[10] a similar risk could exist for people sitting within 1 metre of a TV set or VDU.

It is a common practice in some homes for young children to sit or lie on the floor close to the television set while viewing and this practice may be a contributing factor in the increasing incidence of childhood leukemia. P Landrigan, the chairman of the Department of Community Medicine at Mt Sinai School of Medicine in New York City recently pointed out that the incidence rate of new cases of childhood leukemia has increased steadily in the United States over the last two decades. For acute lymphoblastic leukemia, the most common form of leukemia among children, the cumulative increase in the incidence rate from 1973–1991 was 20 percent.[11]

The magnetic field component for the radio frequency electric fields produced by TVs and VDUs may also have serious health implications. CR Roy and G Elliott from the Australian Radiation Laboratory measured the electromagnetic radiation emissions from 99 different colour and monochrome VDUs, including the RF magnetic fields. The authors report that RF field levels for all the VDUs were well below the International Radiation Protection Association's (IRPA) then recommended upper limit of 3 milligauss, and concluded that the emissions of VDUs do not pose a health hazard to operators.[12]

However, approximately 50 percent of the VDUs tested emitted RF magnetic fields stronger than 0.25 milligauss at a distance of 30 centimetres from the screen corresponding to exposure levels of 15 microwatts per square centimetre, and greater. These values are of the order of 2–75 times the range of calculated RF field exposure levels reported by Hocking (see Chapter 3, page 57) in his study which found a significant association between proximity to TV transmitters and increased risk of leukemia.[13] Given this finding, working with a VDU or personal computer must be considered a considerably greater potential risk for cancer than living close to TV transmitters. The Swedish government voluntary guidelines to restrict emissions from VDUs set a maximum limit for the radio frequency magnetic fields at 0.25 milligauss. This level was set not on the basis of health or other biological research but rather on what was the lowest level technically attainable in order to reduce overall magnetic field exposure in offices.[14]

However, we have already shown even at this level, the radio frequency magnetic fields are much stronger than those found in close proximity to major TV transmitter towers, and of the order of 100 000 times the magnetic fields which occur naturally in the brain[15] (see Chapter 3, page 58).

Higher frequencies in the environment

Today's computer microprocessors work at speeds of 100 million cycles per second (100 megahertz) or more, generating electromagnetic fields at this frequency and at higher VHF harmonies such as 200 or 300 megahertz. Faster microprocessor chips will generate even higher frequencies. These frequencies are absorbed, not only by the human body, but by other radio receivers.

One day in April in 1994, a commuter aircraft took off from Washington DC and the crew set course for Newark, New Jersey. A little while later, air-traffic controllers radioed the pilot that the plane was heading in the wrong direction and was 16 kilometres off course. The plane's instruments said its position was correct even though controllers on the ground knew that it was not.

Navigational instruments on board work out the plane's position using radio signals from beacons on the ground. Unknown to the crew, one of the passengers was using a laptop computer that emitted stray radiation at a frequency that happened to coincide with broadcast frequencies from a beacon. This interference affected the plane's instruments. When a flight attendant noticed the computer and asked the passenger to switch it off, the planes navigational instruments returned to normal.[16]

The International Federation of Airline Pilots reports that there are about 20 electromagnetic interference (EMI) incidents each year, involving computers, mobile phones and portable CD players. The United States Food and Drug Administration has also noted more than 100 EMI incidents where heart monitors and defibrillators have stopped working in hospitals.[17] These examples serve to demonstrate how our proliferation of computers and similar devices are contributing to increasing levels of electronic smog in our environment.

Computer VDU use and miscarriage: Is there a connection?

Since 1979, in a dozen or more office locations, small clusters of miscarriages and birth defects among VDU users have been reported in the media.[18] Because the levels of x-radiation around VDUs had previously been found to be very low, several authorities dismissed the increased incidence of these abnormalities as chance occurrences, while others argued that many of the reported defects could be hereditary. Others refused to recognise the health threat presented by VDUs, urging that the birth defects were non-specific and that some could not, in any case, be expected to be due to radiation.[19]

However in 1982, Delgado and others reported powerful inhibitory effects on chicken embryogenesis produced by low frequency 10, 100 and 1000 hertz electromagnetic fields of 1.2, 12 and 120 milligauss. Some of these fields are sufficiently similar in frequency and intensity to those emitted by VDUs to be of comparative interest.[20] The following year Ubeda and others also observed teratogenic changes to chicken embryos exposed to low intensity pulsed electromagnetic fields of 100 hertz.[21] Significantly, teratogenic effects were observed with very weak magnetic field strengths of about 4 milligauss, with stronger and weaker fields less effective, thus indicating a window effect or a particular magnetic field strength region where changes were most pronounced.

These studies make clear that the potential serious health issues surrounding the use of human-made pulsing magnetic fields in the home and workplace can no longer be dismissed. Although the frequencies studied were slightly different from the domestic frequency of 50 or 60 hertz, the magnetic field strength is typical of that displayed by many household and workplace appliances, including microwave ovens, vacuum cleaners, copy machines, washing machines, electric drills, fluorescent lights, television sets, and VDUs.[22] In the case of the cathode ray tube, a number of additional magnetic fields exist, all of which contribute to a complex low frequency magnetic field environment around these devices.

With approximately half the workforce using VDUs being women of childbearing age, the health implications are enormous. Not surprisingly, these issues were discussed somewhat anxiously at international meetings convened in London in 1984, Geneva in 1985, and in both London and Sweden in 1986. Despite there being sufficient evidence and concern to generate these conferences, it is intriguing that the general consensus from the meetings was that neither ionising nor electromagnetic radiation from VDUs is likely to affect health.[23] Since the conclusions deriving from these meetings became public, there has been a sense of nervous caution from reviewers and researchers in reporting and reinterpreting their findings. McDonald and co-workers, who studied births in the Montreal area in 1984, reported in their preliminary findings, for example, that the rate of spontaneous abortion in 2609 current pregnancies with no VDU use was 5.7 percent, compared with 8.3 percent for 588 pregnancies with a weekly exposure of less than 15 hours, and 9.4 percent for 710 pregnancies with VDU use greater than 15 hours per week. We later find the same authors cautioning that selection and response bias could have affected the results and that further analysis was necessary.[24]

In 1988, Goldhaber and co-workers pointed to the consistent evidence of various studies linking reproductive hazards with VDU work, stating conservatively that the studies provide some basis to suspect that an excess exposure risk could be real. They subsequently found in a case control study of pregnancy outcome:

> a significantly elevated risk of miscarriage for working women who reported using VDUs for more than 20 hours per week during the first trimester of pregnancy compared with other working women who reported not using VDUs.[25]

This increased risk could not be explained by age, education, occupation, smoking, alcohol consumption or other maternal characteristics.

Since Goldhaber's significant finding, other epidemiological studies have been reported which suggest that exposure to electromagnetic fields may be linked to adverse pregnancy outcomes.[26] These findings lend strong support to the VDU–miscarriage connection. In the meantime, as VDU studies continue, some researchers find significantly elevated risks for women working with VDUs[27] while others do not.[28] Given variable research findings, it is certain that the claims for the *safe use* of VDUs are far from settled. This being so, one rational response would be to monitor their growing use as it impacts on health and also, where possible, to use them in ways which reflect the growing body of evidence which suggests that they may not be as safe as we have been led to believe.

Other health aspects of TVs and VDUs

Pulsed electric fields

So far, we have discussed the potential health risks of the electromagnetic fields associated with TVs and VDUs. The comprehensive health picture, however, does not stop here. For example, few people are aware that VDUs can produce a sound or hum which may also impact adversely on health. This mechanical noise is produced by the pulsed electromagnetic fields of the flyback transformer, a device which provides the picture scanning voltage. Depending on the scanning frequency of the TV or VDU, the emissions are either sonic in the range

15–20 kilohertz, or ultrasonic in the range 20–32 kilohertz. One characteristic of these frequencies, especially in the latter range, is that they are inaudible to most adults and hence unsuspected. Measurements of the sound levels at the operator position carried out on 25 models of VDUs gave levels up to 68 decibels.[29]

On the Bel scale of sound intensity, this value is just below the level reported to cause temporary hearing loss, nausea, fatigue, headache and tinnitus to adults in industry.[30] It is thus significant that headache and irritability in VDU operators have been observed.[31] Accurate sound measurements in this particular case revealed a sound energy peak at 16 kilohertz.

The health implications of these noise levels for pregnant VDU operators are at present unknown. However, it is not unlikely that the accumulation of hundreds of hours of exposure to unnatural repetitive noise, much of which is likely to be audible to the foetus, may well serve to contribute to the environmental stress of the developing embryo.

Static electric fields

In addition to *pulsed* electric fields there are also extremely high *static* electric fields produced near cathode ray tubes. These fields are often noticeable when switching on or off a television set, and commonly detected when the hairs on one's hand and arm rise as a result of static charge. One study of 44 VDU models, reported electrostatic field strengths of up to 30 000 volts per metre at a distance of 30 centimetres from the screen.[32] These strong fields are capable of producing effects on the skin, and there a number of documented cases in which rashes have been reported. In each of these cases, the condition of skin irritation was found to subside when the employees were removed from VDU work.[33]

VDU screen dermatitis is part of a newly recognised category of patient problems now surfacing in the literature of dermatology.[34] Symptoms include skin itch, smarting, pain, heat sensation and skin redness, after exposure to VDUs. O Johansson, Head of the Experimental Dermatology Unit at the Karolinska Institute in Stockholm has found major effects in the dermis and epidermis of the skin which suggest that profound biological effects may be present in patients suffering from this type of electromagnetic field exposure.[35]

The powerful electrostatic field on the cathode ray tube screen also causes an imbalance of ions in the air with an excess of positively charged air molecules present in the environment in front of the TV or VDU. Children who often sit close to TV, along with VDU operators,

are continually exposed to these ions. Feelings of malaise and attendant symptoms of lethargy have been associated with extended exposure to high positive ion levels.[36] Once again it is apparent that the
immediate impact of TV and VDUs is to create an electromagnetic
environment which represents a radical departure from that found in
nature. Our view is that such departures may themselves be inimical to
health. Evidence has accumulated to show that large electrostatic fields
exist in the atmosphere, with the ground level field in fair weather
measuring about 130 volts per metre.[37] Not only is this electrostatic
field much smaller than that found near VDUs but the polarity is also
reversed, with the upper atmosphere being positive and the ground
negative. Furthermore, under normal circumstances when the positive
charge near the earth increases, thunderstorms often follow with lightning providing an enormous flow of negative ions back to earth, thus
relieving the build-up of the positive charge. There is considerable evidence that negatively ionised air, for instance, is beneficial to health.[38]

It is apparent that nature provides mechanisms to reduce the build-up
of positive ions in the atmosphere near the earth's surface, and it would
now seem that such natural mechanisms afford significant health benefits.

Dangers to the eye from TVs and VDUs

The purpose of the cathode ray tube is to produce an image to be
viewed by the eye and thereby to provide information from an electronic device capable of transmission to the brain. In order to do this,
the screen itself produces the light radiation which enters the eye.
Reading a book involves light from a source such as sunlight or
fluorescent light, being reflected off the book and entering our eyes.
Similarly, the light entering our eyes as we look at a landscape is
reflected sunlight. We cannot see scenery on a dark night or read a
book in the dark, but we can watch TV or a VDU in the dark. Thus
we can *in the dark* see rocks, flowers and trees on TV and *in the
dark* read a 'page' of type on a VDU screen. The important difference
is that in these situations what we are viewing is a completely
synthetic environment of optic input which is flashing at 25 hertz
(cycles per second).

The stress produced by this unnatural optic stimulus is illustrated
by the occurrence of photosensitive epilepsy, otherwise known as 'television epilepsy'. Television viewing may serve to provoke fits in some
patients with known epilepsy and in others who do not apparently suffer fits except when viewing.[39] While viewing television, epilepsy
patients were found to exhibit spike-wave discharges in their elec-
troencephalograms *even when the set was functioning normally.*[40]

The pineal gland and melatonin production

Of far greater significance, however, is the effect of the unnatural light spectrum from the cathode ray tube on the functions of the pineal gland. This gland influences the hormonal system of the body, in effect controlling much of our entire body chemistry.

That the eye was the channel for light's stimulatory effect on the pineal gland was first demonstrated by F Hollwich at the University of Munster in 1948. Over 30 years later, after decades of research into these effects, Professor Hollwich wrote in the preface of his book, *The Influence of Ocular Light Perception on Metabolism in Man and in Animals:*

> It is the purpose of this book to demonstrate to architects as well as teachers, physicians, and lighting experts that the source of the light rays entering the eye is of great importance. Artificial light may be an optic substitute but it is by no means equivalent to natural light in physiological terms. Too much artificial fluorescent light interferes with the natural development of the child and subjects the nervous system of the adult to inordinate stress. The health of our organism is dependent to a large degree upon the environmental factor of light, that is, upon the entry of natural light into the eye ...[41]

Professor Hollwich goes on to demonstrate the importance of natural sunlight entering the eyes on the health of the human organism. The light produced by the cathode ray tube, whether it be in a TV or VDU, is a type of fluorescent light with a spectrum vastly different from the spectrum of natural sunlight which manifests in the region below 400 millimetres wavelength.[42] We are as a culture increasingly exposing ourselves to a spectrum of artificial light inimical to health, while decreasing our exposure to natural sources of light which can enhance health.

The pineal gland is an active neuroendocrine transducer, converting light into an endocrine signal. Consequently, the pineal gland is implicated in a large number of biological rhythms, including the production of the hormone, melatonin. The melatonin synthesis is controlled by light which reaches the pineal via the superior cervical ganglia.[43]

Melatonin is produced especially during the night. This leads to elevated blood melatonin levels at night. The exposure of humans or animals to visible light at night rapidly depresses pineal melatonin production and blood melatonin levels. Likewise, the exposure of animals to various pulsed static and extremely low frequency magnetic fields also reduces melatonin levels.[44] Melatonin is a potent oncostatic agent and it appears to prevent both the initiation and promotion of cancer.[45] Reduction of melatonin, at night, by any means including exposure to visible light or electromagnetic fields has the potential to increase the vulnerability of cells to alteration by carcinogenic agents. The practice of watching TV late at night or using a personal computer for long hours at night is likely to provide both photic and electromagnetic suppression of pineal melatonin and so possibly increase an individual's susceptibility to cancer.

Adequate melatonin levels are associated with feelings of comfort and well-being; assist the human organism in accommodating stress; help initiate sleep; and are clearly associated with the healthy function of the reproductive system.[46] It is more revealing than it might at first appear that several researchers who found a correlation between VDU exposure and miscarriages and birth defects, but who were reluctant to ascribe electromagnetic fields as a significant cause, suggest stress as the most likely causal factor.[47] It seems that long hours of viewing TV or VDU emissions may not only constitute a stressful situation for the human organism, but may actually stimulate physiological changes in the endocrine system which function to make the body less capable of handling stress.

Since 1980, research has shown that photic stimulation of the retina is responsible for the entrainment of physiological circadian rhythms, regulation of various neuroendocrine parameters and control of metabolism.

When there is insufficient exposure to sunlight during the day, after a time, physiological functions are affected and carbohydrate craving and depression can result.[48] The levels of light emitted by a VDU or TV screen are of the order of 22 times less than level of outdoor light on a cloudy day.[49] Consequently, with long periods of viewing, the body is deprived of what appears to be the essential health-giving photic stimulation otherwise provided by natural sunlight, to reduce the stress placed upon the relevant physiological and neurological mechanisms involved in the viewing process. It is revealing that a 1996 United States study of over 4000 young adults, showed that heavy television viewers (over four hours per day) had significantly higher scores for blood pressure and depression than light viewers (less than one hour per day).[50]

When all is said and done, there is a strange irony in the fact that our success at creating a photoically artificial environment which has come increasingly to dominate our work and play (TV and video entertainment), is really a measure of our failure to relate to the environment and to nature in ways conducive to our well-being. The more we indulge ourselves in our TVs and VDUs, the more we deprive ourselves of the health benefits deriving from natural photic radiation, some of which include lowered blood cholesterol and blood sugar, enhanced immune function, and even lowered blood pressure.[51]

The evidence is persuasive that nature has in many ways provided an environment far more conducive to health than our technological distortions of it. It seems our substitution of the natural by the artificial products of modern technology is leading to an array of health and environmental problems beyond all expectation.

What you can do

The amount of time we spend looking into the cathode ray tube of our VDUs and TVs should be kept to a minimum. Using the TV or VDU as a baby-sitter could have adverse health consequences on the development and growth of children, both physically and mentally.

1. Children and adults should be encouraged to sit as far as practicable from the TV set and VDU to minimise exposure to magnetic and electric fields.
2. When not in actual use, these devices in the home and office should be switched off.
3. We recommend that expectant mothers keep their distance from both TVs and VDUs, and spend as little time as possible watching them.
4. People who must, by way of their work or profession, spend large numbers of hours looking at VDUs or TV monitors, should as often as possible spend time out-of-doors, enjoying scenery and viewing distant objects. These eye exercises are important to balance the eyestrain of long hours of close focusing.

chapter 6

Electricity in the home: Should you be plugged in?

When you settle down for your long winter's nap, you may be getting far more electrical stimulation than you ever dreamt. Our willingness to go to bed with electromagnetic technology in one or other of its household guises is now being blamed for miscarriages, possible birth defects, slower foetal development and even some forms of cancer. Let us try to determine in this chapter the extent of the problem.

Early warning signals

Smitten by the marvels of electrical power, society has been slow either to notice, or to investigate, its harmful effects on the human body. One of the first published epidemiological studies on electrical power frequency fields was carried out by Dr Nancy Wertheimer and Ed Leeper of the University of Colorado.[1] Wertheimer was concerned in this initial study to determine whether a connection existed between exposure to electrical currents generated by high voltage power lines and the increased risk of childhood leukemia. Upon reflection it is perhaps not surprising that emanations from power lines with voltages as high as 230 000 volts (230 kilovolts) might be regarded as hazardous to health, even at some distance from the lines themselves.[2] The capacity of power lines to generate significant electrical fields at quite some distance from the line-wire source is amply illustrated in a well-known photograph of a young boy beneath a high voltage power line, holding a pair of fluorescent lamp tubes which were lit simply because he was standing in the electrical field generated by the line.

Initial research indicated that there was indeed a vague correlation between the location of high voltage power lines and the proximity of the homes of childhood leukemia victims, but considerable research still had to be undertaken until a more definitive pattern was to emerge.

Although the potential effect of the high voltage power lines on human health would continue to prove worrying, Wertheimer quickly became aware that the high voltages emanating from power generating plants and transported through major power lines are soon stepped down to 13 000 volts (13 kilovolts) by large transformers set up at strategic points at a prescribed distance from the high voltage lines. Mapping the locations of the substations and the birth addresses of childhood leukemia victims, the pieces of the puzzle began to come together. In order to make electricity in the 13 kilovolt lines accessible to customers, pole-mounted transformers are used to reduce the primary wire voltage of 13 kilovolts down to the 240 and 120 volt levels required by electrical appliances in the home. The association between the location of the transformers and the vicinity of the birth homes of childhood leukemia victims proved to be statistically interesting, but the case was not as clear-cut as it may have seemed.

Although there appeared to be a correlation between the increased incidence of leukemia among young victims and the proximity of their homes to the transformers, a puzzling aspect of the distribution pattern of the relevant leukemia rates had also emerged, which was inconsistent with this finding. In addition to the fact that the incidence of leukemia was significantly higher for those children living in houses closest to the pole-mounted transformers, a significant percentage of young leukemia victims had lived in *not* the first house away from the transformer, but the *second* house away. What was equally puzzling was that the leukemia rate fell sharply for children in the *third* house away from the transformer and was negligible in respect of the overall population for all the remaining houses on the line.[3]

Having consulted a physicist friend, Ed Leeper, about these unusual findings, it became clear to Wertheimer that there was a factor missing in her original equation about the direct and apparently straightforward relationship between the increased rate of leukemia in children and the distance of the transformers from the childhood homes of leukemia victims. The correlation could certainly not be explained in terms of either the magnetic field generated by the transformer, or by the alternating electric field generated by the wires as a result of the 60 hertz alternating current which passed through them on its path from the transformers to households. The reason for this is simply that the magnetic field given off by the transformer should in principle drop off so sharply that it would be negligible *even at the house closest to it*. Nor could the electric field generated by the alternating current in the lines be responsible for the *varying rates* of leukemia from one house to another, since the voltage upon which these fields would depend does not change in strength according to the wire distance from the transformers.[4] There would be no difference, in other words, in the force of the electric fields as manifest from one house to the other.

The key needed to unlock this final door in the investigation was found when Wertheimer enlisted Leeper's help for the production of a gaussmeter, a device used to measure magnetic field strengths. Testing the device in a neighbourhood whose pole and wire configuration was typical of the distribution network of the electricity system of the city, she was surprised by what she discovered. Beginning at the base of the transformer pole at an alley entrance, the gaussmeter gave off a loud hum, indicating the presence of a strong magnetic field. As she walked up the alley past the first house, however, the hum of the gaussmeter did not subside and strangely continued until she reached the next pole, located at the far corner of the lot of the second house, where the hum from the gaussmeter suddenly ceased. Wertheimer noticed that this was the point at which several wires known as 'service drops' linked up with, and reduced the current load coming from the secondary distribution line fed by the transformer two houses away. She noticed also, that the point at which the first span secondary distribution line finished, coincided with the pronounced decrease in the childhood leukemia rate. On the basis of these initial observations, Wertheimer postulated that the association she was searching for was *not* the proximity of the neighbourhood *transformer* to the homes of childhood leukemia victims, but rather the proximity of the *secondary distribution line* which ran from the transformer past the first two houses on the line to the service drop wires.[5] Since the secondary wire was carrying sufficient current to feed the dozen or so service lines directly supplying the local houses, it was producing the strong magnetic field which seemed to be implicated in the increased rate of childhood leukemia.

Wertheimer followed up her preliminary findings with an extensive investigation of the correlation between the field strengths of first-span secondary wires and the birth addresses of childhood leukemia victims. These follow-up studies revealed that the rate of leukemia for children living in dwellings where first-span secondary wires ran past them was disproportionately higher than for children living in homes away from these wires.[6] Encouraged by these findings, Wertheimer decided to expand her research to determine whether the incidence of other forms of childhood cancer could be associated with exposure to high current wiring of any kind, not just first-span secondary wires.

Several categories of high current exposures were identified and included in the study. Homes situated less than 50 feet (about 15 metres) from first-span secondaries, those within 65 feet (20 metres) of a group of three to five small gauge primary wires, and homes located within 130 feet (40 metres) of either three-phase, large gauge primary wires or of a group of six or more small gauge primary wires, were all regarded as high current risk homes.[7] The results of her research were startling and alarming. As Brodeur puts it:

> During 1976 Wertheimer visited the birth and diagnoses addresses of each of the cancer cases, the birth addresses of each of the controls, and the addresses at which control children had been living at the time their

matched cases had been diagnosed with cancer. She then proceeded to draw a diagram describing the location, size, type, and proximity of the electrical wires and transformers she had observed in the vicinity of each of these homes. Once that was done, she analysed the data and found that her prediction had held up: children who had lived in homes near high-current electrical wires had died of cancer at twice the rate seen in children living in dwellings near low current wiring. The association was strongest among those children who had spent their entire lives in a high-current home. Particularly disturbing was the fact that of six children in the study population who had lived near high current wires coming directly from power substations, all were cancer victims.[8]

Although it had previously been assumed that any harmful effects of electromagnetic fields would ordinarily be cancelled in household wiring, by virtue of the fact that return current tends to balance the supply current, Wertheimer and Leeper argued that such equilibrium is rarely preserved.[9] The problem is that some of the current which should return through the wires, instead tends to flow through the ground. Inasmuch as most household electrical systems are grounded through the plumbing, the return current passes through the pipes of the house to produce a new magnetic field within the home itself. This suggestion was especially disturbing, since the duration of continuous exposure to a magnetic field appears to be equally, if not more damaging to human health than the strength of the field in itself. This being so, exposure to a weak but constant magnetic field could be far more deleterious to human health than was at first supposed. Additional support for their hypothesis came to light in a United States Public Health Service Report which correlated the cause of death for men between the ages of 20 and 65 with their occupations.[10] According to Wertheimer and Leeper's analysis of this data, it was clear that the cancer rate for workers having fairly regular exposures to alternating current magnetic fields was significantly higher than for the overall population. Among some of the workers referred to were telephone and power linesmen, subway and elevated railway motormen, power station operators, electricians, and even welders.

When Wertheimer and Leeper's research was first published in 1979, their results were resoundingly rejected by the medical and scientific community, which criticised the work as shoddy and poorly evidenced. The electric utilities industry quickly joined the ranks of orthodoxy to condemn the findings as heresy and the researchers as heretics.[11] Dismissed out of hand, it was not until 1986, when Savitz and his colleagues were one of the first research groups to announce that they had accumulated sufficient data to replicate and confirm Wertheimer and Leeper's conclusion, that it was recognised that prolonged exposure to low level magnetic fields generated by high-current wires significantly increased the risk of developing cancer in children.[12] According to Savitz's longitudinal study, the risk of developing any of the various types of childhood cancer is increased by more than five times the control population for those children living in homes in close proximity to high-current wires. It is important to note, by the way, that the Savitz study did not include any of the same cancer cases used in Wertheimer and Leeper's study and thus provides a genuinely independent measure of the magnitude of the problem.

Electric razors and electric hair dryers

Once it was established that prolonged exposure to low level electromagnetic radiation could increase the risk of cancer, Wertheimer turned her attention to household appliances in respect of which prolonged exposure is generally a characteristic of their use. The risk factors of relevance here need to be carefully distinguished, and the distinction should not be reduced simply to the difference between the health hazards associated with extremely low levels of electromagnetic radiation and those associated with extremely high levels. In addition to the question of individual hypersensitivity, the concept of dose rate must be included in any risk–benefit ratio regarding the use of household appliances and electrical equipment.[13]

The electric razor is a case in point. Electric current which is used for household appliances is supplied at a frequency of 50 or 60 hertz (see page 39). In basic terms, this means that the current provided for our homes is an alternating current which flows first in one direction and then in the other, generating an electric field. When it does this at a frequency of 60 hertz, it is moving back and forth 60 times per second or generating a 60 hertz field.

In the case of the electric razor the electromagnetic fields produced have been measured as 60 hertz fields with magnetic strengths as high as 200–400 milligauss, about 2 centimetres away from the cutting edge of the blade. Since the blade is often in direct contact with the surface of the skin during the process of shaving, it is clear that the nearby tissues are being exposed to a powerful magnetic field. Since it has been suggested that 60 hertz fields of as little as 3 milligauss are associated with an increased risk of cancer (see pages 46–47), the exposure level from appliances such as electric razors, electric hair-dryers, curling irons and so on, needs to be carefully monitored.[14]

It is of course true that the electric razor is normally used for only a few minutes every day, which raises the dose rate as a factor in assessing the adverse impact on health of magnetic fields. It is generally assumed that if the dose rate is low (exposure for only a short duration), the use of appliances which give off relatively high magnetic fields is safe. We have argued elsewhere that this assumption may not be as uncontentious as some researchers make it seem, but we do accept that the dose rate is an important factor in any analysis of these matters. Let it be said, however, that neither of the authors makes use of an electric razor. The dose rate for electric razors and other such high field strength appliances also varies considerably, so that a general statement about their safe use could be misleading.

The dose rate for hair-dryers becomes equally variable. Many people use an electric dryer to dry their hair at least once per day. The exposure time is generally longer than the dose rate for the electric shaver, and the high strength magnetic field generated is directed most often towards the skull and thus the brain. This brings into play another factor, namely, the extent to which certain organs or parts of the body are susceptible to toxic intervention (for example, the accumulation of mercury or aluminium in the brain) or electromagnetic intervention (for example, the special sensitivity of the reproductive organs to magnetic fields). This issue has special import for those people working in occupations where certain vulnerable parts of the body are more directly exposed to magnetic field sources than others (such as women working at computer display terminals). In regard to the health risks of hair-dryers, the much neglected example is the hairdressing profession, in which the hair-dryer is used frequently throughout the day. Exposure of hairdressers to the relatively high electromagnetic fields generated by the professional hair-dryers used in salons is not only regular, but the hair-dryer is also frequently held in a position reasonably close to the breast, neck or head of the hairdresser. As far as the authors are aware, no study has yet been done of the potential health risks of electromagnetic radiation within this profession, though the questions which arise are of great interest.

The electric blanket and heated waterbed connection

Although the field strengths of electric razors are, as we saw, relatively high, the daily exposure or dose is generally minimal. While this daily dependency upon an electrical device with strong field strengths is worrying, the concern pales in comparison with the potential health risks associated with the use of electric blankets and heated waterbeds.

The field strengths of electric blankets are considerably lower than those associated with electric razors, ranging from 50–100 milligauss or about one-half to one-quarter the field strengths exhibited by razors.[15] The difference in the two cases is that the electric blanket is used for many hours at a time and is maintained as close to the total surface of the body as possible. This means that the accumulated exposure or total administered dose of electromagnetic radiation is considerably higher than the dose level for the electric razor. The fact that the exposure is comprehensive across the entire surface of the body also adds a new dimension to the question of dose impact, the health implications of which need to be explored further.

Although the electric blanket generates a greater field strength than the heated waterbed, whose heating element is located on the underside of the bed, the heated waterbed produces a magnetic field reading of approximately 5 milligauss, the dose to which anyone lying in the bed would be exposed. Since the children involved in Wertheimer's earlier study on childhood leukemia were exposed to magnetic field strengths of about 3 milligauss, the potential health risks associated with heated waterbeds cannot go unattended. In respect of the dose–rate calculation, it is important to appreciate that while electric blankets are ordinarily used only during the coldest months of the year, heated waterbeds tend to be heated year-round.

Why electric blankets generate magnetic fields

The parallel wires in the middle of the electric blanket form an 'S' pattern through which the electric current flows. Because of the configuration of lines, the current flow through the 'S' pattern is balanced by the current flowing in the opposite direction, thus tending to cancel out or prevent the generation of magnetic fields. The heating element within the waterbed is similarly designed to minimise the generation of electric fields. The problem is that at the outer edges of both the electric blanket and the heated waterbed, the current becomes unbalanced. When this happens, the relatively significant electromagnetic fields characteristic of each appliance are generated.[16] Whether electric blankets and heated waterbeds could ever be constructed to avoid completely any current imbalance which might lead to the generation of a magnetic field strength of significance is a moot engineering point. Suffice it to say that, to date, we know of no brand of electric blanket or heated waterbed which can claim its product to be free of relatively strong electromagnetic fields.

The health hazards of electric blankets and heated waterbeds

Because of the problem of having to select a separate control population to establish whether those who use electric blankets or heated waterbeds have a higher incidence of cancer than those who do not, the research in the area has been designed instead to determine the effects of these electrical appliances on foetal development. This study is much simpler since it is possible to compare the effects of magnetic fields on the same population. For example, by comparing the rate of miscarriage during the winter versus the summer months for a single control population, it is possible to assess whether the use of electric blankets and heated waterbeds during the winter increases the rate of miscarriage during that period. Because the study deals with the same group of people, it is possible to eliminate the confounding influence of other factors such as a couple's dietary, smoking or drinking habits, thus isolating the role in miscarriage played by electromagnetic fields.[17]

Once again, it was the pioneering work of Wertheimer and Leeper in the United States which has served to advance the frontiers of knowledge in this area and encourage other researchers to continue their initial investigations.

Utilising a rigorously controlled experiment, Wertheimer and Leeper compared the rate of miscarriages among users of electric blankets and heated waterbeds as they occurred during the summer and winter months of the northern hemisphere. They found that the incidence of miscarriage was disproportionately higher during the months in which these appliances were used than when they were not.[18] They discovered also, that the rate of miscarriage was higher between September and the end of January. The greatest risk of miscarriage was thus observed to occur during the first months after conception and to coincide with the coldest period in the United States, when the temperature setting of the appliances was likely to be set higher, thereby increasing the strength of the magnetic field generated.[19] No such seasonal pattern of miscarriage was reported among non-users. The results of their study also demonstrated a trend towards slower foetal development.

In their first study on this matter published in *Bioelectromagnetics* in 1986, Wertheimer and Leeper admitted that the effects they had documented could also be interpreted as the effects of excessive heat exposure. In other words, the increased rate of miscarriage, slower foetal development, and perhaps even some forms of childhood disability, might be due to the heat radiation generated by the appliances in question, rather than their electromagnetic fields. Excessive heat, for example, is known to have a deleterious effect on both sperm and ova.

In order to distinguish the different effects which each of these two factors might be playing in the increased risk of miscarriage, Wertheimer and Leeper undertook a new study in which the effect of electromagnetic radiation could be separated from the effects of heat radiation.[20] To do this, they now focussed their investigations on ceiling cable heating units, popular in many homes built during the 1960s and 1970s in the Eugene, Oregon area of the United States. They conceived of the ceiling cable units as a kind of big electric blanket in the ceiling. As in the case of the electric blanket, the currents flowing along the edges of the ceiling cable pattern were also unbalanced, thus generating quite strong magnetic fields.[21] The difference was that while the human body would be exposed to the magnetic fields produced by the ceiling cable, it would remain relatively unaffected by the heat radiation. The room would be warmed, in other words, without the body temperature being raised excessively.

With the new experimental situation well defined, Wertheimer now compared the seasonal rate of miscarriage for families living in homes heated by ceiling cable. As in the electric blanket study, the rate of greatest foetal loss was observed in the coldest months, when the magnetic field strength generated by the cable was at its peak.[22] This variation in the seasonal pattern of spontaneous abortion was *not* manifest among families living in homes which relied on non-magnetic field heating sources. On the basis of these results, Wertheimer and Leeper have reaffirmed their original hypothesis with confidence. Enough has certainly been said to show that it would be imprudent to ignore their conclusion. Given that exposure to the electromagnetic fields generated by electric blankets and heated waterbeds can significantly increase the incidence of miscarriage, these household appliances need to be used with caution, if at all. As the cost of home heating rises, the temptation is to make more, rather than less, use of electric heating devices designed to warm our beds. If the result of succumbing to this temptation is that the level of exposure to electromagnetic fields is increased, the temptation is best left resisted.

What you can do

It is clear that exposure to electromagnetic radiation in all of its forms needs to be monitored carefully. Some appliances, such as the electric razor, generate considerable magnetic field strengths and more research needs to be done to establish the acceptable dose exposure, if any. The rule of thumb which we propose is, *where there is a possible health risk and the return is not great, seek a minimal risk alternative.*

1. *In the case of the electric razor,* the safety razor is a relatively safe bet. If you feel you have to opt for the electric razor, try a battery-operated razor.

2. *The electric blanket* case deserves our special attention. In light of current research on the magnetic field effects of electric blankets, the threat of increased risk of miscarriage should be a sufficient deterrent in respect of their use, at least for pregnant women. Woollen blankets and duvets provide a safe alternative, but if you feel you must use an electric blanket, there are ways in which you can minimise the risks. One helpful strategy is simply to heat the bed for an hour or so before retiring. Just before you are ready to go to bed, shut off the blanket and enjoy the warmth. Be careful not to switch the blanket off at the temperature regulator, or to assume that all is well simply because you have switched off the wall switch. A number of electric blankets are capable of generating a magnetic field if they are left plugged in at the wall socket. Unplug the electric blanket from the wall before actually getting into bed.[23] Similar precautions should be taken when you use a heated waterbed. Once you are in the waterbed, your own body heat, coupled with the heat from the bed water, will keep you surprisingly cosy and comfortable.

    Choose an electric blanket of the correct size so that wired areas of the blanket are never tucked in under the mattress. Do not use an electric blanket with a waterbed. If you use a waterproof sheet, cover the electric blanket with an ordinary blanket to absorb moisture below the waterproof sheet. When the blanket is switched on, never pile blankets or clothing on it. If you are not using the electric blanket, it is best to store it in an unfolded, hanging position. Never use pins to secure something to the blanket. Do not dry-clean electric blankets; if laundering is necessary, it is imperative to follow the instructions of the manufacturer. Keep the electrical leads and controls free of the bed at all times. Check the blanket and the controls on a regular basis to ensure that neither has become faulty.

Electric blankets should be replaced every two to five years, depending upon use.

3. We believe that just as electrical inspections of home and workplace wiring is necessary before connection to supply, the measurement of magnetic fields as part of the inspection should be mandatory. Areas of high field intensity should be identified and these areas avoided as much as possible. It is particularly important to locate sleeping areas away from high current installations such as meter boxes and distribution boards.

4. Prevent, whenever possible, members of the family, especially children, from sleeping near electric heaters and other major electrical appliances, such as clothes dryers, washing machines and dishwashers, or on the side of the wall against which these appliances are placed in another room.

PART three

Food technology

chapter 7

Food processing: To eat or not to eat? That is the question

There is little doubt that modern society and the technology to which it has given birth have radically transformed our lives. Much of this transformation is to be applauded, but the obsession with synthesising virtually every type of food should not to be. Although we have—largely through the influence of advertising—come to regard as palatable and attractive the array of synthetic foods that have been fabricated as surrogates for real ones, it is clear that the progressive use of synthetic ingredients as substitutes for nature's own products, has resulted in a progressive reduction in the nutritive value of what we eat. This progress has been accompanied by distinct changes in the health patterns of our society. In the pages which follow we will address some of the subtle and often unexpectedly far-reaching ramifications of modern food production not only for the health of the human organism but also for the entire planet.

Unheeded warnings

One of the world's leading nutrition researchers, HM Sinclair, from the Nutrition Research Laboratories at Oxford, in his address at London University to celebrate World Health Day in 1957, remarked:

> I started to work on human nutrition because I had become convinced that what was and is perhaps the most serious problem in medicine arose from alterations in our diet from the processing and sophistication of foods. My clinical teachers could not answer why the expectation of life in this country of the middle-aged man is hardly different from what it was at the beginning of this century or even a century ago. That means that despite the great advances in medicine—pneumonia almost abolished, tuberculosis comparatively rare, the magnificent advances in surgery, endocrinology, and public health—a middle-aged man cannot expect to live more than four years longer than he could a century ago—and indeed in Scotland the expectation of life is now actually decreasing.[1]

Sinclair concluded his speech by saying, 'nutrition has now become the most important single environmental factor affecting health in every country of the world'.[2]

Forty years later the situation has not changed; if anything it has worsened. At a valedictory address in 1980, Sinclair pointed out that the life expectancy of 50-year-old males in the United States had increased by about 5 percent—from 22–23 years—in the last 150 years. During those 150 years medicine had been making unparalleled progress.[3]

This same theme continues to be echoed by nutrition researchers. In the 1984–85 *Yearbook of Nutritional Medicine,* DO Rudin writes:

A comprehensive survey of epidemiological studies also suggests that the dominating illnesses in modernised societies are new, or have become newly prominent, in the past 100–150 years. Moreover, when traditional societies modernise today, they, too, seem to develop these same 'modernisation diseases' within a few decades. These illnesses include schizophrenia and probably other behavioral disorders, ranging from neuroses to delinquency and criminality; ischemic heart disease and hypertension; major types of cancer, tinnitus and Meniere's disease; cystic fibrosis; celiac disease and probably other immune disorders; arthritis; obesity; appendicitis; possibly irritable bowel syndrome; various endocrinopathies and the drying and scaling dermatoses.[4]

Interestingly, he goes on to say:

The internal consistency of these concurrent findings—chemical, industrial, clinical and epidemiological—suggests that in scaling up to mass feeding, we are now producing an insidious new type of interactive or *synergistic multi-malnutrition* ...[5]

Further evidence comes from the observations that food intolerance appears to be steadily increasing in our society, and appears to be positively linked to modern urban living.[6]

In the opening address for the 1996 Commonwealth and Scientific Industrial Research Organisation (CSIRO) Conference on Human Nutrition, the Chairman of the Australian Food Council, Murray Rogers, told participants that in Australia around 100 000 potential years of life are lost annually due to premature deaths related to poor nutrition.

Up to 50 percent of heart disease, 40 percent of strokes, 50 percent of stomach cancers, 35 percent of bowel cancers and 50 percent of non-insulin dependent diabetes are attributable to bad diet. The economic cost of all diet-related disease in Australia is now estimated to be over $3.2 billion per annum and increasing.[7]

The vested interests of science and technology

The developments of science and technology over the past 150 years have bequeathed to modern society the tools required to reduce food to its basic components and to analyse these components in minute detail. During this same period, societal values have been encouraged which fostered a public preference that foods look pure and perfect. Considerable emphasis has thus been placed on aesthetic appeal coupled with culturally adapted tastes. This being so, it is unsurprising that our science-based technology has taken upon itself the responsibility to provide us with 'perfect' food—allegedly pure, nutritious, sensual and unblemished, but most importantly, sufficiently long-lasting to increase shelf-life and enable marketing and distribution in our growing urban centres.

This has been achieved by refining the basic food, thus making it chemically pure and, hence, predictable from food processing and hygiene aspects. Chemical nutrients such as vitamins and minerals are also added to make the food 'nutritious'. Many foods are chemically coloured to meet our stereotyped expectations, and various other chemicals are added to make the food feel smooth or crisp. Flavours are added to meet our expectations, and the addition of extra salt is usually part of the processing. For example, the much higher levels of sodium found in most processed foods compared with natural foods is shown in Figure 7.1. Finally the inclusion of preservatives enable long shelf-life in warehouses and supermarkets.

Figure 7.1 Sodium content of natural foods vs processed foods

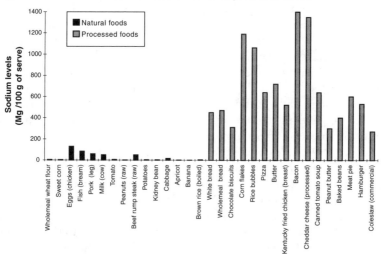

R English & J Lewis, *Nutritional Values of Australian Foods*, Australian Government Publishing Service, Canberra, 1992

This synthesising of food on the basis of scientific knowledge is far more extensive than we often realise. For example, canned peas have been coloured green, and strawberry jam coloured strawberry red for so long now that when a leading British food manufacturer attempted to discontinue colouring these items customers reacted by asking what was 'wrong' and sales fell by 50 percent.[8]

The naive assumption that we can readily improve upon nature is sadly exemplified in some ways by the promotion for several decades of bottle feeding babies with bovine milk. This ignored the possibility of the existence of subtle health factors associated with the composition of human milk which were absent from bovine milk. However, research has shown that human and bovine milk lipids have substantial differences, with human milk lipid being quite unique structurally and nutritionally.[9] We also now know that breast milk plays a major role in the development of the gastrointestinal tract. The unique nutritional composition of breast milk also provides factors which stimulate the developmental changes that occur following birth.[10] Epidemiological evidence now also suggests that breast milk decreases the risk of food allergies.[11]

While it is well known that breast feeding babies protects them from infections, it has recently been discovered that human milk also has a powerful anti-cancer effect.[12] Full-term, breast-fed infants have higher levels of the essential fatty acid docosahexaenoic acid (DHA) in their brains than bottle-fed babies. There are also studies which suggest that children with abnormally low levels of DHA in their brains have slightly, though nonetheless significantly, impaired brain development and score lower in IQ tests.[13] Other studies have found that bottle feeding newborn infants with cow's milk can lead to a much greater chance of the child developing diabetes later in life.[14]

Despite increasing recognition of the adverse effects on the body of processed foods, chemical technology is leading us into a new era of biotechnology and genetic engineering in which the food industry is poised to take full commercial advantage of yet another radically transforming kind of technology.[15]

Since 1992, at least six genetically engineered transgenic food crops have been approved for sale in the United States, including two varieties of tomato with characteristics of delayed ripening; insect-resistant potato; insect-resistant corn; herbicide-resistant soybean; virus resistant squash; a variety of tomato engineered to have a higher solid content for easier processing into sauces; and canola with the oil composition altered to be high in lauric acid which is otherwise normally obtained from coconuts and palm kernels.[16]

In 1995, the Ministry of Agriculture, Fisheries and Food in the United Kingdom cleared three genetically modified (GM) foods for sale; a tomato paste from GM tomatoes, GM canola (oilseed rape) and GM soy beans.[17] The tomato paste was developed from fruit engineered to carry a gene which delays softening, but also contains an antibiotic-resistant gene which was used as a marker. The soy beans have an inserted gene which makes the bean resistant to the herbicide Roundup,™ allowing farmers to spray their crops, killing the weeds, while the soy beans survive. Authorities in Canada and Japan have approved these beans for human consumption while the beans can be imported into Australia for processing. For example, Australia imports approximately 100 000 tonnes of United States soy beans each year for processing into vegetable oil, and approximately 2 percent of the United States soy crop contains the 'Roundup Ready'™ gene. Separate labelling is not required in the United States which means that GM modified beans are sold alongside and are indistinguishable from conventional beans.

Soy bean derivatives of different types are permitted to be used in well over half the processed food and beverage products on Australia's supermarket shelves.[18] Soy bean oil is incorporated as a component of margarines, frying oils and blended oils, which are then used in a variety of products including baked goods (pastries, cakes, biscuits), snack foods and sauces. Soy proteins are used as meat extenders, to produce tofu and textured vegetable proteins, and in the manufacture of soy milks. Another soy product, lethicin, is used in a wide range of food products including cakes, pastries, biscuits, margarine, table spreads, milks, creams, yoghurts, chocolate and confectionery. With modern food technology processes, small amounts of genetically modified soy bean components are therefore being incorporated into the diets of most of the community. While many authorities regard these GM modified foods as safe for human consumption and the safety issues regarding genetically modified foods have been extensively reviewed recently,[19] other researchers remain unconvinced that long-term health and environment problems will not arise.[20]

The 'Roundup Ready'™ soy beans, for example, contain a foreign gene which increases the amount of a particular enzyme protein in the plant. This, in turn, protects the plant from the effects of Roundup™. It is simply naive to rule out the possibility that unanticipated adverse reactions may occur in humans after a time. The herbicide-resistant properties of the plant may also encourage more herbicide use and lead to higher residues of Roundup™ remaining in the bean.[21]

At the beginning of 1997, the European Commission decided to allow genetically modified maize into Europe.[22] For three months prior to this decision, an estimated 4000–5000 tonnes of the maize had been illegally imported into the Community market each week.[23] The maize contains three foreign genes. One introduced gene changes biochemical processes in the plant so that it is resistant to a herbicide, another new gene from a bacterium produces a toxin which acts as an insecticide, and a third gene produces an enzyme which destroys the antibiotic ampicillin. Some scientists are particularly concerned by the presence of the third gene. It seems that the new maize is capable of making not just one, but 600, copies of the ampicillin-destroying enzyme gene. This being so, there exists the serious possibility that conditions could arise in which this gene could be transferred, from corn to bacteria in an animal's intestine and from there spread widely to other living things in the environment.[24]

At the present time in the United States, hundreds of field trials of trangenic plants are now being carried out each year and for most of them the United States Department of Agriculture (USDA) requires only that the researchers give notification. It does not usually require a permit, nor does it request data showing a probable lack of harmful environmental effects.[25]

If genetically modified crops are grown in areas where they have not been adequately field tested, there could be unexpected harmful effects such as the spread of genes to wild plant relatives or the creation of new weeds or new viruses from virus-resistant plants.[26]

China is now cultivating thousands of hectares of genetically modified rice which produces its own insecticide.[27] Whether insects will, in due course become resistant to the crop is undetermined, as is the long-term effects on human health of an engineered rice containing its own insecticide.

Other scientists are concerned that genetically engineered crops such as the soy bean and corn will lead to more intensive agriculture and increasing destruction of the soil in the long term. They point out that between 1990 and 1995 the USDA has spent about US$640 million on biotechnology research and only about US$28.5 million on sustainable agriculture research.[28] It is the view of the authors that sustainable agriculture should be considered one on the most vital areas of government research. At the present time, we clear oxygen-producing forests to produce grazing land for cattle, just as we pollute our waterways with agricultural fertilisers, pesticide, and other contaminants. We have still to learn to grow crops in such a way that we minimise, if not avoid, topsoil loss altogether. Without fresh air to breath, clean water to drink and good soil in which to grow our food, the goal of community health will remain progressively elusive.

Biotechnological developments have also recently been extended to included the use of growth hormones. For example in Australia, pigs may be given daily injections of growth hormone to increase their productivity.[29] In the United States since February 1994, cows may be injected with a synthetic growth hormone called bovine somatotropin (BST), one of the most controversial drugs yet produced by biotechnology. This drug increases the quantity of milk produced by dairy cows by as much as 10 and 15 percent.

What we are not told is that cows receiving the drug are likely to suffer more than 20 ailments including digestive disorders such as bloat and diarrhoea, along with increases in ovarian cysts and a significantly higher risk of mastitis.[30]

Researchers cautious of the use of BST point out that the increased incidence of mastitis in BST-treated cows is likely to lead farmers to use more antibiotics, residues of which could stay in the milk and eventually undermine the effectiveness of antibiotics prescribed in the case of human illness. Notwithstanding these concerns, in February 1994 the United States Food and Drug Administration (FDA) issued guidelines to State regulatory bodies recommending that they rule that milk producers should not have to tell customers when their milk comes from cows treated with BST.[31] This suggests to the public that BST-produced milk is the same as naturally produced milk, even though in this case, it contains higher than normal levels of an insulin-like growth factor.[32]

Dr B Mephan from the University of Nottingham and Dr P Schofield from the University of Cambridge, point out that the ten-fold higher concentrations of this growth factor induce a proliferation of mucous cells in the intestinal tract of test animals.[33] Dr S Epstein from the University of Illinois Medical Centre in Chicago, asserts that the growth factor is absorbed across the wall of the gut, especially in children, and may be a risk factor for causing undesirable breast cell transformations, which may later lead to breast cancer. A very potent form of this growth factor has been estimated to be present at levels up to 3 percent in BST milk. Dr Epstein suggests that prenatal and infant human breast tissue may be particularly susceptible to these hormonal influences, which may in turn also increase the vulnerability of the breast to other cancer risk factors later on.[34]

Given our obsession with ever-increasing technology, the question of nutritional requirements for the metabolism of these bioengineered foods in relation to the complex biochemical pathways in our bodies and in the environment is barely being addressed.

The complexity of human digestion, assimilation and the detoxifying processes associated with these subtle mechanisms is appreciated by all too few. L Gamlin has pointed out that research suggests that the enzymes needed to detoxify many synthetic chemicals are deficient in a number of people. She concludes her findings with the thought, 'Food intolerance could simply be a sign that the human body is not fully adapted to the novel environment we have created for ourselves'.[35]

It has been argued that what we eat is controlled by the 'foodmakers'. However, it is salutary to remind ourselves that the food companies themselves are primed to respond to consumer demand. Consequently, we are constantly seduced, by way of media advertisements, to believe that the products of food technology are better for us than whole food, particularly in respect to taste and convenience.[36] Advertising defines, reflects and energises our culture in the developed world. In 1997, United States food companies are expected to spend $20 billion to ensure that they are part of the American culture.[37] Let us now look more closely at some of these highly technologised foods we take for granted.

Bread: Basic food or staple poison?

The modern refining of wheat flour for bread-making began about 200 years ago. The advent of steam power and the accompanying growth of technology was applied to the milling industry to produce pure white flour as a result of many grindings, siftings (boltings) and separations. In 1870, the middlings purifier came into use. This device used strong air currents to remove remaining particles of lint and bran. By the early part of this century the refining process was such that the best white flour may have gone through 180 separations.[38]

In the 1920s, the bleaching agent known as agene or nitrogen trichloride was introduced in flour processing as a bread 'improver'. This produced a 'strengthening' of the flour, or more precisely the gluten, which resulted in a bigger loaf volume. Later, other powerful oxidising or bleaching agents such as chlorine dioxide and chlorine gas were used to treat the flour. White flour produced this way also had excellent keeping properties, one of the reasons being, in the words of a former President of the American Chemical Society and world-renowned nutrition researcher, RJ Williams, 'it will not support the life of weevils'.[39] However, by the late 1930s research had accumulated evidence that was to be the beginnings of what to some must be re-garded as a nutrition horror story.

Around 1900, Christian Eikman, a Dutch medical officer, noticed that a beriberi-type symptom was produced in chickens fed polished rice. He and a co-worker, G Grijns, demonstrated that these symptoms could be prevented or cured by feeding rice bran or polishings. A decade later, Dr RR Williams became interested in the concept that the then very prevalent beriberi was caused by a nutritional deficiency. In 1933, Williams isolated 450 milligrams of the anti-beriberi factor and in January 1936, after 25 years of research, discovered the structure of the compound which was called vitamin B1 (thiamine). Six months later, Williams and co-workers produced synthetic thiamine in their laboratory. Between 1936 and 1938, RA Petes and co-workers showed that thiamine was essential for carbohydrate metabolism.[40]

The implications of this research on the nutritional quality of refined white flour were soon realised and, from 1941, white flour was enriched with the addition of the vitamins thiamine, riboflavin and niacin, and the mineral iron.[41]

This step epitomises the philosophy of food technology which prevails to this present day. A natural food is reduced and purified to basic components and modified to meet industrial processing requirements. Certain synthetic nutrients are then 're-added' according to the prevailing nutritional science and consumer demands.

However, our story has only just begun. Research by HM Sinclair in the early 1950s showed that vitamin B6 levels were also reduced by the flour refining or extraction process, as were the levels of essential fatty acids (EFAs) and vitamin E which protects the EFAs. Ironically, the remnants of these latter nutrients which manage to survive are subsequently destroyed by the bleaching agents. Sinclair's research led him to the view that deficiencies in these nutrients were contributing factors to diseases of civilisation such as atherosclerosis and heart disease,[42] and there is now substantial evidence that vitamin B6 is an important, yet mostly overlooked, key to preventing heart disease.[43] Sinclair's admonitions have gone unheeded by the milling and baking industries, despite the fact that his concerns were re-echoed by RJ Williams in 1971.

To demonstrate this point, RJ Williams performed experiments in which:

> Young weaning rats of four different strains, 128 in number, were placed on two bread diets. One group of 64 was placed on commercial enriched bread still produced essentially in accordance with the practices of about 30 years ago. A matched group of 64 was given the same bread which had been supplemented in accordance with more up-to-date knowledge by the addition of small amounts of minerals, other vitamins, and one amino acid, lysine.[44]

The two breads were indistinguishable in appearance, and the added nutrients at wholesale rates would cost only a fraction of a cent per loaf.

That the rats on the commercial enriched bread would not thrive was to be expected. After 90 days, about two-thirds of them were dead from malnutrition and the others' growth was severely stunted. The rats on the improved bread did surprisingly well; most of them were alive and growing at the end of the 90 days. Their growth and development on the improved bread was on the average seven times as fast as on the commercial bread.

Williams rightly contends that '… the milling and baking industry should have been doing experiments like this for decades', and that 'enriched flour' on the basis of present knowledge be called 'deficient flour'.[45]

Williams goes further to point out:

> … the complacent acceptance of the limited 'enrichment' with thiamine, riboflavin, niacin, and iron, has held back the nutritional improvement of all other cereal products. When corn or rice or breakfast foods are 'enriched', the same four nutrients are added, as though these constituted an exclusive 'sacred quartet', and no other nutrients could be of concern.[46]

This same attitude persists in the food manufacturing industry to the present day even though, for example, flour produced by modern milling contains 75 percent less vitamin B6 than did the original whole wheat.[47]

Furthermore, the bran and embryo (germ) contained within the grain, which are sources of many essential nutrients, are discarded during flour processing and termed 'offal' within the food industry. Yet, whole wheat is a nutritious source of essential fatty acids, a large number of vitamins and a rich range of essential trace minerals which are now known to play key roles in many enzymatic and biochemical pathways. An electron microscopy study of vertical sections of a whole wheat grain revealed a distribution of the essential minerals sodium, magnesium, phosphorus, sulfur, potassium, calcium, manganese, selenium, cobalt, chlorine, silicon, arsenic, iodine and molybdenum. The researchers concluded that 'wheat was found to be a good source of trace elements'.[48] Modern wheat flour refining destroys this source and deprives the human organism of essential nutrients. And fortification is not the answer. Even W Mertz, chairman of the United States Department of Agriculture Nutrition Institute, conceded in 1979: 'Although cereal products in the United States have been enriched with iron for about 40 years, a disappointingly high incidence of iron deficiency still exists in some parts of the population'.[49] Furthermore, artificially fortifying foods with iron may contribute to an increased risk of heart attack in men and post menopausal women, and also contribute to the risk of getting cancer.

Consider, for example, a five-year Finnish study of over 3000 men which found that subjects with blood ferritin (iron) levels above 200 micrograms per litre had more than twice the risk of heart attack compared to men with normal ferritin levels of around 165 micrograms per litre.[50] Recent research suggests that a key factor in the progression of heart disease may be the breakdown of certain fats in the presence of iron, which subsequently accumulate on arterial walls.[51]

Two studies of human populations, one Finnish and one American have also reported an increased risk of cancer in men and women who possess high levels of iron in their bodies. The American study led by Dr R Stevens at the Pacific Northwest Laboratory in Washington, examined data from 8500 men and women. Of those studied since the early 1970s, 720 developed cancer. It was found that people with an iron level of twice the normal level had twice the risk of cancer.[52]

The Finnish research, led by Dr P Knekt of the Social Insurance Institution Research and Development Centre in Helsinki, studied data on 40 000 men and women over a period of 14 years, during which time 2500 people in the group developed cancer. It was found that those people whose iron levels were greater than twice the normal range had three times the risk of developing rectal and colon cancer.[53] Further evidence for the latter association comes from a recent report which found a connection between high iron intake and increased incidence of colorectal polyps. The authors of the report cautiously suggest that the use of iron supplements and the consumption of iron-fortified foods should be avoided by people with high plasma iron levels.[54] In the light of growing evidence against the widespread iron fortification of foods, especially breakfast cereals, it is the view of the authors that such practices need urgently to be reviewed and reassessed outside the context of vested interests.

RJ Williams emphasises that 'in order for our cells and tissues to have a proper environment, we need to have adequate supplies of a team of nutrients, and the team is as strong as its weakest member'.[55] Unfortunately, in flour processing not only are we removing essential compounds, but we are also adding some which are proving to have an adverse impact on human health.

After 25 years of using bleaching agents, in the mid-1950s Sir Edward Mellanby made the discovery that when bread made with flour treated with the bleaching agent agene was fed to dogs, it produced 'canine hysteria' or 'running fits'. When the findings were confirmed, the use of agene was phased out. The attitude of health administrators in this case is revealed by Wickenden, who reports the words of Dr Paul Dunbar, Commissioner of the Food and Drug Administration (FDA), '... just as soon as a new maturing substance was found, the food standards were amended so as to outlaw nitrogen trichloride'.[56] The new maturing substance was chlorine dioxide.

Chlorine is still used especially to denature the gluten in high ratio sponge cake flour. Yet, chlorinated organic compounds are linked with cancer and other disease.[57] Research, using traceable radioactive chlorine-36 to chlorinate unbleached flour, has revealed that 27–33 percent of the total chlorine 36 was incorporated into the flour lipids. A further 20–26 percent of the total chlorine-36 was incorporated into the non-fat portion of the wheat. Only 40–50 percent of the total chlorine used to bleach flour is reduced to harmless chloride, the remaining 50–60 percent of deliberately added chlorine forms potentially extremely harmful organochlorine compounds in the flour itself.[58]

Modern bread-making flour contains many other chemicals, including potassium bromate. This chemical has been established as a known carcinogen in animals.[59] Yet its use continues, presumably on the assumption that during the baking process it is all reduced to presumably harmless potassium bromide. However bromide itself has no known function in human nutrition[60] and at certain levels is known to produce depression, emaciation and at higher levels mental deterioration.[61] The effects of regular long-term exposure at low levels is unknown, as is the extent to which bromine is incorporated into the flour lipids.

The picture is clear. Through modern technology we have turned one of nature's oldest and most important whole foods, wheat, into a precursor for disease.

Butter versus margarine: What are the facts?

Fats are an important and essential part of our diet and are present naturally in many foods. Fats are a source of the fat-soluble vitamins, namely vitamins A, D and E, and the essential fatty acids (EFA). In Australia in 1985, almost 20 percent of the average total fat intake came from butter or margarine [62] both of which are humanmade modifications of natural foods.

Butter is mostly made from bovine milk. Milk is a water-based solution in which about 3.5 percent fat is incorporated as very fine globules dispersed in the water. Using modern technology the milk fat (cream) is separated and pasteurised (which inhibits natural enzyme activity as well as bacteria). The cream is then churned, which tends to agglomerate the fat globules into granules of butter, with a watery solution, buttermilk, separating out. After the buttermilk has been drawn off, the granules of butter are washed with water, which is usually chlorinated, and salt is mixed in, which serves the 'technological purpose of inhibiting infection by yeasts and moulds and by proteolytic and lipolytic bacteria'.[63] Further processing is used to reduce the moisture droplet size which has been found to enhance the keeping properties.

Thus we take a nutritionally balanced whole food, milk, which has been shown to lower blood cholesterol levels in certain conditions,[64] to contain components which tend to protect against gastrointestinal infection,[65] dental caries,[66] and possibly stomach cancer[67] and breast cancer,[68] and remove these beneficial components, convert what remains to butter, now linked with heart disease[69] and atherosclerosis,[70] and add unnatural amounts of salt which is linked with cardiovascular hypertension.[71] It is now believed that full-fat milk may prevent a number of common cancers including melanoma, leukemia and cancers of the breast, ovaries, prostate and colon. The protection appears to be associated with a particular fat component in the milk referred to as conjugated linoleic acid (CLA). Researchers believe the CLA works by mopping up oxidant chemicals that can induce cancer-causing mutations in cells.[72] A recent study has also found that men who drink milk regularly are less prone to strokes.[73]

Publicity since the 1960s on the possible saturated fat–coronary heart disease link led to the promotion of polyunsaturated margarine as a more healthy form of dietary fat. However, a substantial literature has now accumulated to suggest that this dietary change has been equivalent to 'jumping out of the frying pan and into the fire'.

Margarine was invented by a Frenchman, Meges-Mouries, in 1870. It was originally made by melting down and clarifying various animal fats and then churning with a proportion of cream to simulate the making of butter. In 1912, food technologists developed methods for the catalytic addition of hydrogen to polyunsaturated oils to produce solid, stable fats that could compete with butter and lard. Today, most table margarines are made by chemically hardening various vegetable oils with hydrogen gas. The composition of common edible oils and fats are shown in Figures 7.2 and 7.3.

Figure 7.2 Composition of common edible oils

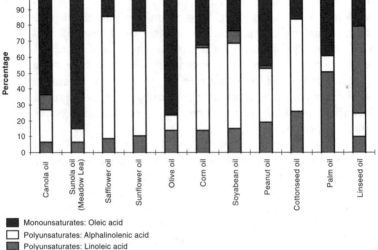

JF Ashton, University of Newcastle, NSW, 1996

Figure 7.3. Composition of common edible fats

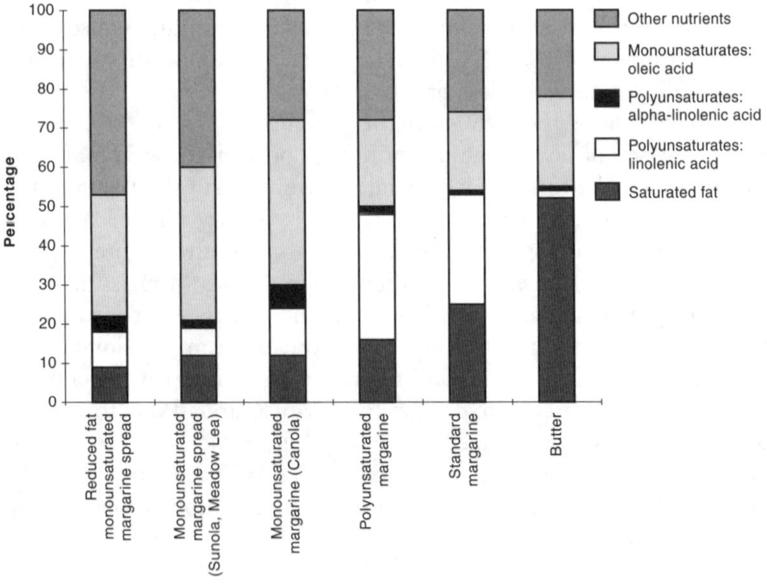

JF Ashton, University of Newcastle, NSW, 1996

The vegetable oils used for margarine, and also as sold as cooking oil, are highly refined. Initially the oils are removed by a combination of mechanical crushing of the seeds and extraction with an organic solvent. At this stage, various amounts of free fatty acids and other substances are still present in the oil. This crude oil is refined further by washing with caustic soda to remove the free acids and heated and specially filtered to decolourise and deodorise the oil. The hardened oil is blended with a water solution, usually made from 'ripened' skimmed milk to impart flavour. Emulsifiers, colouring agents and synthetic fat-soluble vitamins are also added.

In the process of hardening or hydrogenation, double bonds between atoms of the carbon chains which constitute unsaturation, are changed to single bonds between carbon atoms constituting saturation. Thus polyunsaturated refers to oils containing a number of double bonds. Saturated fats contain no double bonds. The more saturated an oil or fat, that is, the fewer double bonds, the more viscous or solid it becomes. Thus a controlled chemical process is used to produce a product which has a similar texture to butter yet is more unsaturated than butter, and therefore deemed to be better for us.

However, let us look more closely at this synthesised highly modified fat. To begin with, the initial refining of the oils removes most of the nutritive factors present originally in the seed, including a range of B vitamins, zinc, magnesium and copper. Out of the 45-or-so known essential nutrients, at least 29 are involved in fat metabolism.[74]

It is of interest to note that olive oil is the only edible oil which is generally cold pressed rather than chemically processed prior to consumption. LA Cohen points out:

> In this regard, it is intriguing that in Mediterranean countries such as Italy, Greece and Spain, where olive oil consumption is high, and as a consequence red cell and platelet levels of oleic are high, breast cancer mortality rates are intermediate with regard to those of the higher-risk northern European countries and low-risk Japan ... in the light of the experimental and epidemiological findings and the question of processed versus unprocessed oil, further investigation into the special effects of olive oil are merited.[75]

In the decade since Cohen's observation, substantial evidence has accumulated to show that olive oil, and in particular cold pressed 'extra virgin' oil, is rich in polyphenolic compounds and may serve as antioxidants which have been shown to be protective against cancers. Dr. D Trichopoulos from the Harvard School of Public Health reports that there is strong evidence that olive oil consumption may reduce the risk of breast cancer, whereas margarine intake appears to be associated with an increased risk for the disease.[76]

The incidence of breast cancer in Spain has risen 100 percent since 1960, particularly among women in the 35–64 age group. Spanish researchers studying this problem have noticed that significant changes in the Spanish diet have occurred during this time and one of the changes has been the increased consumption of polyunsaturated oils rather than olive oil. The Canary Islands, where there is a high consumption of polyunsaturated sunflower oil, registered Spain's highest incidence of breast cancer.[77]

The hydrogenation process, as well as increasing the degree of saturation, adds the hydrogen in such a way that it bonds in an opposite 'trans' position rather than an adjacent 'cis' position, thus forming what is termed the 'trans isomer' of the fatty acid chain or 'trans fatty acid'. These particular isomers are rare in nature, and in 1984 Krause and Mahan pointed out that:

> Unfortunately there is little information on these isomers, and they have generally been accepted as safe when in fact this may not be true.[78]

It is now known that trans fatty acids, which are present in most margarines in substantial amounts, will block the utilisation of the cis fatty acids and therefore should be considered an anti-nutrient. They have also been linked with breast cancer and do not even lower plasma cholesterol concentration in the way all cis polyunsaturated fatty acids do.[79]

In May 1994, GV Mann wrote to *The Lancet* suggesting that the epidemic of coronary heart disease in the western world followed the introduction of partially hydrogenated fats in food.[80] He proposed that the trans fatty acids in these fats interfere with normal fat metabolism processes in such a way as to cause increased cholesterol levels, blocked arteries, obesity and insulin-resistant type diabetes. That same month Dr Walter Willet from the Harvard School of Public Health reported that there was now very strong evidence that the trans fatty acids in margarines were worse for the heart and blood vessels than saturated fats. Willet estimated that as many as 30 000 heart disease deaths per year in the United States could be caused by diets high in trans fats.[81]

Another aspect of the hydrogenation process to be considered, is the health implications of the small amounts of nickel catalyst which remain in the margarine. Analysis of 16 samples of margarine collected from food retailers in 1978, revealed nickel at levels ranging from 0.22–1.94 parts per million (ppm), with five samples containing more than 1 ppm.[82] The possibility of this nickel-promoting atherogenesis (the build-up of degenerative fatty deposits within blood vessels) in the body needs further investigation, especially in the light of recent research with rats which revealed that levels of nickel in drinking water exceeding 0.05 ppm should be regarded as a risk factor.[83] The environmental implications of the disposal of spent nickel catalyst need also to be considered.

The medical basis of increasing polyunsaturated fats as a preventive measure against heart disease has been questioned for decades. In 1959, in a letter to *The Lancet* Sinclair pointed out:

> There is still no proof that alteration in the quality or quantity of dietary fat will affect the incidence of atheroma or of ischaemic heart disease, but there is abundant circumstantial evidence that the incidence of both is increased by a relative deficiency of certain highly unsaturated fatty acids, usually referred to as essential fatty acids.[84]

Three decades later Professor WE Stehbens has concluded:

> There is so much conflicting evidence in the dietary claims for the lipid hypothesis of atherogenesis that in the absence of more reliable scientific input, considerable caution is advisable before recommending desirable proportions of saturated/unsaturated fats in the diet. Reappraisal of the role of serum cholesterol and LDL [low density lipoproteins] in human atherogenesis is urgent and essential.[85]

Passwater has also argued strongly that polyunsaturates are not the answer[86] and recent findings from the Research Institute of Public Health in Finland has found that a high intake of linoleic acid from polyunsaturated cooking oils is associated with increased oxidation susceptibility of LDL cholesterol and hence risk of plaque build-up in the arteries.[87]

Perhaps the most serious and yet mostly overlooked aspect of margarine production is the destruction of essential fatty acids (EFAs). These are the three polyunsaturated fatty acids called linoleic (omega-6 type), linolenic (omega-3 type) and arachidonic. Arachidonic can be formed in the body from linoleic acid but vitamin B6 is required for this to happen.

Vegetable seed oils are in many cases rich sources of EFAs but during hydrogenation most of the EFA content is destroyed. For decades, Sinclair attempted to draw attention to the need for an adequate intake of the EFAs, a deficiency of which he persistently believed was associated with diseases of civilisation such as atherosclerosis and heart disease. Deficiencies in EFAs have also been linked with a number of skin disorders.[88]

It is now established that EFAs play a major role in the health of the human organism. EFAs are required for the metabolism of both cholesterol and fats. They are required for brain development and brain function, especially omega-3 types. They are required for the manufacture of healthy cell membranes in our body and serve to stimulate many metabolic processes. EFAs are now known to play a vital role in the production of prostaglandins. These hormone-like substances regulate a large number of biochemical functions including platelet stickiness, hence their involvement in the clotting diseases. They also play a role in the maintenance of muscle tone which, in turn, involves blood pressure regulation; sodium excretion; inflammatory response; and immune function.[89] U Erasmus has also noted that a number of research reports show that EFAs, particularly omega-3 types, retard the rate of cancer development.[90] Not surprisingly, Erasmus also observed that as the EFA content of our diet decreased, the consumption of synthetically hydrogenated vegetable oils with their high trans fatty acid content has increased. These high trans fat types now constitute 15 percent of all the fats we consume. He concludes that:

> Trans fatty acids show better statistical correlation with increased cancer incidence over the last 70 years than does our consumption of vegetable oils or our total fat consumption.[91]

The food industry is responding with research into hydrogenation which produces cis fatty acids rather than trans types. However, one process, which has been developed uses chromium carbonyl catalysts, and leaves various toxic chromium species in the oil which are difficult to remove.[92]

In the late 1970s, Passwater pointed to the evidence linking polyunsaturates with cancer and premature aging due to the release of high levels of free radicals by these fats.[93] This greatly increases the requirements of the free-radical scavenger, vitamin E, low levels of which are also linked with an increased risk of heart disease and cancer.[94]

The health risk of artificially high levels of polyunsaturates is at last being recognised by health officials,[95] although official statements are veiled and carefully worded. A 1989 report from the United States National Research Council Committee on Diet and Health, now recommends that individual consumption of polyunsaturated fatty acids should not exceed 10 percent of total calories. The committee said it is prudent to recommend that polyunsaturated fatty acid intake not be increased above current average levels due to the lack of information about the long-term consequences of polyunsaturated consumption at high levels.[96]

Another health factor just coming to light is the possible connection between increased margarine consumption with increased skin cancer. Some decades ago, Sinclair showed that essential fatty acid (EFA) deficient animals were much more sensitive to ultraviolet light, with mild doses producing complete necrosis of the irradiated paw of EFA-deficient animals.[97] Almost 25 years later in 1980, other researchers proposed a more general connection between polyunsaturated fat consumption and malignant melanoma.[98] In 1988, Chedekel and Zeise suggested a possible 'peroxy radical' mechanism linking polyunsaturated fatty acids and skin cancer[99] and that same year V Reeve and co-workers from the University of Sydney published their findings, that test animals fed 20 percent saturated fat in their diet were almost completely protected from UV tumorigenesis.[100] Since then there has been a steady accumulation of evidence linking the use of polyunsaturated margarines with an increased susceptibility to sunlight-induced skin cancer. An important part of our defence against cancer involves our T-cell mediated immunity, a controlling factor in the promotion of UV initiated tumors. Furthermore, it has been found that polyunsaturated fats impaired T-cell based immune function.[101]

V Reeve and co-workers have demonstrated that in mice the promotion of UV induced tumours requires a dietary source polyunsaturated fat. In mice fed saturated fat, the tumours are initiated but remain latent in the skin. They can be reactivated by refeeding with a polyunsaturated fat diet. In one experiment, a series of diets containing increasing amounts of polyunsaturated sunflower oil mixed with hydrogenated saturated cottonseed oil were fed to UV radiation-exposed animals. The skin cancer response increased in severity as the polyunsaturated content of the mixed dietary fat was increased whether measured as tumour incidence, tumour multiplicity, progression of benign tumours or reduced survival.[102]

In another study, the Sydney researchers observed that when mice were fed polyunsaturated fat during a ten-week regime of UV radiation exposure but changed to a saturated fat diet post-irradiation, there was not only protection for tumorigenesis but also a tendency for tumours which had appeared to regress. The researchers went on to study these dietary effects in more detail by examining the effects of four different fats: predominantly polyunsaturated sunflower oil and margarine, and predominantly saturated butter and clarified butter. The latter product consisted of butter with a reduced no-fat solids content. It was found that there was protection by clarified butter against UV suppression of the immune system. This protection increased linearly as the saturated fat content of the diet increased. There was no difference between butter and clarified butter as a source of saturated fat and there was no difference between margarine and sunflower oil as a source of polyunsaturated fat. The researchers concluded that feeding mice butter or clarified butter provided significant protection against suppression of the immune system by UV radiation compared with sunflower oil or margarine.[103]

Thus the picture emerges that saturated fats protect against skin cancer whereas the unnatural increase in refined polyunsaturated fats in the diet from margarines and chemically extracted polyunsaturated cooking oils has actually increased the risk of skin cancer.

The butter versus margarine controversy illustrates the erring mentality underlying food technology and nutrition. Namely, that there are simple single factors relating diet and disease. The cholesterol and saturated fat versus polyunsaturated fat story illustrates this. Cholesterol is essential for normal body processes. However, we now know that there are a number of different types of cholesterol. For example, there is low density lipoprotein cholesterol (LDL) and very low density lipoprotein cholesterol (VLDL), both of which are associated with the build-up of degenerative fatty deposits or plaques within blood vessels, and are atherogenic. There is also high density lipoprotein cholesterol (HDL) which plays a protective role.[104]

Saturated fats increase the blood levels of the atherogenic cholesterols and total cholesterol. On the other hand, polyunsaturated fats decrease total cholesterol levels and the atherogenic types, but also decrease the protective types of cholesterol. However, studies have shown that the group of fats called mono-unsaturated actually lowered total cholesterol and atherogenic cholesterol without lowering the protective HDL.[105] It is also interesting that rich sources of these mono-unsaturated fats are olives and avocados, both of which require minimum processing (simple squashing or peeling). As observed earlier, a number of recent studies have now concluded that olive oil and in particular cold pressed extra virgin olive oil, offers substantial protection against heart disease.[106] (The composition of olive oil is shown in Figure 7.2, page 106.)

Again, the paradox is that we apply sophisticated modern technology to extract and manufacture unhealthy fats, while nature provides healthy ones in a readily available form. Another aspect of the synthetic fat story remains to be told.

In January 1996, the United States Food and Drug Administration (FDA) approved a synthetic fat substitute Olestra, for use in foods such as potato crisps and other snack foods.[107] Olestra mimics fat but is not absorbed by the body. In other words it tastes like fat, behaves like fat, but doesn't make you fat. Natural fat consists of a molecule of glycerol attached to three molecules of fatty acids. In Olestra the glycerol has been replaced by sucrose and six, seven or eight fatty acids attached. These fatty acids prevent the body's digestive enzymes from breaking down the sucrose, so this synthetic fat substitute is not digested.

The development of synthetic fat substitutes such as Olestra is a response by food manufacturers to criticisms of the high fat content of many processed foods, particularly fast foods and snack foods. This new product which takes processed foods yet another step further removed from that provided by nature is estimated to be worth $1 billion in sales in the United States alone within ten years.[108] While there is the potential to generate enormous profits for food companies there is also the potential to create additional health problems in the community. In trials of the product some people suffered from stomach cramps, bloating and diarrhoea while others had the problem of 'passive oil loss' where anal leakage of the undigested Olestra occurs.[109]

Another concern involves the possible loss of fat-soluble nutrients. The vitamins A, E, D and K normally attach to fat molecules and are absorbed by the body when the fat is digested. These nutrients also attach to Olestra and pass straight out of the body with it. Although Olestra's manufacturers will saturate their product with vitamins so that it will not capture vitamins from other foods, it cannot be saturated with carotenoids which are precursers to vitamin A and which are believed to reduce the risk of cancer.[110] From this it follows means that people who eat foods containing Olestra will have caretenoids flushed out of their bodies with the Olestra, thereby resulting in reduced levels of these protective natural compounds in their system.

Dr Walter Willet, chairman of the nutrition department at the Harvard School of Public Health, has expressed concern that an entirely new food is being allowed into the American diet when there is a question mark over its safety.[111] Dr Henry Blackburn, professor of epidemiology at the University of Minnesota in Minneapolis is concerned that the 22-member advisory board to the FDA was made up mostly of people who had past or present ties to the food industry. He points out that decisions on Olestra's safety were based on studies of a few hundred people conducted over three weeks. These studies are so limited that they have almost no predictive power.[112]

A number of other fat substitutes have already been developed.[113] While these empower the consumer to eat more high 'fat' foods without getting 'fat', this outsmarting of nature in one area is likely to lead to a serious and detrimental imbalance in another. The role of fats in the diet is very important to health, and one aspect of this which is only beginning to receive the importance it deserves, is the role of the essential fatty acids (EFAs) in health.

Fats essential for health

Primitive diets consisting primarily of fresh fruits, leafy vegetables and wild animals, provide a relatively equal balance of omega-6 and omega-3 essential fatty acids (EFAs). Over the past 150 years, however, changes in the composition of the food supply of Western societies has resulted in a substantial increase in consumption of omega-6 EFAs and a corresponding decrease in intake of omega-3 EFAs. Today, the ratio of omega-6 to omega-3 consumption is estimated to be in the range of 10:1 to 25:1.

Modern food production methods have been responsible for the decreased omega-3 fatty acid content of many foods. Wild animals, once free to feed on wild vegetation, were leaner and had significantly more omega-3 fatty acids in their tissues than farm-raised commercial livestock today. Domestic beef which has little access to green leafy plants, which are good sources of omega-3 fatty acids, have little or no detectable amounts of omega-3 fatty acids in their meat. Similarly, modern aquaculture produces fish that contain less omega-3 fatty acids than those that grow naturally.

The industrial revolution introduced vegetable oil technology which, in turn, led to the popular use of cooking oils such as sunflower, peanut and corn, all of which are good sources of omega-6 EFAs. The use of infant milk formulas has increased since the Second World War. In contrast to human milk, most milk formulas are devoid of the essential omega-3 fatty acid docosahexaenoic acid.[114]

A startling example of omega-3 type EFA deficiency is illustrated by the following case study.

> A young girl on total parental feeding who was given all of her EFA as safflower oil (0.7 percent omega-3 EFA), developed a variety of semi-acute neurological problems over a period of four months—paresthesias, neuralgias, unsteady gait and visual clouding. These problems all disappeared when she was placed on soybean oil (7 percent omega-3 EFA) as sole source of EFA. Serum analysis established that this was a pure omega-3 EFA deficiency.[115]

Analysis by Rudin suggests that there is a growing body of literature implicating EFA deficiencies and imbalance in a number of the illnesses of modernised societies.[116] For example, he suggests that:

> The subtleties of the emerging distinctions between the omega-3 and omega-6 EFA families also explain certain discrepancies in 'official' dietary recommendations. For example, the public is now advised both to *increase polyunsaturate consumption to prevent heart disease and to decrease polyunsaturate* consumption to prevent cancer.[117]

It is revealing that the two groups of people, Cretans and Japanese, with the lowest rates of heart disease in the world have diets which contain minimum amounts of processed food and have high dietary intakes of omega-3 fatty acids. The Cretans, who have the second lowest rate of heart attacks have a diet rich in grains, vegetables and fruit together with a moderate amount of fish, cheese and olive oil. They have a high intake of alpha-linoleic (an omega fatty acid) from their use of walnuts and purslane, a salad vegetable. Japanese living on Kohama Island have the lowest incidence of heart disease, having a diet rich in alpha-linolenic acid as a result of their high consumption of soya beans.[118]

In one experiment carried out by the French National Institute for Health and Medical Research in the early 1990s, the progress of 606 patients who had been admitted to hospital in Lyon after their first heart attack, was monitored. Half the patients were put on a typical Cretan diet and were given a solid spread with similar composition to olive oil but enriched with alpha-linolenic acid, as a substitute for butter or margarine. The second group of patients followed the American Heart Association (AHA) diet which aims to lower blood cholesterol levels by reducing the intake of saturated fat and increasing the consumption of polyunsaturated fat.

Over a period of 27 months, 16 patients on the AHA diet had died from heart attacks, four had died from other causes while 17 had survived another attack. Of the patients on the Cretan diet, three had died from a heart attack, five from other causes and only five had had another non-fatal attack. The benefits of the Cretan diet were so striking that the trial was stopped and all the patients were told to follow the omega-3 rich Cretan style diet.[119]

The evidence indicting processed food, particularly processed fats, in the diseases of modern civilisation is enormous, further confirming the potential health hazards which results from disrupting the balance of nutrients which nature provides in whole foods.

Refined foods: Diet for disease

The modern refining and processing of foods affects the dietary patterns and human nutrient intakes around the world to an extent appreciated by few. The vegetable oils we consume are refined, our sugar is refined, and our major cereals (wheat, corn and rice) are refined. The size of the food refining industry is enormous. For example, in India, the second largest producer of rice in the world, approximately 75 percent of the 80 million tonnes produced is refined by removing the bran. This bran, which constitutes nearly 8.5 percent of the rice grain, is actually highly nutritious being rich in natural oils (including essential fatty acids), protein, minerals and vitamins. Yet much of this bran is further refined with organic chemical solvents used to make some edible oils for margarine manufacture, with the larger portion being used for soap production and other related products.[120] From this one example alone, it can be seen that nearly 60 million tonnes of a whole food has been split up into its components with important balance trace nutrients removed and largely destroyed. Furthermore, this processing involves the use of thousands of tonnes of organic chemical solvents which eventually have to be discarded into the environment.

Another aspect of the processed food problem is the huge range of everyday foods (and particularly the so-called 'fast foods') which are simply combinations of refined cereals, refined fats, and refined sugar with water, flavours, colours and processing chemicals added. An examination of the 15 largest multinational food companies in the world reveals that the majority of their products are based on combinations of the refined food groups mentioned above. In 1989, the sales from these 15 companies alone amount to more than 1000 billion dollars annually.[121] Yet this is but a small fraction of the income derived around the world from the sale of nutritionally deficient refined foods.

In the United States, the food processing industry has an enormous impact on the economy with the sum of direct, indirect and induced effects giving a total industry output, reported in October 1996 as being greater than $1.8 trillion annually.[122] In Australia, the food and beverage industry is one of the largest manufacturing industry sectors with a turnover of $35.6 billion and employing 867 000 people in the 1992–1993 financial year.[123]

Fast foods: Costly to our health and the environment

In Japan, western processed foods such as hamburgers and other 'fast foods' are replacing traditional diets particularly in highly industrialised areas such as Osaka-Kobe. These dietary changes, in turn, appear to be having a seriously adverse impact on the health of the Japanese people who have adopted them. In 1984, 46 percent of people in the Osaka-Kobe region passed physical checkups, compared with only 21 percent in the 1994 health survey.[124] Across the country, the Japan Hospital Association has found that the number of people with kidney and cholesterol problems also doubled during the ten year period.[125]

Western fast food companies are also changing India's eating habits, affecting not only the health of those who can afford the foods but undermining the nation's agriculture. Maneka Gandhi, a former minister for the environment in India's parliament believes that the intensive farming of poultry and other livestock which fast-food restaurants encourage will have serious implications for India's ability to feed itself.

In a paper submitted to India's parliament in 1995 she warned that using food crops such as maize to feed chickens was a poor use of resources. Grain could feed five times as many people if consumed directly rather than being given to animals to be transformed into meat, milk and eggs. In addition, she pointed out that around 1700 litres of water was consumed in the modern commercial production of half a kilogram of chicken meat. This amounted to 20 times the amount of water 'a normal Indian family is supposed to use in one day-if it gets water at all'.[126]

In many western countries, children face a daily barrage of advertisements for processed junk foods on television. A 13-country survey carried out by Consumers International found that children in Australia, the United States and the United Kingdom see more than 10 ads per hour during children's television viewing times.[127] Heavily processed and refined foods such as sweets, breakfast cereals and fast foods accounted for more than half the food advertisements in the survey. See Figure 7.4.

Figure 7.4 Food advertisements during children's TV viewing time

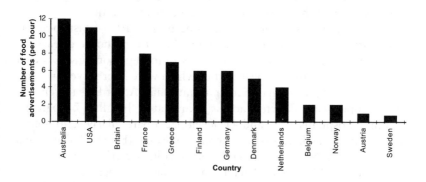

A Coghlan, *New Scientist*, 30 November 1996

On British television, nearly 95 percent of the foods advertised were high in fat, sugar, salt or fixtures of these ingredients, and advertising for non-processed foods such as fruit and vegetables was virtually nonexistent.[128] In Sweden and Norway, where television advertising directed towards children is banned, young people rarely see food advertisements.

Consumers International fear that this type of advertising may undermine advice on healthy eating. A survey of fast foods by the Australian Consumers Association found that fast foods tend to be high in fat, low in complex carbohydrate and high in sodium. This is not the recipe for a healthy diet. A well-balanced diet is low in fat and sodium, and high in complex carbohydrates and fibre.[129]

Not only is food technology being used to add economic value to food that is often nutritionally poor, but part of the financial return from the sale of these foods is used, in turn, to promote other processed foods which are nutritionally inferior to the ones provided by nature. The public are lured into buying more and more processed foods which in turn generates more revenue for commercials which encourage more sales, thereby setting in motion a vicious circularity which spirals towards ill health. Refined foods are favoured by food companies because their properties are more predictable in manufacturing processes. The technique of separating basic foods into components and then recombining them with flavours and colours and manufactured shapes also gives endless possible combinations for new products. It is known that consumers will tend to consume more of a product, for example, if it is produced in a variety of shapes or colours.[130] Thus the potential for increasing sales via new products is almost unlimited. However, the point that has been lost amidst the fray is that very few of these manufactured foods are beneficial for health, and some, as we have seen, may be dangerous to health.

Modern, large-scale food processing particularly of fast foods, is also increasing the risk of major outbreaks of food-borne disease. In the United States, a multi-state outbreak of food poisoning was traced to the use of pre-frozen meat patties which were later served in hamburgers at a chain of hamburger restaurants. According to reports, 583 people were affected across four states. In all, 171 patients required hospitalisation, and 41 developed a particularly serious condition known as haemolytic uraemic syndrome (HUS).[131] The illness was produced by an *Escherichia coli* (E.coli) bacterium described as serotype 0157.H7. This bacterium can produce permanent illness including kidney failure, brain damage and diabetes. Dr K Bettelheim from the Victorian Infectious Diseases Reference Laboratory in Australia believes the multistate outbreak in the western United States occurred only because the meat patties were prepared in bulk in one plant and shipped to many restaurants. The meat for these patties came from a variety of abattoirs, which drew their supplies from many farms. Since meat offcuts are used in hamburger meat production and modern machines produce tonnes of thoroughly mixed mince meat at a time, it is possible theoretically for a single ham-burger patty to contain meat which has come from thousands of different animals. Dr Bettelheim suggests that:

> The extensive development of fast-foods and the sale and storage of frozen foods, and the use of microwave ovens to heat foods rapidly may all contribute to the provision of new niches for potential pathogens to enter or survive in the food chain.[132]

The interconnection of human food production, with other human activities such as sewerage treatment is illustrated by Bettelheim's suggestion that:

> A simple path for the spread of a pathogen like E. coli 0157 to a new environment could be that an individual carrying it jets to the new areas, his/her faeces enter the sewerage, which is inadequately treated before discharge into the sea. The E. coli infect the sea birds. When the weather at sea is rough, they feed and defecate on pastures where cattle and sheep graze. The route is then short from the pasturing animals to human infection.[133]

Strong support for Dr Bettelheim's hypothesis has come from a British study published in March 1997 which found that three percent of 700 samples of fresh gull faeces collected from beaches where they feed contained E. coli 0157.[134]

At the annual meeting of the American Society for Microbiology in May 1997, members were told that dangerous viruses are being spread around the world through the sewage held in aircraft. A study by researchers from the University of North Carolina found that from 40 samples of sewage pumped from international flights landing at United States airports, 19 contained infectious viruses that had survived exposure to disinfecting chemicals in the aircraft's sewage tanks. The leader of the study, Mark Sobsey, commented that the range of illnesses that could be transported by the world's airlines is 'quite worrisome'. Furthermore, even when aircraft sewage is pumped into local municipal sewage treatment works, only about 90 percent of viruses are irradicated. Ten percent of the viruses survive and are discharged into the environment.[135]

A recent survey of pollution levels at several popular Sydney beaches, which are not far from where treated sewerage is pumped into the sea, found very high levels of pathogens. For several of the beaches including famous Bondi Beach, 98–100 percent of water samples taken over a two-year period had faecal coliform counts greater than 1000 coliforms per 100 millilitres.[136]

In January 1995, Australia experienced its first outbreak of hemolytic uremic syndrome (HUS), which was associated with the consumption of contaminated fermented sausage in the state of South Australia. Twenty-three children were hospitalised with HUS and one died.[137]

The intensive farming of cattle where protein concentrates sometimes containing meat offal are fed to animals which are normally vegetarian is also linked to the spread of disease. Since the early 1980s, sheep processing wastes were incorporated into animal feed concentrates for cattle. In 1986 'mad cow disease' or bovine spongiform encephalopathy (BSE) was first diagnosed in British cattle. The sheep offal (brains and spinal cord) contained scrapie, a long-standing disease among sheep. By 1996, at least ten Britons aged from 18–41 years had died from a type of Creutzfeldt-Jakob disease (CJD) which was probably caused by eating beef contaminated with BSE.[138] A total of 40 cases of CJD had been diagnosed altogether. Because modern mass production processes were used in producing this unnatural high protein animal feed, thousands of cattle were infected and, by 1996, 158 000 cattle had died.[139] Despite widespread publicity about the problem, contaminated high-protein animal feed was also being fed to pigs and chickens up to March 1996, even though a feed ban for cattle was introduced in 1988.[140] Evidence that cows can pass the disease to their calves has also led to some fears that women might pass on the deadly disease to their children before showing the symptoms themselves.[141]

Modern meat production has introduced other possible disease exacerbating factors. For example, many intensive farming methods may tend to reduce the proportion of available polyunsaturated fatty acids in food and increase the proportion of saturated fats.[142] Modifying the fatty acid profile of animal products through feedlot technology alters the fatty acid profile in the blood of humans consuming these foods which may in turn affect their susceptibility to heart disease and cancer.[143] Since the 1940s, many farmers have given their livestock antibiotic growth promotors. It was found that antibiotics improved the animal's feeding efficiency so that the animals need less food to reach marketable weight. There is now growing concern however that this practise could promote the emergence of bacterial strains which have genetic resistance to antibiotics. The hypothesis is that antibiotic-resistant bacteria could then cause diseases in humans that would be untreatable with conventional antibiotics.

In 1983, there was an outbreak of food poisoning in the United States which involved a strain of the bacteria Salmonella that was resistant to a number of antibiotics. The strain was linked to hamburgers made from the meat of cattle fed on low doses of the antibiotic chlortetracyline.[144]

Not everyone is agreed about the best way to resolve the problem. Some researchers believe, for instance, that if the use of low-dose antibiotics in animal production is stopped, there will be an increase in the incidence of diseases in farm animals which require other drugs to control, or alternatively farm animals will simply die. Supporters of this view cite, for example, the situation where following the ban on the use of all antibiotic growth promotors by Sweden in 1986, an extra 50 000 pigs died of 'scours' or diarrhoea.[145]

Part of the problem is that antibiotics treat only the symptoms, not the causes of diseases. The National Research Council in the United States in its comprehensive review *Alternative Agriculture* correctly reminds us that many of the animal diseases that farmers currently confront are a direct consequence of the intensive animal breeding measures now used by farmers, which forces animals to be raised in highly unnatural and confined conditions. The National Research Council refers to a number of studies which have shown that proper hygienic care of animals with access to fresh air, sunlight, clean pasture and water together with proper waste disposal substantially reduces disease in farm animals and the associated dependence on antibiotics.[146]

Nature's blueprint for health applies to animals as well as to humans. As we recognise that the earth's resources are limited, the use of animals for food must be urgently reviewed in favour of more sustainable forms of food production,[147] along with food production techniques such as permaculture.[148]

In the United States, about 200 million tonnes of maize are grown annually,[149] the bulk of which is fed to livestock to produce beef for hamburgers, and plump chickens for fast foods. In the 1980s in Central America and Amazonia, cattle ranchers burned at least two-and-a-half million hectares of forest each year, mainly to clear land to raise beef for lucrative export markets in the developed world.[150] Much of the land is now incapable of sustaining even a crop of corn.

In Australia, clearing land for grazing together with overgrazing, have led to massive destruction of soil structure as well as extensive salinity problems in some areas.[151] Given the destructive nature of western livestock corn production on the soil,[152] together with its high fertiliser and pesticide/herbicide requirements,[153] it is clear that meat and especially beef should be considered a luxury food in terms of the environmental cost required to produce it, not an expendable for fast foods.

Furthermore, some observers have noted that modern production methods may mean that up to 100 000 litres of water need to be used to produce 1 kilogram of beef compared to say 2000 litres to produce 1 kilogram of soy beans.[154]

Nature's own fast foods, fruits and nuts, not only contribute in most cases to a sustainable environment (trees build up the soil, prevent erosion and contribute positively to the oxygen–carbon dioxide balance) but are in harmony with nature's own blueprint by providing optimum nutrition.[155]

Health implications of refined foods

In general, there are two important aspects of refined foods that have a direct bearing on health. Firstly, there is the removal of essential nutrients during refining, and secondly the subsequent addition, in most cases, of chemical additives, flavours and colours to make the 'new' food more desirable. Both these aspects are inextricably interwoven with the diseases of civilisation.

Let us now consider some of these health implications in more detail. We now know that there are at least 14 essential trace elements required for human nutrition in amounts ranging from a few micrograms to milligrams per day.[156] The elements perform unique functions, often acting as catalytic or structural components of larger molecules which are essential for specific biochemical pathways. For example, chromium is required in microgram quantities for proper metabolism of glucose or sugars.[157] White sugar and white flour, which are ultimately a source of glucose for the body, are refined to the extent that they no longer contain chromium. Furthermore, chromium deficiency is now implicated as a significant risk factor for cardiovascular disease,[158] illustrating simply one of the many mechanisms linking refined foods with heart disease.

Another example is the element zinc. More than 90 zinc-containing enzymes have now been identified, including at least one in every major enzyme classification.[159] Moreover, zinc appears to have a major role in nucleic acid metabolism, protein synthesis, normal development of the outer horny layer of skin and sexual maturation. Zinc deficiency can also cause impaired immune response.[160] The nutritional requirements for this metal are particularly high in children[161] and deficiencies have been linked with foetal abnormalities.[162] However, substantial quantities of this element are removed by the refining of sugar cane and the milling of wheat.[163]

In 1929, *The American Journal of Diseases of Children* published a case where:

> A boy aged 12 had been termed by psychiatrists and psychologists a 'nervous hyperkinetic child'. To use his own words, 'I have to keep on

going and be doing something ... '. His diet was meagre in meats and vegetables and rich in candy and cookies.[164]

Since that time there has been a lively debate linking refined sugar to hyperactivity[165] and antisocial behaviour,[166] although the latter claims are especially controversial. What is arguable, however, is that foods and food constituents can adversely affect human health and behaviour[167] which raises very serious implications and new responsibilities for the food-processing industries.

For example, we know that nutrients are intimately involved in the regulation of brain function through their cofactor and precursor roles. However, as CE Leprohon-Greenwood and GH Anderson at the University of Toronto Faculty of Medicine point out:

> The role of vitamins and minerals in metabolic processes in most body cell types has been well described. However, their role in the nerve cell, particularly in the process of neurotransmission and in view of current developments in neurochemistry, requires considerable elucidation. Furthermore, there has been relatively little examination of the effects of marginal deficiencies on brain metabolism and function.[168]

They also write:

> In addition to this relative uncertainty of the biochemical role of vitamins and minerals in neurotransmitter metabolism, it is uncertain whether the brain has any special ability, relative to other tissues, to preserve its metabolic and neurochemical activity during nutrient deficiencies.[169]

The refining of carbohydrate foods such as sugar cane and cereals involves not only the removal of vitamins and minerals, but also the removal of dietary fibre. More than 20 years ago Burkitt and Trowell presented a powerful case from the available research directly linking refined carbohydrate foods and diseases of civilisation—such as bowel cancer, ischaemic heart disease (IHD), diabetes, gallstones, constipation, appendicitis, diverticular disease, hiatus hernia, dental caries, periodontal disease and obesity.[170] Their pioneering work had been inspired by the seminal ideas of a British naval doctor, TL Cleave, who in 1956, upon reflecting on his naval experience, was led to conclude that most of the diseases listed above could be attributed to the excessive consumption of refined carbohydrate foods and of refined sugar in particular.[171] In the meantime, evidence has continued to accumulate establishing the protective role of whole food components and certain dietary fibre in heart disease, cancer and diabetes.[172]

One of the most recent studies confirming the importance of dietary fibre, and particularly whole foods, was the *Health Professionals Study* by Dr Walter Willett and colleagues at Harvard University. The researchers found that men who ate more fibre, especially fibre from cereals, suffered fewer heart attacks.[173] The dietary fibre from oats,[174] rice[175] and legumes (beans, peas and lentils)[176] have been found significantly to reduce blood cholesterol levels, a major risk factor in heart disease, which accounts for 44 percent of all deaths in Australia every year.[177]

Dietary fibre appears to protect against colon cancer by increasing stool bulk and/or decreasing intestinal transit times and thereby reducing exposure of the colon to carcinogens.[178] Sufficient levels of fibre in the diet also enhances the production in the colon of one important short chain fatty acid, butyric acid, which is believed to inhibit the growth and proliferation of certain tumours.[179] Sufficient fibre in the diet may also serve to reduce the risk of breast cancer in premenopausal women by lowering the circulating levels of oestrogen. Other evidence suggests that it may be that the oestrogen-lowering effects are due to other components of whole foods such as lignands or isoflavonoid compounds which occur in very close association with dietary fibre.[180]

A newly emerging aspect of our modern refining and processing technology is the often lower glycemic index (GI) of processed foods. The GI of a food is a measure of how quickly a food is broken down during digestion into simple sugars and in particular into glucose. Fibre present in foods slows down the absorption of food components and the foods remain in the digestive system longer. Reducing the dietary fibre content of a food by milling and processing increases the GI value (see Figure 7.5).

Figure 7.5 Effect of processing on the glycemic index of some foods

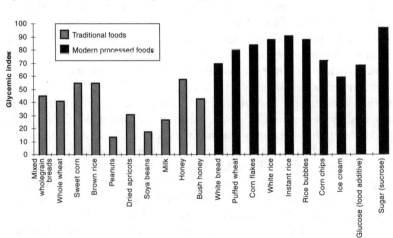

J Brand-Miller, *American Journal of Clinical Nutrition*, October 1995

High GI foods are believed to be a significant factor in the development of mature onset diabetes, one of the modern diseases of civilisation.[181] In many countries including Australia and the United States, the number of people with non-insulin-dependent diabetes mellitus has reached epidemic proportions. In the early 1990s, there were an estimated 700 000 diabetics in Australia and about 10 million diabetics in the United States where the estimated cost of the health care to the economy was US$90 billion.[182] Traditional foods in their least processed state, generally have the lowest GI values. In contrast, highly processed foods such as white bread, cornflakes and rice bubbles have high GI values.[183] It is revealing that unprocessed natural bush honey has a lower GI than commercial processed honey (see Figure 7.5).

The growing epidemic of diabetes and the economic cost to the community of this one of many health problems associated with processed foods, casts serious doubts on the perceived value to our economy of the food processing industry.

Another aspect of food refining and food processing which has been essentially ignored by both food technologists and nutritionists is the destruction of the original masticatory properties of the original whole food. As machines and chemical processes are used to extract food components from the original whole food matrix, the need for chewing in most instances, especially for any length of time, has been done away with. Yet the act of chewing provides mechanisms for maintaining good health.

After reviewing the literature relating to dental caries and periodontal disease, AK Adatia found that unrefined food is chewed for a longer period, and protective factors which are extracted during mastication remain in the mouth in a high concentration for a relatively longer period.[184] Furthermore, there is evidence to suggest that during compression of food between the cusps of teeth water-soluble protective factors were incorporated into enamel. Adatia goes on to suggest that vigorous mastication may help to resist dental caries and periodontal disease in other ways by pointing out:

> Foods which require to be chewed for a long period are cleared from the oral cavity more thoroughly than those not requiring much chewing. Refined carbohydrate foods such as white bread are retained for longer than wholewheat bread ... Secondly, mastication not only encourages salivary flow, which would have a flushing effect on teeth, but also saliva of a high buffering power ... Thirdly, vigorous mastication would wear and flatten the cusps of teeth, reduce the depths of the fissures and eliminate potential stagnation areas on the teeth ... Finally, vigorous mastication of firm fibrous foods would remove debris from

the exposed areas of the teeth and help to maintain the gingival tissues in good tone.[185]

Mastication of firm, fibrous foods also stimulates optimum growth of the jaws. For example, children who regularly chew sugar cane not only usually have caries-free teeth but have perfect orthodontal development as well. The consumption of refined foods also tends to promote over-nutrition, particularly excess consumption of calories which ultimately leads to obesity and many associated diseases. CW Binns wrote in 1989:

> Since these early days [of European settlement] the nutritional status of the colonising peoples has improved dramatically to our present situation where overnutrition and imbalance of nutrient intakes are the major problems.[186]

GB Gori, a deputy director at the United States National Cancer Institute, had suggested a decade earlier that:

> it is plausible that nutritional deficiencies and/or excesses influence metabolic processes that, after many years of insult, result in the appearance of certain forms of cancer. [187]

Thus K Heaton, after reviewing the relationship between food ingestion and disease, concluded that:

> ... carbohydrate refining makes the consumption of starch and sugar quicker, easier and less satisfying. As a natural obstacle to food ingestion, fibre adds effort to eating. Nutritionally, the role of fibre may be to act as a natural preventive of overnutrition.[188]

Unfortunately, these properties of whole foods are rarely, if ever, considered by food technologists when refining and recombining foods. Thus nature's intermeshed mechanisms to sustain health are thrown into disarray by the ignorant processing and refining of foods.

This nutritional vandalism does not stop here. Numerous food additives are combined with the refined remnants of food. Many of these are inimical to health, often in ways which are subtle or indirect. These additives include colours, flavours, sweeteners, preservatives, emulsifiers, thickeners and hundreds of specific food-processing chemicals. There are so many food chemicals used that, in the words of Arnold Bender, then Professor of Nutrition at Queen Elizabeth College in London, 'The trouble is that we 're using so many of these additives that you couldn't possibly test them all—not in the next 100 years'.[189]

Space does not permit us to discuss the varied health implications of the multitude of food additives commonly used, but reviews of progress in the evaluation of the health aspects of these substances which are generally recognised as safe are published regularly.[190] There is one more group of additives which should be looked at. These are the sweeteners.

Artificial sweeteners

Sweeteners are components of natural foods or synthetic chemicals that produce the taste sensation of sweetness. Sugar from sugar cane, called sucrose, is the most commonly used sweetener. There are many other sugars, including glucose sometimes called dextrose, fructose sometimes called levulose, lactose or milk sugar and maltose, to name a few. Glucose is produced commercially in large quantities by hydrolysing starch such as corn starch. Fructose is produced commercially by the enzymatic conversion of glucose. Other natural sweeteners have protein-type structures. These include Thaumatin and Monellin, which are found in certain African fruits and are 2000 to 3000 times sweeter than cane sugar.

Another group of sweeteners are humanmade chemicals. These include saccharin, which was developed in 1879, cyclamate in 1937 and aspartame in 1965. Saccharin and cyclamate are not metabolised and therefore do not add calories to the diet when used. Aspartame is metabolised but, being about 200 times sweeter than cane sugar, it is used in much smaller amounts, thereby contributing negligible calories to the diet.

Seventy years ago, the most common sweetener was cane sugar which was incorporated into the diet mainly by the homemaker. Today, processed foods contribute the largest portion of sugar to the diet and this amount is increasing. In the United States per capita cane sugar consumption deriving from beverages alone had increased from 2 kilograms in 1909 to 10 kilograms in 1980.[191]

This increase mirrors the general increase in refined food consumption, with the associated tendency to over-nutrition, particularly the over-consumption of calories. Increased community awareness of the need to avoid being overweight for both health and appearance aspects has prompted a demand for low-calorie foods, particularly beverages. This has resulted in a rapid increase in the use of synthesised, chemical, non-caloric sweeteners. Per capita consumption of these sweeteners in the United States in 1950 was 1.3 kilograms, rising to 3.2 kilograms in 1978 and more than doubling to 7.2 kilograms in 1984.[192] The rise in the use of these sugar substitutes is especially worrying given that both cyclamate and saccharin have been indicted as carcinogens.[193]

Cyclamate which serves as a co-carcinogen has been banned in the United States since 1970 but is currently used in about 40 other countries, including Australia.[194] In a number of tests and studies, saccharin has been found to cause bladder cancer in test animals. However, because of public and vested interest protest against a proposed ban, saccharin continues to be widely used in low-calorie beverages, canned fruits, desserts, salad dressings, jams and baked goods.

In 1981, after eight years of controversy regarding its safety, the United States Food and Drug Administration (FDA) approved the use of the new low-calorie sweetener aspartame for virtually unrestricted use in dry products and carbonated soft drinks. Marketed under the name Nutra Sweet™, it has found widespread acceptance with an annual sales turnover reaching US$700 million in 1988.[195] Aspartame is promoted as a healthy lifestyle alternative to sugar, being low in calories, non-cariogenic and metabolised naturally.[196] This last claim requires more careful scrutiny than perhaps it has been given.

In the body, aspartame is hydrolysed to aspartic acid, the amino acid phenylalanine and methanol. Methanol is a particularly poisonous alcohol and, in 1984, W Monte warned that aspartame may release toxic levels of methanol into the bloodstream.[197] There have been reports of brain damage to babies where the mother consumed aspartame while pregnant or when aspartame was fed to the babies during the first six months after birth.[198] However, in 1985 the American Medical Association concluded, after an investigation, that aspartame was safe. Their conclusions were based on the assumption that the methanol levels produced are not significant and that, since a number of natural proteins metabolise to phenylalanine, the production of this compound in the body is not a problem except to a very small minority who are sensitive to it.[199] The conclusions of the AMA may have been premature.

In 1987, TJ Maher and RJ Wurtman revealed a number of fascinating aspects of the aspartame picture. Firstly they found from a number of studies reported since 1985, that doses of aspartame which are within the range actually consumed by some people can affect the chemical composition of the brain, and may thereby contribute to particular central nervous system (CNS) side-effects, including headaches, inappropriate behaviour responses, and seizures.[200]

Secondly, although rats and mice are often used to test the safety of food compounds, there are major metabolic differences between rodents and people. Maher and Wurtman state that:

> The major biochemical effect of aspartame, in humans, is to raise blood and, presumably, brain phenylalanine levels; in contrast, its main effect

in rodents is to raise blood (and brain) tyrosine levels and tyrosine is often the antidote to phenylalanine's effects on the brain.[201]

They comment that these findings underscore the necessity for human studies before conclusions can be drawn that a food or compound is risk-free. This is an echo of the warnings of Sinclair three decades earlier, that certain chemicals added to foods 'believed to be innocuous are so classified because they do not kill rats or mice. Man is a different animal, as is sometimes forgotten'.[202]

Thirdly, Maher and Wurtman observe that:

The effects of aspartame, and of certain other food additives, like caffeine, on the nervous system are sometimes not of such a nature as to allow their detection by the standard neurotoxicological tests used to assess the safety of food additives, inasmuch as these effects need not be associated with cell death, nor with other visible manifestations of neuronal damage. Rather, they involve more subtle biochemical changes, as well as functional consequences that are demonstrable only in specially treated animals (and possibly, by extrapolation, only in especially vulnerable people).[203]

Fourthly, they point out that consumption of dietary phenylalanine in the usual way as a constituent of protein does not elevate brain phenylalanine levels. In contrast, consumption of aspartame with protein at a meal, elevates plasma phenylalanine levels without elevating those of the other large neutral amino acids (LNAA). This causes marked elevations in the plasma phenylalanine ratio with a parallel rise in brain phenylalanine levels. They point out, 'It should be noted that aspartame is probably the only phenylalanine-containing food that man has ever eaten which elevates this ratio'.[204]

Furthermore, if an aspartame-containing beverage is consumed along with, for example, a carbohydrate-rich, protein-poor dessert food, its effects on brain phenylalanine are doubled. Once within the brain, excess phenylalanine can also inhibit enzymes needed to synthesise the neurotransmitters. The researchers go on to report that elevated levels of phenylalanine circulating in the blood can lower the production of important brain chemicals such as catecholamines and serotonin. This in turn can adversely affect a number of physiological processes including the body's response to stress.[205]

The significance of these findings is enormous and Maher and Wurtman comment reflectively in their discussion: 'It is unfortunate but perhaps not surprising that questions about aspartame's phenylalanine-mediated neurologic effects arose after the sweetener was added to the food supply.'[206]

Dr J Olney at the University of Washington believes that aspartame may be linked to brain cancer. In animal studies, 12 out of 320 rats fed aspartame developed brain tumours while none of the controls developed cancer. Dr Olney analysed the National Cancer Institute data and found that it showed a 10 percent jump in brain cancer incidence amongst humans in the three years after the approval of the use of aspartame.[207] At a press briefing in November 1996, Dr Olney said the rise was most-likely due to an environmental carcinogen and that the evidence points to aspartame.[208]

The widespread introduction into the food supply of a chemical producing a new and unique brain chemistry has the implied potentiality for a major health disaster. Ironically, this situation has arisen as a result of ignoring the importance of nature's inbuilt calorie limiting components, such as fibre, while substituting synthesised chemical substitutes in their place.

There is yet another side to the sweetener story. Even seemingly harmless sugars, such as glucose, can be inimical to health when used in combinations not intended by nature.

Glucose syrups (corn syrups) can substantially impart mouthfeel to beverages or a glassy appearance to canned fruits or help retain the colour of certain jams. They also cost 10 to 40 percent less than cane sugar. There has thus been a steady increase in the use of glucose as an added sweetening agent in a wide range of foods including beverages, jams and ice cream. Yet nature does not provide free glucose in foods except in combinations with fructose and mostly with fructose in approximately equal proportions or in excess. Research has shown there is good reason for this.

In 1981, VA Koivista and co-workers demonstrated that glucose ingestion, especially prior to exercise, results in the release of large amounts of insulin into the bloodstream and subsequent reduction in blood sugar levels.[209]

Further studies indicated that fructose ingestion before exercise maintains stable blood glucose and insulin concentrations and appears to lead to the sparing of muscle glycogen.[210]

It is revealing to consider that the most concentrated sweetener provided by nature, honey, is a unique combination of fructose and glucose with a little sucrose and minerals such as iron, copper and manganese at levels large enough to be significant nutritionally.[211] In a study of the use of honey for infant feeding, performed over four decades prior to the VA Koivista work, FW Schlutz and EM Knott, reported that:

> Honey appeared to have special advantages for infants because, with the exception of dextrose, it was absorbed most quickly of all the sugars

tested during the first 15 minutes following ingestion, yet did not flood the blood stream with exogenols sugar. It also maintained both a steady and a slow decrease in blood sugar until the fasting level was again reached. The response of honey is presumably due to the unique combination of the two easily digested monosaccharides, dextrose and levulose. Honey is quickly digested and absorbed into the body because of its dextrose content, while the levulose, being somewhat more slowly absorbed, is able to maintain the blood sugar. Honey has an advantage over sugars which contain higher levels of dextrose since it does not cause the blood sugar to rise to higher levels than can be easily cared for by the body.[212]

Again, use of apparently harmless food additives, such as glucose, in ignorance of nature's blueprint for health can lead to nutritional imbalances which predispose us to poor performance and disease.

Fibre and fish oil fads: Looking for a quick cure

In the 1960s, the phenomenon of the hippy revolt against the orthodox establishment coupled with the warning statistics of increased cancer and heart disease mortality rates, awakened public concern for health and the environment. Since that time we have experienced a number of health fads, as people attempt to protect against or remedy the health problems they have inherited. But curiously we find that the reasoning behind the fad cure is often underpinned by an identical philosophy of health or the same interpretive framework which produced the problem in the first place. The current fads of fibre and fish oil serve to illustrate this point.

Steadily increasing publicity has been given to the cholesterol–heart disease link, and numerous public health authorities have implemented programs to reduce cholesterol intake. However, in March 1987, Robert Kowalski published his book *The 8-Week Cholesterol Cure: How to Lower Your Blood Cholesterol by up to 40 percent without Drugs ar Deprivation* (Bantam, New York). By 1988 the book was the number one non-fiction book on the *New York Times* best-seller list and the publisher had sold over 1.2 million hard-cover copies.

The book essentially focused on the cholesterol-lowering properties of soluble fibre and oat bran in particular, which had been reported in medical literature, and the author's 'personal fight against—and triumph over—the Western world's number one killer'. A number of further medical studies followed[213] and in April 1988 the American Medical Association released its confirmatory findings.[214] The result was, in the words of James Tindal, the executive vice-president of the

United States Foods Groups, that 'demand [for Quaker™ Oat Bran] went right through the roof'.[215]

The food industry rapidly responded to the growing consumer trend with hundreds of new products, new in the sense that they now contain the additive oat bran. In similar terms, the grain 'offal' once fed to animals is now sold to us at ten times the price.

The fibre fad expanded to rice bran[216] which has also been shown to lower cholesterol levels.[217] The cholesterol lowering properties of oat bran seem to be related to its high soluble fibre content, whereas in the case of rice bran the lowering effect appears to come from the rice bran oil. The high oil content of rice bran necessitates that the bran be stabilised soon after separation as it becomes rancid in a matter of hours. Defatted rice bran is also available but does not appear to significantly lower cholesterol in animal tests.[218] Let the consumer beware!

However, many common whole foods without special additives, also reduce cholesterol levels when incorporated into the diet. The cholesterol lowering properties of whole rolled oats were reported by AP de Groot and co-workers as far back as 1963[219] and have been confirmed in numerous recent studies.[220] Whole rice also lowers blood cholesterol levels,[221] as do many other whole foods. In 1968, KS Mathur and co-workers reported that mature leguminous seeds such as chick peas and mung beans reduced serum cholesterol levels in humans.[222]

Since that time, literature has accumulated confirming that a wide variety of whole foods,[223] including the common 'baked bean',[224] peas and lentils,[225] soy beans,[226] nuts including walnuts, macadamias and almonds,[227] mushrooms[228] and flax seed (linseed)[229] significantly reduce cholesterol levels.

Much publicity, has been given to the possible role of fish oils in preventing heart disease. Populations such as the Greenland Inuit (Eskimos) and the Japanese, that consume a large proportion of fish in their diet are well known to have a low incidence of coronary artery disease.[230] The beneficial factors are believed to be the omega-3 essential fatty acids (EFAs) and their derivatives,[231] which are found in the fish oils. These oils, particularly those from cold-water fish, can contain up to 30 percent derivatives of omega-3 EFAs, making them rich sources of these nutrients. Consequently there has been an influx of products containing fish oil onto the market.

Often fish oils have been used only in a hardened or hydrogenated form to prevent the occurrence of fishy off-tastes and smells. New developments in the fats and oils industry has enabled 'novel' refining techniques to be used to produce fish oils that can be added to foods without effecting the flavour profile of the food. Another technique microencapsulates the fish oil so that it forms a powder that can be added to foods.[232]

The special refining process for fish oils suitable for food applications involves neutralisation of free fatty acids and removal of pigments. Special techniques remove peroxides, aldehydes and ketones which play an important role in the development of off-flavours. Antioxidants are then added to stabilise the final product.[233] Recent research suggests however that some of these natural oxidation products have cancer-inhibiting properties and may contribute to possible protective roles that fish oils appear to play in protecting against diseases such as breast cancer.[234]

Given that the highly refined food additive form of fish oil which is stabilised with added antioxidants is thus quite different to the natural unprocessed fish oil found in fish, from which it was derived, the use of fish oil supplements as an alternative to eating whole fish gives rise for concern, especially when there is strong evidence that whole fish does protect against breast cancer. In a 1993 Norwegian study, the health records of 533 276 women aged 35–64 years were categorised by reference to the occupation of their husbands over a 15-year period. The wives of fishermen, who were found to consume twice as much fish as other women, were found to have the lowest incidence of death from breast cancer.[235]

Furthermore, while eating fish oil supplements or whole fish produced comparable blood fat and cholesterol changes, only the consumption of whole fish improved factors. The researchers suggested that the greater health benefit of whole fish may be related to the presence in the fish flesh of other compounds that could complement the action of the omega-3 fatty acids.[236]

Clearly the addition of 'health additives' to foods has the potential to transfer the manifestation of disease in the human organism from one area to another rather than be the answer to disease produced by an unbalance diet in the first place.

Whole foods: The long-term answer to human and environmental health

The nutrient interactions that maintain a healthy balance of the intricate biochemical pathways of the human organism are both complex and subtle. The conventional reductionist approach to nutrition and the associated food technology fails to provide an understanding sufficiently complete to safely alter to any great extent the combinations of nutrients provided by nature. The failure of technologists to recognise this fact has resulted in many of the products of food technology becoming stepping stones to disease rather than a pathway to health.

Conventional nutrition education has exacerbated the situation by persistently emphasising a simple component model as a basis of understanding desirable food combinations. The nutrition biochemist RJ Williams emphasised the same point two decades ago when he wrote about nutrition education, 'If we simplify matters too much, we do violence to the truth'.[237] Yet there is a simple approach that may be taken. That is the utilisation of whole foods, which, by way of an intrinsic balance of nutrients within themselves are able to provide the same balance to the human organism.

We have seen, for example, that whole foods such as wheat contain a balanced array of nutrients and other factors such as fibre, which together provide the requirements for health. Even whole foods such as eggs which have been maligned in the past have now been vindicated by research.[238] Professor Stehbens writes:

> The effect of eggs on serum cholesterol levels in humans is variable, often misrepresented, and without the dire consequences so often alleged. Much depends on other dietary constituents and food additives.[239]

An important, yet often overlooked, value of whole foods is their ability to enhance the health mechanisms of the human organism to the extent that disease processes produced by consuming adulterated and refined foods are actually slowed and in some instances even reversed. The protective role of fresh fruit and vegetables against the development of stomach cancer is one of the most consistent findings.[240]

Onions and garlic have been shown to have a very significant protective action against high serum-cholesterol and blood coagulation changes which occur after fat ingestion.[241] Both these foods are also effective in controlling the hyperglycaemic effect of eating glucose, and onions have been observed to lower blood glucose levels in clinical diabetes.[242] Findings such as these prompted J Carper to catalogue and synthesise the growing evidence that many whole foods are as effective as prescription drugs, yet without harmful side-effects and at a fraction of the cost.[243]

The failure to recognise the provisions of nature for health through food has led to a carelessness in how the food is produced. The artificial feeding of our plants and animals has led to alterations in the compositions of these foods which are also inimical to health as well as contributing to the degradation of our environment. There are also the harmful effects of the hormone, pesticide and antibiotic chemicals which are deliberately administered to plants or animals during growth.

Over four decades ago H Sinclair, in his World Health Day address remarked, 'We feed hens on concentrates that are deficient in the unstable fatty acids, and the eggs in consequence are deficient'.[244] This same issue was raised again by M Denton and RW Lacey in 1991[245] and in 1996 researchers from Australia's Commonwealth Scientific and Industrial Research Organisation (CSIRO) showed that changing the fatty acid content of animal feeds changed the cholesterol levels in humans consuming those animal products.[246]

Even before this time, experts such as Dr McCarrison, a former director of nutrition research with the British Government in India, and Sir Albert Howard, reported that organically grown foods were more effective in restoring health and preventing disease than foods grown with artificial fertilisers and chemicals. McCarrison in his early experiments found that rats raised on a deficiency diet, when fed compost grown grain rapidly regained their health, whereas rats fed grain grown with artificial fertilisers continued to lose weight. These results were later confirmed by Rowland and Wilkinson in their Knightsbridge laboratory, yet chemical analysis revealed no difference in the major constituents between the compost-grown grain and that grown with artificials.[247]

A Guest, a former BBC broadcaster, reports the words used by Sir Howard to describe his findings:

After 15 years in horticultural colleges all that I have learned was not worth a candle ... I then turned away from these things to study nature's way and how she feeds her crops. Then I started composting from vegetable and animal refuse without the aid of the horticultural scientist or chemist or any artificial fertilisers, insecticides, germicides, spraying machines or any of the expensive tackle of the modern research station. I soon found the key to good crops—crops of quality and free from disease—was a fertile soil and that such fertility could only be brought about by the stopping of the use of artificial fertilisers and putting into the soil sufficient humus got by composting vegetable and animal refuse. I soon found out that plants grown with soil treated with compost had a great measure of immunity from disease and when this produce was given to animals it passed on to them the same immunity from disease, for I actually saw my own cattle rubbing noses with cattle that had foot and mouth disease and my own cattle were none the worse and did not take the disease.[248]

These experiences, which defy the simple reductionistic model of plant nutrition and which have provoked scorn and antagonism from many food technologists, are now being realised by an observant few as indicating the direction which food production must take if we are to restore the health not only of the human race but of our planet. J Reganold, Associate Professor of Agronomy and Soils at Washington State University, in a 1989 article titled 'Farmings Organic Future' urges:

> Land farmed without fertilisers or pesticides can be just as productive as land pumped full of chemicals. More important still, organic farming preserves the land's most precious commodity, the soil.[249]

He goes on to point out:

> Organic matter has a profound impact on the quality of the soil; it encourages mineral particles to clump together to form granules, improving the structure of the soil; it increases the amount of water the soil will hold and the supply of nutrients; and the organisms in the soil are more active. All in all, organic material makes the soil more fertile and productive.[250]

A large body of evidence has accumulated which indicates that modern intensive tillage practices associated with continuous monoculture or short rotations may make soils more susceptible to erosion.[251] Conventionally cultivated soil with continuously intertilled crops almost always experiences a decline in organic matter and the ability to retain moisture (see Figure 7.6). On the other hand, organic farming techniques such as the use of manure and legumes produce soils which tend to have higher levels of overall organic matter. Organic matter improves soil quality by increasing granulation, water infiltration, nutrient content, soil biota activity and soil fertility and productivity.[252]

Figure 7.6 Effect of agricultural practices on soil organic matter

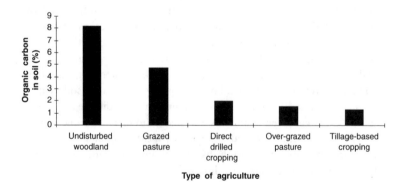

State of the Environment Advisory Council, 1996

A 1982 inventory by the United States Department of Agriculture showed that about 44 percent of arable land is eroding faster than is considered tolerable. A similar percentage applies to England and Wales, according to 1986 data.[253] In Australia, soil erosion on average amounts to about 14 billion tonnes or about 19 percent of the total soil moved each year globally, even though Australia occupies only 5 percent of the world's land area.[254]

We find that 'technological advances' in the form of new fertilisers, pesticides and plant varieties have masked, for the time being, the decline in productivity as a result of erosion. But these 'technological advances' in general only have the potential to increase yields on deep, relatively uneroded topsoils. They have little application for the areas with existing poor or eroded soils, and as good soils erode further, these so-called 'advances' will progressively fail to increase yields.

We earlier referred to the enormous value of the sales of refined and processed foods. By replacing these altered foods with whole foods we may also provide economic opportunity to replace conventional farming by organic farming. The cost of technology could become the cost of conservation.

Reganold shares with us this thought:

> We worry about whales and sea otters, and rightly so. But, in general, we do not concern ourselves with soil, because most of us live in cities, unaware of the inherent value of soil. We rearrange and restructure it, pump it full of chemicals, and otherwise abuse it. Our obligation as farmers and as society is to put soil before profits.[255]

What is needed to achieve this shift of emphasis is a new consciousness in which our role as stewards of the earth takes precedence over our desire to rule the earth. We must recapture the value of nurturing the earth rather than contriving through our greed, to secure its rape.

What you can do

W e have seen from this chapter that much of the food we eat these days is pulled apart in the name of technology and reconstituted for the purpose largely of maximising the food processing industry's profits. We have been educated to desire what the food industry can produce—without comprehending what it destroys to produce it. We degrade our soil, our water, our air, the fragile economies of agricultural regions and, of course, our own bodies. However, we all can play a part in minimising the destructive impact of this vicious circle.

1. We can rediscover the rich variety and wonderful tastes of whole foods. That is, we can buy and cook unprocessed whole grains, legumes, fruits and vegetables, and other whole foods available to us. In order to do this efficiently it may be necessary to seek out the wide variety of 'wholefood' cookbooks. Vegetarian and ethnic cookbooks are often a good source of these recipes.

2. We can avoid buying refined foods such as white bread, flour and sugar and replace them with natural wholemeal bread, wholemeal flour and raw sugar or, better still, honey. We can avoid reconstituted wholemeal flour and reconstituted fruit juices. The latter have lost important flavours and components and may contain synthetic flavours and additives (which may not be listed on the label).

3. We can choose not to use highly refined products such as margarine and vegetable cooking oils. We can choose to replace these foods with cream or cold pressed olive oil and other natural fats such as those found in avocados. Olives and avocados are tree crops, and these help reduce soil erosion and help slow down the 'greenhouse effect'. Instead of being seduced by the desire for refined 'smooth' foods, we can deliberately select chewy natural foods. These foods will help to control weight and appetite, as well as aid digestion and protect against dental caries.

4. We can replace repetitious meals of nutritionally fortified food (added vitamins and minerals) with a wide variety of foods at meals, particularly fresh fruits and vegetables.

5. We can, where possible, avoid buying commercial fast foods and choose instead nature's own fast foods: fruits and nuts.

6. If possible, we should buy organically grown produce. Not only is this likely to be more nutritious but you will also be supporting farming methods that assist conservation of the soil and the maintenance of fragile rural and labour intensive industries.

7. We can buy foods grown in as close as possible conditions climate-wise to where we live. This is illustrated by the example of sunflower seeds which when grown in a hot climate produce a much lower level of polyunsaturated oils compared with when they are grown in cold latitudes. There is some evidence to show that polyunsaturated oils may make us more susceptible to skin cancers from sun exposure in hot climates. Here is just one example of how nature has finely tuned our food supply to harmonise with our environmental needs. In hot climates saturated fat appears to help protect against skin cancers, and at the same time the increased sunlight entering our eyes lowers blood cholesterol levels ensuring optimum lipid metabolism for health.

We can each optimise our own personal health by taking full advantage of the whole foods nature has provided. Remember, variety is the spice of life, not the food additives we sometimes substitute for spice.

chapter 8

Aluminium in food: Is it a factor in Alzheimer's disease?

The common industrial and household metal, aluminium, epitomises the ability of modern science and technology to extract and use for man's convenience the materials stored in the earth's crust. Today this element, the most abundant metal in the earth's crust, is used to clad our houses and vehicles, to frame our windows, to cook, process and package our food and drink, and as a material for general construction and ornamentation. Compounds of this metal are added to our water supply, to our foods and to our medicines.

Now, just over 100 years since the independent discovery of the modern electrolytic process of extracting aluminium by Paul Heroult in France and Charles Hall in the United States, a vast literature has accumulated which indicates that aluminium is toxic at very low levels to both plant and animal life, depending on the bioavailability of the aluminium ingested.

In this chapter we propose to show that when the adverse health effects of aluminium are viewed together with the environmental impact of aluminium production, the implications for the health of the inhabitants of this world are enormous.

Aluminium: A health hazard?

The connection between aluminium and health is set in the background of the belief by most scientists, even in recent times, that aluminium compounds are harmless to the human body except when taken in very large doses.[1] This belief arose because aluminium is a major component of rocks, clays and soils and consequently is found in plants, despite the fact that there is no known biological function of aluminium.[2]

It was also widely held that aluminium compounds were not readily absorbed by the gastrointestinal tract and thus excreted almost entirely in the faeces.[3] Standard texts on toxicology reflect these views.[4] As recently as December 1988, experts were not concerned when 20 tonnes of 8 percent aluminium sulfate solution were accidentally dumped into the water supply of a small English town, causing water aluminium levels to rise from the usual 0.1 parts per million (ppm) to 40 ppm, an increase of 400 times.[5]

On the basis of this assumed low toxicity, aluminium compounds have been widely used as chemicals in water purification, food and pharmaceutical industries. Aluminium sulfate and alums are commonly used to treat municipal water supplies which have a high colloidal or suspended matter content. Treatment of other water supplies after heavy rain is common. Aluminium sulfate is also used as an acidifier in some baking powders.

Aluminium ammonium sulfate is a buffering agent used to maintain acidity during food processing. Aluminium calcium silicate is a common anticaking agent added to prevent certain food powders such as table salt from compacting. Highly soluble aluminium compounds are also used as miscellaneous food additives. A specific example would be the commercial addition of alums to frozen strawberries, maraschino cherries and pickles to improve their appearance.[6] Aluminium salts are sometimes added to processed cheeses and beer. Sodium aluminium phosphates are common emulsifiers for processed cheeses comprising up to 2.6 percent by weight of the cheese.[7] This compound may also be present in baked goods as a leavening agent at levels up to 2.5 percent.[8]

Even table salt commonly contains 1 percent sodium silico aluminate. The presence of aluminium is ubiquitous—aluminium compounds are found across the entire spectrum of processed food from table salt to cakes. There is also the aluminium which, as we shall see later, is subsequently leached into food when cooked in aluminium containers.

Aluminium compounds also find extensive use in pharmaceuticals including intravenous fluids,[9] antacid preparations, anti-diarrhoeals, buffered aspirin tablets, vaginal douches, and as formulation aids in diphtheria, tetanus and pertussis (whooping cough) vaccine. These vaccines are commonly administered to immunise newborns and young children when they are two, four, six and eighteen months, and four to five years of age.[10]

Even the alternative medicine industry has not escaped the aluminium contamination of health supplements, with some herbal tablet manufacturers using up to 2 percent aluminium sodium silicate as a lubricant. Another area of aluminium usage is in personal care agents. For example, personal deodorants and anti-perspirants can contain up to 15 percent aluminium chloride, a highly soluble form of aluminium, which as we discuss later, can be absorbed into the body.

The aluminium toxicity controversy

The toxicity of aluminium compounds has not always gone unquestioned. In 1893, EE Kellogg (wife of the famous Dr JH Kellogg from Battle Creek) wrote in her book *Science in the Kitchen*, concerning the addition of alum-containing baking powder to wheat flour: 'Such adulteration is exceedingly injurious'. She then went on to describe a method for detecting the presence of alum in flour.[11]

Just after the turn of the century, aluminium cooking utensils increased in popularity and in 1913 staff from the medical journal *The Lancet* did tests to determine the extent to which aluminium entered foods cooked in aluminium pots and pans. Measurable aluminium residues were detected when the following foods were prepared in aluminium cookware: beef steak fried in an aluminium pan; tomatoes, butter, salt and pepper fried in an aluminium pan; soup prepared in an aluminium saucepan; and brussel sprouts boiled in an aluminium saucepan.

The Lancet researchers dismissed the small quantities of aluminium entering the food as being harmless.[12] During the late 1920s, a controversy over the possible poisonous effect of the use of alum in baking powders led to a number of animal studies on the gastrointestinal absorption of aluminium. After feeding the animals foods which contained alum most researchers reported finding elevated levels of aluminium in tissue, along with indications of aluminium absorption from the gut. As has been pointed out, however:

> the effects of aluminium absorption were generally dismissed as insignificant because the bulk of the orally administered aluminium could be found in the faeces, and the aluminium absorbed from the GI tract was only a small, and often immeasurable fraction of the oral load.[13]

Given the admitted elevated levels of tissue aluminium, the conclusion that dietary aluminium was harmless seemed to be premature. Fortunately, it was not a conclusion accepted by everyone. In 1939, the American herbalist Jethro Kloss urgently warned his readers against using aluminium cooking utensils and referred to the animal experiments with alum.[14] In 1951, American health writer Julius White included a major review of the adverse effects of aluminium in his book *Abundant Health*.[15]

In the post-war era the toxicity of aluminium cookware was raised again,[16] and in 1974, JRJ Sorenson and co-authors from the University of Cincinnati in Ohio, published a 92-page review of over 800 research papers on aluminium. These authors concluded: 'As in the past there is still no need for concern by the public or producers of Al or its products concerning hazards to human health derived from well established and extensively used products'.[17] Since then, the issue has been kept alive essentially by the natural health and alternative medicine proponents and certain nutrition writers. For example, the *Nutrition Almanac* (1975) states:

Foods cooked in aluminium utensils may absorb minute quantities of the mineral. Small quantities of soluble salts of aluminium present in the blood cause a slow form of poisoning characterised by motor paralysis and areas of local numbness, with fatty degeneration of the kidney and liver. There are also anatomical changes in the nerve centers and symptoms of gastrointestinal inflammation. Symptoms result from the body's effort to eliminate the poison.[18]

In 1980, S Levick at Yale University School of Medicine wrote to the *New England Journal of Medicine*:

> I wish to call attention to a possible public health hazard: aluminium-induced organic mental syndromes from corrosible aluminium cookware. Large numbers of people in our aluminium-using society may be the victims of slow aluminium poisoning from several sources. Corrosible aluminium cookware may be a nontrivial source. This hypothesis needs systematic study.[19]

His hypothesis was dismissed by JH Koning of Riverside General Hospital who argued that most ingested aluminium was recovered in the faeces and that much more aluminium was ingested when taking antacid medications than one could ever leach from an aluminium pot.[20]

Aluminium and disease

Although the debate continues, there is no doubt that since the early 1970s the evidence has mounted to show that soluble aluminium ions are neurotoxic.[21] In 1973, DR Crapper and co-workers reported that aluminium may be an aetiological factor in Alzheimer's disease[22] and AC Alfrey reported that the dialysis dementia observed in some kidney disease patients may be caused by aluminium ions.[23] It is now recognised that aluminium toxicity figures prominently as a major cause of disease and death in kidney disease patients who are unable to detoxify and eliminate aluminium as effectively as healthy individuals.[24] The aluminium–Alzheimer's disease link is currently controversial with an active debate in the medical literature.[25]

In 1993, the eminent epidemiologist Sir Richard Doll reviewed the evidence linking Alzheimer's disease and environmental aluminium.[26] He points out that while a number of researchers have reported above average amounts of aluminium in the brains of people who have died of Alzheimer's disease, a recent analysis of unfixed, unstained frozen sections of senile plaques failed to detect any aluminium. This has led some observers to speculate that earlier findings may have been due to contamination of samples or stains with dust containing aluminium.

While Doll does not believe that there is sufficient evidence to show that aluminium in the environment is the cause of Alzeimer's disease, he affirms that given a number of unanswered questions, the debate must be kept open until the controversial evidence is resolved. He concludes that the combined clinical and epidemiological evidence suggests, however, that whatever the link to Alzheimer's, aluminium is neurotoxic to humans.[27]

Not all researchers agree that Doll's paper adequately addressed the biochemical evidence implicating aluminium as a causal factor in Alzeimer's disease. They point out that acute exposure to aluminium such as in dialysis patients, and chronic exposure to aluminium such as from acidic drinking water, both result in cell death. Admittedly, the pathologies are different because the biochemical mechanisms alter as the concentration of aluminium changes.[28] Nonetheless, the biochemical evidence implicating environmental aluminium in Alzheimer's disease continues to accumulate[29] though the exact connections remain to be elucidated.

Alzheimer's disease now affects 5 percent of Australians over 65 years of age and in 1993, more than 4 million Americans had been diagnosed as having the disease.[30] Current research is steadily adding weight to the aluminium cookware–disease link. Recent and more rigorous studies have demonstrated clearly that substantial levels of aluminium are leached into solution when acidic foods are cooked in aluminium utensils. Aluminium concentrations of 15 ppm[31] and 200 ppm[32] have been recorded when citric acid solutions (pH3) were boiled for ten minutes. N Ward at the University of Surrey has found levels of leached aluminium ranging from 600–1000 ppm in water adjusted to simulate cooking acidic foods.[33] Wine and apple juice stored in aluminium containers for two years contained levels of 1.5–3 ppm and 9.4 ppm respectively.[34] Soft drinks sold in aluminium cans can have up to 6 times as much aluminium when compared to their glass-bottled counterparts.[35] Even cooking oatmeal and carrots in aluminium utensils increases their aluminium content measurably.[36]

Accurate analytical data has revealed that by cooking beef in aluminium, the aluminium content of the beef is increased, on average, by a factor of 38 times[37] (see Figure 8.1).

JAT Pennington from the United States Food and Drug Administration has published a major review of the levels of aluminium leached into various foods as a result of cooking in aluminium utensils.[38] She concludes from her data that compared with leached aluminium, aluminium food additives are the main source of dietary aluminium (see also Figure 8.2).

Figure 8.1 Uptake of aluminium into beef after cooking in aluminium pan (mean values for 10 replicate analyses)

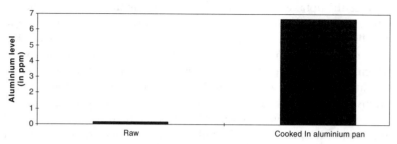

DM Sullivan et al., *Journal of Association of Official Analytical Chemists*, vol. 70, no. 1, 1987

Figure 8.2 Estimate daily aluminium intake from various sources

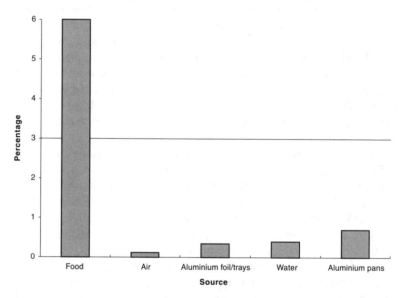

Note: Aluminium may have different bioavailability to the human body, depending on its source. For example, the aluminium in drinking water may be more easily absorbed than the aluminium in certain foods.
Choice, January 1992

There is now substantial evidence that the addition of aluminium to municipal drinking water poses a risk of disease. Using an artificially produced light aluminium isotope from the Los Alomos National Laboratory in the United States, and an ultrasensitive analytical technique called accelerator mass spectrometry, Dr J Walton from the Australian Institute for Biomedical Research and co-workers, were able to trace the uptake of very low levels of aluminium into the bodies of laboratory rats. A minute 0.1 micrograms of radioactive aluminium in 4 millilitres of water (that is a concentration of only 25 parts per billion, much less than the aluminium levels in treated drinking water which is often at several hundred parts per billion levels), was given to the rats. Measurable amounts of this aluminium were later found in the brains of the animals. Despite the persistent scientific hypothesis that in principle their kidneys have the ability to excrete much higher levels of aluminium, it is clear that the kidneys are less than fully effective at eliminating all the aluminium entering the body. This research showed that for rats, aluminium even from very low level soluble forms can, despite the body's excretory mechanisms, accumulate in the brain.[39]

The uptake of aluminium into the human brain is likely to be similar to this uptake into a rat's brain. If this is indeed the case, then there is need for grave concern over the addition of soluble aluminium compounds to our drinking water and food. Dr Walton and co-workers calculated that over a period of seven or eight decades of drinking aluminium-treated tap water, a microgram of aluminium would enter the human brain. Aluminium even in this small amount would be toxic to brain cells. Whatever the outcome of the Alzheimer's debate, it is evident that aluminium-induced brain cell death could manifest itself in old age by mental senility or dementia.

An English study has shown that the risk of Alzheimer's disease is 1.5 times higher in districts where the aluminium levels in water exceed 0.11 ppm, compared with districts where the water contained less than 0.01 ppm.[40] A similar Canadian study published in 1996, found the risk of Alzheimer's disease was 2.5 times higher for people who had lived for ten years or more in a region where the water contained more than 150 parts per billion of aluminium.[41]

It has now also been established that the aluminium ion is absorbed in the gastrointestinal tract and that absorption is enhanced in the presence of certain ligands (bonding molecules) such as citric acid which is a common food additive.[42]

Certain foods also increase the bioavailability aluminium from drinking water. One detailed study published in 1994 by the Urban Water Research Association of Australia, found that orange juice, coffee and wine were found to significantly increase the uptake of aluminium.[43] Fresh orange juice gave the largest increase with the absorption of aluminium being increased by a massive 1700 percent (see Figure 8.3). This raises serious doubts about the practice of reconstituting orange juice from concentrates prior to resale. Natural orange juice contains relatively little aluminium. However if the concentrate is reconstituted using town water containing aluminium, that aluminium is now likely to be highly bioavailable.

Figure 8.3 Affect of foods on uptake of aluminium from drinking water

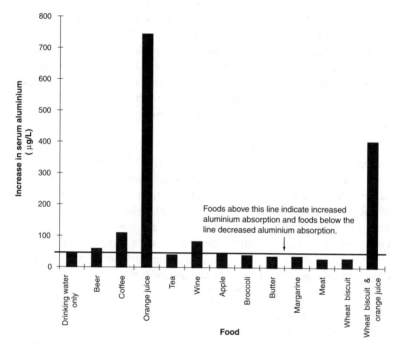

JR Walton et al., Urban Water Research Association of Australia, Research Report no. 83, 1994

It is well established that the aluminium ion is neurotoxic. PO Ganrot points out that normal and lethally toxic brain concentrations of aluminium ion are documented in over 15 research papers and that: 'These investigations are in good agreement and imply that the lethal concentration exceeds the normal level by only a factor of three to ten'.[44]

That such small values can make so great a difference reveals the existence of a very fine line between the states of health and disease. Lamentably, this is a point which has frequently been overlooked in the literature. Once taken up by the brain, aluminium cannot be eliminated, and thus contrary to the conventional wisdom, can have a cumulative effect which is devastating. After reviewing 959 research papers of relevance to aluminium metabolism, Ganrot concludes that aluminium ion is a causal factor in amyotrophic lateral sclerosis (a degenerative disease of the central nervous system) and Parkinson's disease.[45]

The association of aluminium uptake with aging is not new. Soon after DR Crapper and co-workers at the University of Toronto discovered that a high localised accumulation of aluminium occurs in the brain, research was begun by N Kharasch and others at the University of Southern California, to determine the significance of the formation of aluminium DNA complexes. In 1979, a connection was postulated between aluminium and the aging process, with the aluminium DNA complexes affecting genetic information transfer.[46] In 1980, SJ Karlik and associates published research findings which demonstrated that aluminium bound directly and toxically with the phosphate in DNA. In closing the discussion of their paper they cautiously suggested that the cross-linking between DNA and aluminium could lead to 'deleterious biological effects'.[47]

Additional evidence in support of the connection between aluminium and aging has continued to mount. NC Bowdler and others from Michigan State University, published results showing that contrary to the orthodox view, orally administered aluminium from soluble salts and aluminium hydroxide is absorbed in subjects with normal renal function. Their study demonstrated further that elderly people with high aluminium serum levels display a greater incidence of neuropsychiatric defects, including poor memory and impaired visual-motor coordination.[48] Ganrot contributes further to the aging–aluminium link with his observation:

> The structure best known to accumulate Al^{3+} is the nuclear chromatin [the substance of chromosomes] and especially the heterochromaiin [genetically inactive portion]. This accumulation is extremely interesting for several reasons, foremost perhaps, the possibility that it might theoretically cause an increasing heterochromatisation. According to an often-held opinion, heterochromatsation might, namely, be an essential process in cellular aging but no reliable explanation for it has been proposed. Hence, the possibility seems to be at hand that accumulation of Al^{3+} in chromatin actually could be one of the fundamental aging mechanisms, especially in long-lived cells.[49]

For almost a quarter of a century up to this time, J Bjorksten had argued in several articles that aluminium ion could well be an important factor in the aging process. He found, for example, that the aluminium content of the brain steadily increased with age[50] (see Figure 8.4). He even dared to predict that when cancer and arteriosclerosis can be controlled in the future, accumulation of aluminium ion could then become the foremost cause of human disease.[51]

Research continues to support this hypothesis with the discovery that aluminium is the active constituent of a strong cardiotoxic agent.[52] A Sydney University study has also linked aluminium intake with hip fractures. In a study of men and women aged over 65 years, it was found that those who had used the older type aluminium pots for cooking at age 20 were twice as likely to have a hip-fracture incident. It is believed that the older type aluminium pots leached more aluminium into food than do modern ones and that this may explain the higher rate of hip fracture.[53]

Figure 8.4 Aluminium concentrations in the brain

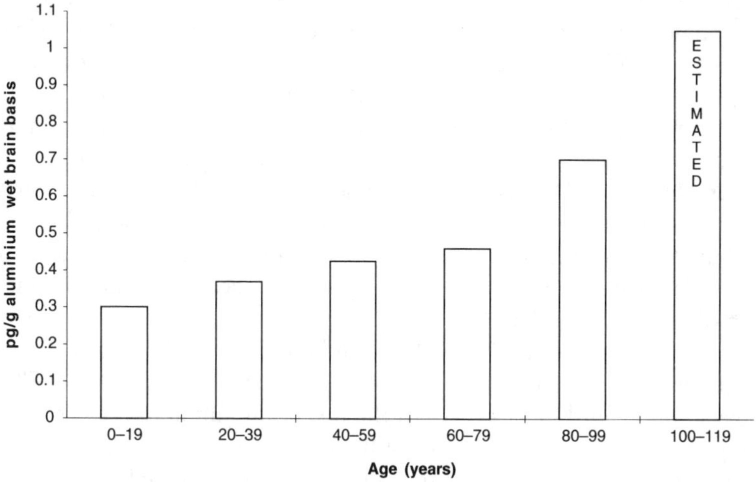

J Bjorksten, *Environmental Health Perspectives*, vol. 81, 1989, p. 241

Of perhaps greater concern is the use of aluminium salts in vaccines which are given to infants. A Lione reports that a small number of patients who receive vaccines containing aluminium salts develop nodules at the site of injection which when excised have been shown to contain aluminium. Thus in these patients, the aluminium salts from vaccines are localised at the site of injection for extended periods of times. More commonly, however, aluminium salts are assumed to be absorbed from the site of injection over a period of time to undergo systemic distribution and possible excretion.[54] Aluminium is added to vaccines such as the common Diptheria–Pertussis–Tetanus (DPT) vaccine, to promote a raised immune repose which, in turn, results in a more effective vaccination because more antibodies are formed.[55] A Lione points out that on the basis of the aluminium content of commercially available products the common course of DPT injections may result in the administration of between 0.75 and 2.50 milligrams of aluminium during the first six months.[56] Aluminium allergy in association with vaccination has been reported[57] and if some aluminium can be rapidly absorbed from the site of vaccine administration, the neurotoxic properties of aluminium could contribute to the onset of post-vaccination encephalopathy (degenerative brain disease).[58]

In 1991, the United States National Academy of Sciences Institute of Medicine released the findings of a committee studying the serious adverse events associated with certain vaccinations. The committee found evidence of sufficient quantity and quality to conclude that it is likely that a causal relation exists between DTP vaccine and acute encephalopathy.[59] Given the neuro-toxicity of aluminium, the possible role of aluminium in these and other post-vaccination adverse effects would appear to have received less attention than it deserves.

The widespread use of the aluminium chloride type anti-perspirants has also been linked as a contributing factor in increased risk of brain disease. In a United States study a significant dose relationship was found between the use of such anti-perspirants and Alzheimer's disease.[60]

Fumes from aluminium welding have also been implicated in a dose-related way to neurotoxic effects such as decreased motor function observed in welders.[61]

It can be seen that the evidence indicting aluminium as a major health hazard is enormous. The question must then be asked: Why have so many authorities rejected the notion of aluminium toxicity in the past and why is it that many still continue to do so?

Part of the explanation for this reluctance is due to the fact that aluminium is such a ubiquitous element. It is the most abundant metal in the earth's crust (about 9 percent) and all igneous rocks contain silicates of aluminium. Consequently aluminium silicate compounds are common in soil, with clay simply being impure aluminium silicate. Without traces of alumina, the quartz sands of the seashore would dissolve in the oceans.[62]

Just as aluminium is widespread in rocks and soils, it is also widespread in the plant kingdom but at much lower concentrations. Seeds and leaves of plants have the highest concentrations of aluminium with tea leaves (*Camellia sinensis*) containing up to 20 000 ppm in mature leaves.[63] However in plants, aluminium appears to be bound in insoluble forms which are not readily absorbed by the body. In animals, the aluminium levels are usually lower.[64] Because of the varying factors which affect aluminium levels in food, it has been estimated that the daily intake of dietary aluminium ranges from around 5–10 milligrams in Britain,[65] and from 7–9 milligrams in North America[66] (see also Figure 8.5). With such high levels of aluminium commonly taken up daily in normal food and with no toxic effect, many authorities deemed it unlikely that the small extra uptake from cooking utensils, or food and water additives, would be significant, especially since relatively large amounts of aluminium are taken orally as antacids and antidiarrhoeals without apparent ill-effect. In addition to the foregoing complexities there is also the chemical analysis problem. Trace levels of aluminium are extremely difficult to analyse chemically. These and other factors have encouraged the reluctance to admit the argument for aluminium toxicity. And it is this same reluctance that has led to a failure to recognise the natural mechanisms which protect plants and animals, from the toxicity of aluminium.

Figure 8.5 Contribution to aluminium uptake from various food groups

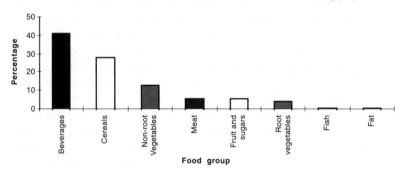

Note: Values are based on average daily intake from food.
Choice, January 1992

Although aluminium is ubiquitous, the point that is missed by the conventional interpretation is that aluminium is present in rocks and soils in very insoluble forms, usually as aluminium silicates, or hydrated oxides of aluminium. The extremely insoluble and unreactive properties of these aluminium compounds have made it one of the most difficult elements to win from the earth. It was not until as recently as 1886, with the discovery of an electrolytic process for producing aluminium from a molten salt mixture, that this most abundant metal was available for our use. This binding of aluminium in the rocks has meant that only extremely low levels of aluminium occur in a soluble form in the environment. This fact is further demonstrated by the extremely low levels of aluminium found in the oceans. The levels of aluminium found in the water of oceans and seas are in the order of only 10 parts per billion (0.01 ppm).[67]

Aluminium is toxic to plants and to nitrogen-fixing bacteroids,[68] and increased soluble aluminium levels reduce plant productivity.[69] Yet here again we observe that nature has provided mechanisms in the roots of most plants to prevent the uptake of aluminium. This protective mechanism usually depends on the formation of insoluble phosphate complexes[70] with subsequent inhibited absorption in the roots.[71]

Australian studies with wheat have shown that the root tip (just the final 2–3 millimetres) is the part of the plant most sensitive to aluminium. If this region is exposed to aluminium ions in the soil, it rapidly becomes damaged and stops the new root growth, thus leaving the plant unable to take up sufficient water and nutrients. Some wheats, however, are tolerant to higher levels of aluminium in the soil and it has been found that these plants release the chemical malic acid at their root tips in the presence of aluminium ions. This chemical binds up the aluminium and protects the root tip from aluminium damage.[72]

Nature has also provided a plan for certain plants to be even more tolerant to aluminium and thereby grow in soils high in aluminium such as the leached acid soils found in high rainfall tropical areas. Most of these acidophilus plants contain strong organic buffer systems in the plant sap. Two plants that have been studied in this regard are mustard and tea. In these instances there seems to be no mechanism in the roots to inhibit aluminium absorption but instead the aluminium is detoxified by being tightly bound as strong organic complexes within the plant.[73] In the case of tea it has been hypothesised that there is actually a phosphate pump which transports aluminium from the roots to the leaves.[74]

Similarly in the animal kingdom, nature has provided a protective mechanism which results in low gastrointestinal absorption.[75] This protective mechanism relies on aluminium generally being bound up with phosphate in an insoluble form which is poorly absorbed. Nature goes further in her protective mechanism in that most, if not all, of the small amount of aluminium that is absorbed is eliminated from the body by the kidney.[76] In newborn infants, the permeability of the gastrointestinal tract is high and the kidney function is immature. Consequently aluminium is absorbed and retained.[77] Again, however, we see that nature provides mechanisms to protect the infant from aluminium intake, as evidenced by the fact that breast milk contains among the lowest levels of aluminium of all foods, with reported values in the range of 4–20 parts per billion.[78] Some researchers have suggested that any values above this range probably reflect contamination from talcum powder or deodorants or from storage in glass containers which may have leached minute amounts of aluminium.[79] The death in 1985 of two uraemic (suffering from kidney toxicity) babies in the United Kingdom from the effects of aluminium toxicity, resulting from the accumulation of aluminium from an infant-feeding formula[80] highlights just how finely balanced these protective mechanisms are.

Evidence is steadily accumulating that many modern scientific and technological practices are directly opposed to the patterns of nature. The deliberate addition of aluminium compounds to the water supply, foodstuffs and pharmaceuticals are, as has been suggested above, obvious examples. A less obvious example is the use of apparently harmless food additives such as citric acid with its highly soluble sodium, potassium and calcium salts. These additives are commonly and extensively used in the food-processing industry. However, under the acid conditions which are found in the stomach, insoluble aluminium phosphate complexes in foods are converted, in the presence of citrate ion, to soluble citrate complexes, thus nullifying nature's protective mechanism. As Martin puts it:

> Studies of Al^{3+} ingestion that do not measure or control the amount of citrate have overlooked a significant variable that may affect the conclusions drawn.[81]

Birchall and Chappall also raise the citrate point, concluding that the solubilisation of aluminosilicate species by citrate may be relevant in aluminium transport.[82] Thus the scientifically reasoned use of apparently harmless citrate in the food industry ironically interferes with nature's mechanisms to protect us from aluminium toxicity. This being so, citrate becomes a subtle but significant factor that increases the bioavailability and hence the uptake of dietary aluminium.

The use of aluminium cooking utensils is another obvious human-made source of aluminium in the diet which is very likely highly bioavailable. However, little attention has been paid to the dietary aspects arising from the fact that aluminium is a very powerful reducing agent, being more reactive than many other metals. This means that in the presence of a freshly cleaned or exposed aluminium surface, the bioavailability of these essential elements such as iron, cobalt, manganese, chromium, copper and zinc in foods would be lowered or destroyed by reduction to the metallic form by the aluminium metal.

Environmental impact of aluminium production

The health aspects considered so far are only part of the aluminium picture. There remains to be discussed the health aspects arising from the environmental impact of the production of the metal itself, and the industrial uses of aluminium. For example, the impact of the aluminium-containing effluents from the paper processing, tanning and dying industries, and the use of aluminium compounds in pesticides and fertilisers. Even brief reflection makes clear that the production of aluminium presents a huge health problem. For example, an estimated 30 000 tonnes of perfluoromethane and perfluoroethane gases are released into the atmosphere each year as a waste product of aluminium production worldwide.[83] The long-term effect of the these greenhouse-type gases on the global environment is not known. Fluoride- and aluminium-containing dusts are also emitted. Acid rain is produced by power stations used to produce the electric power required to win the aluminium. Here again we see nature's protective mechanisms being undermined, as aluminium is leached from the soil by acid rain, making it more bioavailable. Dr Neil Ward from the University of Surrey has found that tap water from areas of acid rain showed dramatic increases in aluminium levels.[84] Another aspect to be considered is the health impact of the high voltage power lines carrying the enormous amounts of power required for aluminium production. Yet another aspect is the environmental impact of aluminium consumables such as the thousands of tonnes of cans and foil wrappings which are discarded. A recent Australian survey found that aluminium cans and foil made up more than 13 percent of beach litter.[85]

It is time to take stock and consider what is the real cost to our health and the environment of the convenience of the aluminium can which we throw away.

What you can do

Now that we have discovered the evidence of the long-term toxicity of aluminium there are a number of things we can do.

1. We can check to determine whether aluminium is added to our water supply. If it is added we can install a water purifier, collect rainwater or buy water low in aluminium for drinking and cooking.
2. We can also choose not to buy beverages in aluminium cans.
3. We can recycle our aluminium pots, pans, pressure cookers and other aluminium cooking utensils, replacing them with stainless steel or glass cookware. This will significantly lower our dietary aluminium intake.
4. Aluminium in our diet can also be reduced by avoiding foods which use aluminium in packing, processing or in the form of additives. Aluminium additives are easily recognised from additive codes on the packages. For example, in Australia the aluminium additives sodium aluminium phosphate, sodium aluminosilicate and calcium aluminium silicate have code numbers 541, 554 and 556 respectively.
5. Aluminium compounds are widely used in medicines. When a drug is purchased, check with your pharmacist to discover whether aluminium is used in the preparation, in the base or as a tablet lubricant. If aluminium is used, enquire whether an aluminium-free alternative is available.
6. Another way to minimise our daily bodily uptake of aluminium is to avoid foods to which citric acid or its salts such as sodium citrate or calcium citrate have been added. Food additive codes can be used to identify this common food acid and its salts in foods. For example, the Australian food codes for several citrate compounds are as follows: citric acid (330), sodium citrate (331), potassium citrate (332), calcium citrate (333) and ammonium citrate (380). As we have seen earlier in this chapter, citrates interfere with nature's mechanisms designed to minimise the absorption of aluminium in the digestive system. For similar reasons, ideally, citrus fruit and vegetables should not be eaten at the same meal, and lemon juice and pineapple juice should be used sparingly in savoury and vegetable recipes. Reconstituted fruit juices and beverages should also be avoided unless the manufacturer can guarantee that aluminium-treated water has not been used in the reconstitution process.

The aluminium levels in foods are also determined by the amount of available aluminium in the soil. This is increased by acid rain fallout. Food grown as an introduced species in soils which are naturally very high in aluminium, such as near bauxite deposits or in high acid rain fall-out areas, should be avoided. Details of these locations can be obtained from the nearest university or government department of agriculture.

Aluminium salts are often added as fertiliser on tea plantations so herbal teas and other more healthful beverages could be substituted.

7. It is best to avoid or minimise the use of personal deodorants and anti-perspirants containing aluminium chloride. Aluminium-free products are usually labelled as such, and are commonly sold in health food stores. By applying the simple guidelines outlined above, we can make substantial reductions in our daily aluminium uptake and thereby make an important contribution in the maximising of our own health.

chapter 9

Food irradiation: Towards health or hazard in the 21st century?

Given that some 25 percent of food produce in many areas of the world is lost after harvesting due to the inability to store and transport it,[1] appreciation of the role which food irradiation can play in rectifying this problem has become understandably politicised. Hailed by some as a panacea for feeding the hungry, there seems an almost certain sense of obligation in adopting a technology such as irradiation, which claims to provide a safe and efficient way of increasing the availability of uncontaminated foods for all. About 37 countries have already permitted the use of irradiation as a method of preserving more than 40 food groups for public consumption.[2] Food irradiation is not permitted in Australia at the present time; however, sterilisation of packaging material by gamma radiation is allowed. Other goods, commercially sterilised by irradiation include cosmetics, wine bottle corks, hospital supplies and medical products.

Commercial food irradiation began in the United States in 1992, and irradiated strawberries, tomatoes and citrus have been marketed in some states such as Florida and Illinois. In 1995, the United States Food and Drug Administration (FDA) announced its proposed policy to accept irradiation as a quarantine treatment of fresh fruits against fruit flies. Fruits from Hawaii including papaya, rambutan, lychees and cherimoya have been marketed at the retail level in several American states since 1995.[3]

In China, irradiated apples have been marketed at the retail level in Shanghai and other cities since the early 1990s. Several countries including Belgium, France, Hungary, Japan, the Netherlands and the former USSR countries are irradiating grains, potatoes and onions on an industrial scale to control insect infestation, moulds and premature germination. Considerable amounts of frozen seafoods, as well as dry food ingredients are irradiated in Belgium, France and the Netherlands to control food-borne diseases caused by bacteria such as Salmonella. France also irradiates frozen poultry products for this reason.[4]

Spices are being irradiated to kill bacteria in many countries including Argentina, Brazil, Denmark, Finland, France, Hungary, India, Indonesia, Israel, Norway, the United States and Yugoslavia[5] (see Figure 9.1).

Food-borne diseases pose a widespread threat to human health. For example, between April and September 1993, a nationwide outbreak of salmonellosis food poisoning occurred in Germany and more than 1000 people were made ill. The cause was identified as salmonella-contaminated paprika used to flavour potato chips that had subsequently been sold across the country. This proved to be the largest documented outbreak, to date, of food poisoning due to contaminated spices.[6]

Figure 9.1 Global irradiation of spices and vegetable seasonings

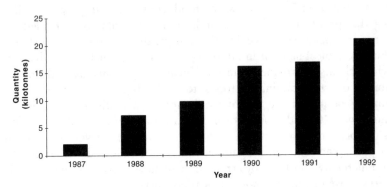

P Loaharanu, *Food Technology*, May 1994

Studies by the United States Center for Disease Control in the early 1990s showed that even in the United States, food-borne diseases caused by pathogenic bacteria, such as Salmonella and Campylobacter, and by Trichinae and other parasites, claim an estimated 9000 lives annually and cause an estimated 24–81 million cases of diarrhoeal disease. The economic losses associated with such food-borne diseases are estimated to be between US$6.5 billion and $33 billion.[7]

It is important to remember, however, when considering the food-borne disease situation that this is an emerging problem which relates largely to our mass production of food.[8] Intensive farming techniques together with modern food handling and processing technologies have contributed to the maintenance of infection reservoirs. The centralisation of food manufacture together with modern transportation, means that any contamination of manufactured foods can potentially infect many people over wide areas, even internationally.[9]

The use of food irradiation to reduce the food-borne disease problem is tantamount to using yet another technological fix to correct a problem largely resulting from our technological distortion of nature in the first place.

It would be naive to deny the range of possible benefits associated with the deployment of food irradiation as a new method of food processing, but it would be equally naive to neglect the array of possible health problems which such a powerful technology may also create. Considerable efforts are being made to convince the consumer public that food irradiation is safe, though some of these efforts are themselves misleading. The United States symbol for a food that has been irradiated, for example, is similar to the 'Radura' label (the emblem of quality) used in the United Kingdom, the Netherlands and South Africa. Indeed, the stylised flower depicted in the centre of the seal gives the impression of a health food rather than one which has been irradiated. It has been pointed out that even the Environmental Protection Agency has noticed the similarity of the irradiation seal to its own logo. Despite reassurances that irradiation of food does not involve health risks, a number of health concerns persist upon closer reflection on the arguments and experiments intended to establish the technology as safe. The burden of what follows is to address these health concerns in a way which neither underplays nor exaggerates political, factual and social-ethical issues which arise from them.

What is food irradiation?

Food and irradiation have been associated ever since humans began using heat to cook food. The term 'radiation' is commonly used to cover a wide spectrum of energy. The heat radiation which emanates from the grilling element of a cooker or from the coals of a fire or from the sun is infrared radiation and has a relatively long wavelength.

In microwave cooking, food is subjected to even longer wavelengths of microwave radiation. Microwave radiation is absorbed selectively by water in the food, causing the water to heat up which, in turn, cooks the food. Inasmuch as nematodes, pathogenic bacteria and other parasites are killed by sufficient heat, heating food will help to sterilise it. Cooking is also believed to enhance the flavours of some foods and, physiologically speaking, can improve the digestibility of certain foods, such as oats.

In these cases, the forms of radiation being used are all of a wavelength longer than visible light, and the radiation has been used simply to heat the food, through a process which involves its absorption at the molecular level. The term 'food irradiation' refers to a process of energy transfer at the atomic, rather than at the molecular level. In the case of food irradiation, the very short wavelengths of radiation are so energised and penetrating that they actually knock electrons out of the atoms of the foods they bombard. When this happens, the integrity of the molecular structure of the food is disrupted, often leaving positively and negatively charged particles called ions. Radiation which leads to the formation of ions is thus called 'ionising radiation'; two examples of this are gamma rays and x-rays.

In certain situations, these ions dissociate into 'free radicals' which are molecules having an unpaired electron. This unpaired electron makes free radicals extremely reactive chemically, and they usually last for only a fraction of a second before reacting with some other molecule, which, in the case of food or a living organism, may be a biologically important molecule. Biological structures and molecules known to be affected by ionising radiation include chromosomes, proteins, enzymes, nucleic acids, lipids and carbohydrates. Even water, a fundamental constituent of all biological cells, is ionised by ionising radiation. J Doull and others report:

> The cytopathologic changes [characteristic of disease in cells] observed in cells following radiation are numerous, varied, and complex and are similar to those seen after other types of cellular injury. The response of individual cells and types of cells to radiation is variable depending on such things as the cell cycle and the oxygenation status of the cell. Many of these responses may result in cell death, which, if extensive, may result in death of the organism. Other cells may undergo either total or partial repair. These changes may also be expressed at later times by tumours or mutations.[10]

If food containing damaged cells and damaged biological structures is assimilated into our own bodies, we need to know much more than we do at present about the way variable food structures affect us. We shall see that the whole area is fraught with health worries.

The history of irradiation

The concept of irradiating food to kill bacteria and thereby preserve it, probably originated just over a hundred years ago with Mink, who in 1886 found that the newly discovered x-rays could kill bacteria.[11] The idea became official in 1905 with the issue of two patents covering the use of radiant energy to preserve food by eliminating spoilage microorganisms, and the treatment was tested in Sweden on strawberries in 1916. The use of x-rays to kill *Trichinella spiralis* nematodes in pork was patented in the United States in 1921 by Schwartz , and just under a decade later, Wust was granted a broad patent for the use of x-rays to kill all bacteria and so preserve food indefinitely.[12] The first commercial use of food irradiation, however, was not introduced until

1937 in Germany, when spices used in the manufacture of sausages were sterilised by irradiation technology.

It was not until just after World War II that research into food irradiation was begun on a broad base in the United States, not by the Department of Health, but by the United States Atomic Energy Commission. The war and immediate post-war periods had seen a very rapid growth of nuclear technology. Nuclear reactors produce an excess of high energy neutrons. Many stable elements when subjected to these high energy particles are converted to an unstable atomic structure known as a radioactive isotope which subsequently emits radiation, often in the form of gamma rays, a radiation of even shorter wavelength and greater penetrability than x-rays. It seems that one of the objectives of the United States Atomic Energy Commission 'was to see if accumulated radionuclide material, the products of a number of years of activity associated with nuclear power, could be put to beneficial use in the food and agricultural industries'.[13] Public acceptance of the program was promoted under the title 'Atoms for Peace'. The United States Army Quartermaster Corps was also involved, and the program was announced in 1953 by President Eisenhower.

In Canada and Europe a similar history is revealed. The development of food irradiation emerges symbiotically with the growth of the nuclear industry. During this time, the highly penetrating gamma rays from radioactive sources such as cobalt 60 replaced the x-rays of the original patents.

Since the 1960s, a joint initiative to fund and promote research on food irradiation was undertaken by the World Health Organisation (WHO), the Food and Agriculture Organisation (FAO) and the International Atomic Energy Commission (IAEC). Under the auspices of these organisations key reports were produced in 1976 and 1980 in which the wholesomeness of irradiated foods was affirmed. Based on the recommendations of this joint committee, the Codex Alimentarius Committee of the United Nations formulated international guidelines on food irradiation, urging that food irradiated up to a dose of 1 million rad presents no health risk. The rad is a measure of the radiation energy absorbed by the food. One hundred rads is equivalent to one gray (Gy) the unit now most commonly used. One million rads corresponds to 10 kilograys (KGy) where 1 kilogray equals 1000 grays. In energy terms, 1 gray represents 1 joule (J) of energy absorbed by one kilogram of food being irradiated. The United States Food and Drug Administration (FDA), has recommended that general irradiation should not be permitted beyond one-tenth of the Codex level, or 100 000 rad (1 kilogray).

Food irradiation and the risk of mutations

The production of mutations by ionising radiation was first discovered in 1927 by HJ Muller at the University of Texas using x-rays.[14] Two decades later in 1947 and also at the University of Texas, WS Stone and colleagues found that mutations were produced in *Staphylococcus aureus* when cultured on broth irradiated by ultraviolet light from a mercury vapour lamp (that is ultraviolet c-type radiation). They concluded their report with the statement, 'Natural radiation is much more important in the mutation process if it can *induce mutation by an effect on the food* [our italics] as well as on the organism'.[15]

Fifteen years later, as commercial food irradiation was being introduced in a number of countries, MS Swaminathan and co-workers raised the matter of irradiated food causing mutations. Their research showed that an increase in mutations in *Drosophila melanogaster* (fruit fly) was observed when the flies were reared on a medium that had been irradiated with a sterilising dose (1.5 kilograys) of cobalt 60 gamma rays.[16]

While these researchers expressed explicit warnings of the potential and very serious, yet random, health risks associated with food irradiation, standard texts on food preservation reassured the reader that irradiated foods had been fed to rats, chickens, dogs, monkeys and human volunteers and that no 'untoward physiological reaction' had been uncovered.[17]

In retrospect, it is now clear that these latter conclusions were made ignoring the possibility that in larger organisms the harmful effects may not become manifest for a decade or more, and may also be dependent on the extent to which other environmental mutagens and carcinogens are taxing the organism's natural protective mechanisms against radiation.

It should be emphasised that mutation from irradiation has proved to be a random process. The probability of mutations happening, however, increases as the dose of radiation becomes stronger, and it is important to see that the radiation used in food irradiation is indeed very strong. The dose of 10 kilograys used to kill vegetative bacteria and yeasts is 1000 times the lethal dose for higher animals and man.[18] When we think in terms of subjecting hundreds of thousands of tonnes of commonly consumed basic foods such as wheat and potatoes to powerful ionising radiation the potential in the long-term for a significant widespread health problem is scaled up enormously.

Some researchers, however, have recognised the weakness of short-duration experiments on well-nourished subjects to test the safety of food irradiation. C Bhaskaram and G Sadasivan from the National

Institute of Nutrition in India proposed to test the hypothesis that adverse health effects are most likely to develop in malnourished subjects and therefore more likely to be detected in short-term trials. They fed freshly irradiated wheat to children suffering from protein calorie malnutrition and found that the children developed polyploid cells (these cells have an abnormal number of chromosomes) and certain abnormal cells in increasing number as the duration of feeding increased. An associate researcher also found that rats fed freshly irradiated wheat were observed to be affected by a 'dominant lethal mutation effect, as well as reduced germ survival'.[19]

Since this finding was reported a number of other laboratories have not been able to reproduce these effects and an investigative committee appointed by the government of India in 1976 concluded that the available data failed to demonstrate any mutagenic potential of irradiated wheat. Scientific committees in Australia, Canada, Denmark, France, United Kingdom and the United States also came to the same conclusion.[20]

By 1979 however, J Barna of the Hungarian Academy of Sciences had identified scores of adverse effects on animals due to the ingestion of irradiated food. Barna's review of the food irradiation literature revealed 1414 adverse effects, 185 beneficial effects, and 7191 neutral effects.[21] Among the deleterious effects manifest by animals fed irradiated food were increased incidence of tumours, reduced growth rates, lower birth weights and reduced survival rates for offspring, kidney damage, changes in white blood cells and genetic damage.

Can DNA survive food digestion?

PM Cullis and co-workers at the University of Leicester have confirmed several earlier studies that DNA attached to proteins is damaged by gamma irradiation.[22]

The current view is that DNA in food is digested and destroyed, therefore any damaged DNA in irradiated food should not pose a threat to the health of consumers of that food. However at the International Congress on Cell Biology in San Francisco in December 1996, researchers from the University of Cologne presented findings which showed that DNA fed to a mouse can survive digestion and invade cells throughout its body. In the study, researchers fed a bacterial virus to a mouse, with the surprising result that sections of viral genetic material large enough to contain a gene survived to emerge in faeces. DNA from the virus were also found in cells in the mouse's intestine, spleen, white blood cells and liver.[23]

Should these preliminary findings be confirmed there could be serious implications for the safety of large-scale food irradiation, as altered genetic material may be incorporated into bacteria and cells which changes their functionality. This being so, it is quite possible that a dysfunctional cell or group of cells could well affect the body's capacity to control disease.

Changes to food caused by irradiation

The public is generally reassured by authorities that 'The irradiation process produces very little change in food', and that 'None of the changes known to occur have been found to be harmful or dangerous'.[24] Given that the food irradiation process causes sufficient changes to the biochemical processes of bacteria and insects to kill them, it is not surprising that significant changes to the major components of foods, such as proteins and starches, are, in fact, produced by low to medium food irradiation.

LA MacArthur and BL D'Appolonia reported in 1983, that the gamma radiation of wheat affects the physical properties of the dough and its baking characteristics.[25] SR Rao and co-workers found that irradiation of flour at 10 kilograys impaired loaf volume and crumb grain.[26] In 1984, O Paredes-Lopez and MM Covarrubias-Alvarez found that at medium doses of radiation (1–10 kilograys) the overall bread quality of wheat was greatly reduced, and that at radiation doses greater than 5 kilograys, irrespective of the baking formula used, loaf volume and baking quality deteriorated.[27] Recent 1996 studies by H Koksel and co-workers, have now confirmed that irradiation of wheat changes the physical characteristics of wheat flour, such as stickiness, firmness and bulkiness.[28] Irradiation of grain also causes problems in noodle and spaghetti quality. Studies have found that irradiation of wheat at low levels (0.2–1.0 kilograys) can result in noodles with increased losses or breakdown during cooking and inferior scores in sensory analysis.[29] These physical changes strongly suggest that significant alterations to the protein and starch components of the wheat are occurring as a result of irradiation. Some of these changes may have subtle and yet unknown affects on health. For example, it is now known that changes in the type of starch in cereals can alter the glycemic index of the food which, in turn, can affect the extent to which that food may contribute to the development of diabetes.[30]

Another area of concern is the destruction or alteration of micro components in foods by irradiation. For example, the vitamins A, E, C,

K and B1 (thiamine) are relatively sensitive to radiation.[31] Some of these vitamins, particularly A, E and C are the antioxidant vitamins which help protect against cancer and heart disease.[32]

But these are not the only antioxidants in foods, and in particular fruits and vegetables, which confer protection against disease. It is now known that fruit and vegetables have many other antioxidant-type components which protect against disease, such as the various carotenoid and flavonoid compounds.[33] For example, researchers have recently discovered that the compound lycopene (a carotenoid antioxidant found in tomatoes) appears to decrease the risk of prostate cancer.[34] Very little is known about the effects of food irradiation on these important food components, but their antioxidant nature suggests that they too would be sensitive to ionising radiation and that their concentrations in irradiated food would be significantly reduced. Tea, spices and whole grain cereals such as rice contain similar polyphenoic antioxidant-type compounds, many of which are now being found to have important anti-disease properties.[35]

By irradiating food we reduce the levels of the antioxidant compounds and vitamins which protect us against deleterious diseases such as cancer and heart disease that the consumption of such foods was intended to prevent. To add food irradiation to the food processing chain is simply another step towards the destruction of the very nutrients we intend to preserve.

There is yet another side to the vitamin aspect. Because irradiated food will tend to be stored longer, the losses of vitamins of all kinds will in any case be greater. This is one reason why the labelling of irradiated food is so important. Consumers have a right to know whether the foods they are eating have been processed or frozen so that they can make informed choices, and ensure that projected nutritional losses can be compensated for with fresh produce. Since fresh foods which are irradiated give the impression of greater freshness and more nutritional value than they actually have, consumers could be fooled into thinking that they are balancing their diets when in fact they are not. As the food irradiation industry grows and expands in the hope of feeding Third World countries, the dream of providing the poor with the nutrition they require could turn into a nightmare.

Space has permitted us to discuss only some of the far-reaching health and environmental implications of food irradiation. We have not examined, for example, the health and environmental problems associated with the production, transportation, and location of the powerful ionising radiation sources used for food irradiation. Nor have we addressed the socio-environmental issues of a technology which encourages the nuclear industry. Our limited analysis of the food irradiation controversy, however, does bring into bold relief the need to reassess and reconceptualise the guiding principles employed to formulate decisions on health policy. To achieve the appropriate answers to the food irradiation issue, we must first recognise what constitutes appropriate questions.

What you can do

It is important to realise that we do not need irradiated food. You can choose simply not to buy irradiated food.

1. Where possible try to buy locally grown produce and avoid imported foods which may have been irradiated. For example, in some countries, wheat, potatoes, onions and seafood may have been irradiated on a commercial scale, but processed foods manufactured from these ingredients would not be labelled with regard to the irradiated ingredients.
2. Support organic farming producers and buy traditionally prepared foods rather than mass produced processed foods.
3. It is important to lobby governments to ensure that irradiated food components and food packaging are also declared on food labels.

The irradiation of food-packaging materials is increasing and this is another area in which there may be potential health risks from packaging–radiation interactions.

chapter 10

Cadmium: The newest toxic link in the food chain

In a world of space exploration and the marvels of the microchip, it is astonishing that the situation has arisen in which farmers are now being paid to stop farming their land because of intolerable pollution from the very chemicals originally designed to make agriculture more efficient.[1] Admission of this state of affairs is an indictment of the unbridled application of scientific technology, with little or no regard for the disruptive impact of such intervention upon nature's finely balanced systems. Quick to applaud the success of technology, but slow to concede its indiscretions, we find ourselves in a seemingly hopeless predicament.

Our faith in the capacity of scientific technology to resolve the very problems it inevitably creates has led to new problems more difficult to resolve. The continuing saga of the European farming community referred to above is a case in point. In the belief that the problem of chemical pollution could satisfactorily be solved by a massive and rapid conversion of polluted farmland into virgin forestland, attempts to renew the European countryside have been made. However, these have brought about yet another unanticipated ecological crisis. Contrary to simplistic expectations, the conversion process seems to have exacerbated the pollution problem by enhancing the acid rain release of toxic heavy metals in the soil.[2] Substantial research studies have now shown that the growing trees are absorbing and accumulating acidic pollutants in the air. These pollutants are subsequently washed into the soil by rain, often increasing the acidity of the adjacent soils by ten- to twenty-fold. The increase in soil acidity tends to dissolve the heavy metals which were bound in the soil from previous chemical contamination of farmlands. When this happens, the heavy metals become mobile and free to spoil municipal and general water supplies, along with neighbouring croplands. Once again, the lack of appreciation of the fundamental interconnectivity among the living and non-living components of nature, coupled with naive technological intervention, has provided yet another point of entry for heavy metals into the food chain.

Among the wide array of heavy metals which are now being released in the soil, cadmium is regarded by some researchers as the most immediately dangerous.[3] In 1993 at its spring meeting, the International Agency of Research on Cancer determined that cadmium is carcinogenic to humans.[4]

Cadmium is a metal which is in many ways similar to zinc. It is used as an anticorrosion plating agent on nails and in plumbing fittings, and as a component of the now familiar rechargeable nickel cadmium batteries. It is also used in pigments, fungicides, phosphors, ceramics and as a stabilising material for polyvinyl plastics. (Figure 10.1 shows the major manufacturing sectors using cadmium.) Cadmium occurs naturally in the environment but at very low levels and in highly insoluble forms. For example, the earth's crust is estimated to consist of 0.15–0.2 parts per million (ppm) of cadmium, with uncontaminated soils being of the order of 0.4 ppm. Uncontaminated water contains even lower levels, with most fresh water containing less than 0.001 ppm. Sea water averages a very low 0.00015 ppm.[5]

Figure 10.1 Global cadmium consumption 1998 (total: 18 500 000 kilograms)

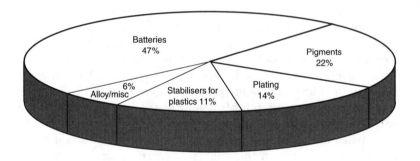

JF Ashton & RS Laura, *Search*, January/February 1992

Over the past century, however, the cadmium content of the environment, and particularly soils, has been steadily increasing as a result of industrial and agricultural processes. Cadmium constitutes about 0.3 percent of important zinc ores and, since it is more volatile than zinc, it is often released into the atmosphere during the smelting process. In 1968, an estimated 2300 tonnes of cadmium metal were released into the air over the United States.[6] Cadmium is released into the atmosphere as a result of burning wastes such as plastics containing cadmium and sewage sludge. It is also released by steel industries from steel scrap reclamation,[7] and even from car exhausts.[8] The proportions of cadmium used in various industrial processes is shown in Figure 10.1. The fall-out from these processes eventually reaches the soil. The major source of direct land contamination is waste disposal and the proapplication of phosphate fertilisers to soils.[9]

Impact of fertiliser usage

The cadmium level of phosphate fertilisers such as superphosphate can commonly be as high as 100 ppm.[10] Given the acidic nature of these fertilisers, the high level of cadmium they contain is mobilised to make them an important source of contamination in soils and subsequently in plants. Wheat and rice are two crops in particular which concentrate cadmium from the soil. In one extensive study of wheat plots grown with farmyard manure, compared with NPK artificial fertiliser, the cadmium content of the manure- grown wheat ranged from 16–50 micrograms per kilogram, while the phosphate-fertilised wheat ranged from 63–90 micrograms per kilogram.[11] A 1995 study of the dietary cadmium intake among different socio-economic groups in south India found that rice was a major source of cadmium among rural populations and the economically deprived classes.[12] In another study involving potatoes, long-term application of artificial phosphate fertilisers has led to cadmium levels of 230–320 micrograms per kilogram compared with 100 micrograms per kilogram in controls where lime only was added.[13] Furthermore, it was discovered that in the case of cadmium contamination, the levels of concentration are not significantly reduced by peeling the potatoes.[14]

Another plant which can concentrate cadmium from the soil is the sunflower. A 1995 American study found that production of non-oilseed sunflower on certain soils, yields kernels with cadmium levels way in excess of international market limit. Average cadmium concentrates of 310–1340 micrograms per kilogram were found in the study of 200 sunflower varieties. Soil properties were found to play an important role in cadmium uptake and accumulation.[15]

Some alcoholic beverages have also been found to contain high levels of cadmium. A 1996 report from University of Granada researchers in Spain, showed that mean cadmium levels in wine samples tested ranged from 0.1–15 micrograms per litre. In other alcoholic beverages (cider, brandy, rum, whisky, gin, and other spirits) the mean cadmium concentration ranged from 'not detectable' up to 11 500 micrograms per litre.[16] The authors of the study suggest that since alcoholic beverages are widely consumed, they contribute greatly to cadmium intake. They also advise that there should be strict control of the cadmium levels in alcoholic beverages.

The now common practice of using acid treated, cadmium containing, superphosphate-type fertilisers in food production is reflected in a market basket survey carried out for the National Health and Medical Research Council (NHMRC) of Australia in 1988. Nine out of 22 samples (or 41 percent) of potatoes and 5 out of 19 samples (or 26 percent) of carrots were found to have cadmium levels above the maximum permissible concentration (MPC) of 50 micrograms per kilogram. Five out of 21 samples (or 24 percent) of wheat analysed also exceeded the MPC. The survey also revealed that 2 percent of beef products and 4 percent of lamb products had unacceptably high levels of cadmium.[17]

A subsequent National Residue Survey in 1991 and 1992 found that 8 percent of offal samples exceeded the established maximum permissible concentrations (MPC).[18] More than 10 percent of samples in the 1992–93 Victorian Clean Agriculture Produce Monitoring Program had cadmium levels above the MPC. Violations were detected in carrots, potatoes, silverbeet and safflower.[19] The estimated typical daily intake of cadmium from dietary sources is illustrated in Figure 10.2.

Results from a survey in Queensland, Australia, reinforced concerns about excess cadmium levels in beef and dairy cattle, and the long-term use of superphosphate fertilisers containing cadmium is suspected as the cause of the contamination.[20] Artificial animal feeds can be another source. Phosphate rock which usually contains cadmium as a trace element is sometimes mixed with feed to make pigs and cattle retain water and gain weight quickly.[21]

Recent concern about cadmium has been such that Australian government authorities and industry have recently taken steps to eliminate the feeding of phosphate supplements containing high levels of cadmium to cattle and by banning the sale of offal from aged sheep and cattle likely to contain elevated levels of cadmium.[22]

In France studies have revealed that cadmium uptake in animals from plant feeds and diet supplements can result in offal meat cadmium concentrations of 480–630 micrograms per kilogram in the kidneys and 130–800 micrograms per kilogram in the liver. The same study reported finding 160 micrograms per kilogram cadmium in commercially produced sliced bread.[23]

Sewage sludge source

Sewage sludge is another important source of cadmium in the food chain. In a general sense, the recycling of sewage would seem to be a process in harmony with nature's finely tuned equilibrium. However, if the sewage being recycled also contains common industrial wastes, the sludge becomes an environmental poison. Vegetables grown using these sludges often accumulate much higher levels of heavy metals than is permissible. In one recent study cucumbers grown on sludge-amended soils were found to have cadmium levels ranging from 100 to 500 micrograms per kilogram.[24]

Animals grazing on pasture which has been fertilised with industrial sewerage sludges can also accumulate high levels of cadmium, particularly in the liver and kidneys. In one recent British study, the average level of cadmium in sheep livers was 1240 micrograms per kilogram of dried tissue for animals which had grazed on heavily treated pasture. This was eight times higher than the levels found in the livers of sheep grazing on pasture which had not been fertilised with sludge.

In sheep kidneys, the levels in animals that had grazed on treated pasture averaged 2570 micrograms per kilogram of dried tissue which was six times as high as the level in the organs of sheep that had grazed on untreated pasture. The level of cadmium in the soil of the sludge-fertilised fields averaged at just under 5000 micrograms per kilogram of dried soil.[25]

Atmospheric cadmium fallout

Soil is not the only source of cadmium for plants. Research has shown that for plants grown in the field, 28–36 percent of atmospherically deposited cadmium is translocated into the plant itself.[26] Atmospheric fall-out is even known to contaminate the production of honey. One European study found cadmium levels of 40–52 micrograms per kilogram in honey collected from an urban area.[27] The estimated weekly intake of cadmium from food is shown in Figure 10.2.

Figure 10.2 Estimated cadmium intake from foods by Australian males

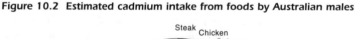

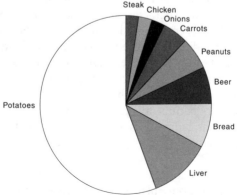

J Simpson & B Curnow, *Cadmium Accumulations in Australian Agriculture*, Bureau of Rural Resources, Canberra, 1988

Another important source of cadmium in the air we breathe is cigarette smoke. In 1972, GP Lewis and co-workers reported that the body burden of cadmium in cigarette smokers was more than double that of non-smokers.[28] Subsequent studies found that blood cadmium levels are five to 15 times higher in smokers than in non smokers.[29] Other studies have showed that sidestream tobacco smoke actually contained considerably more total cadmium than mainstream smoke.[30] 'Mainstream' smoke is the smoke taken into the body when the cigarette smoke is directly inhaled. Since mainstream smoke is inhaled into the body through the filter, some of the toxic substances the smoke contains are removed. 'Sidestream' smoke, on the other hand, refers to the smoke which wafts out from the end of the cigarette and can be inhaled via the nostrils of the smoker. Since sidestream smoke is unfiltered, it contains a number of toxic substances, harmful not only to the smoker but to those who are obliged to breathe the smoke produced by someone else's cigarette. Since up to 50 percent of respired cadmium is absorbed[31] this adds yet another dimension to the well-known health risks of inhaling tobacco smoke.

Health effects

The highly toxic nature of increased cadmium levels in the human diet was first noticed in Japan after World War II. Rice grown on paddy fields contaminated with cadmium from a nearby zinc processing industry resulted in a severe outbreak of the itai-itai ('ouch-ouch') disease in post-menopausal women. This disease produced a variety of skeletal disturbances, including back and extremity pains, along with difficulty in walking. The neuromuscular signs and skeletal effects corresponded to the symptoms characteristic of a series of cadmium poisoning cases which had been reported some years earlier in 1942.[32] Later studies of itai-itai disease showed specifically that kidney function is also reduced.[33] Since that time, environmental cadmium pollution, particularly of rice and wheat crops, has been brought into bold relief as a serious problem for human health in Japan.[34]

As health researchers became aware of the cadmium problem, further studies revealed that cadmium is an extremely toxic heavy metal, adversely affecting a variety of body processes and producing a range of deleterious health effects.

It now appears that the kidney is the organ most sensitive to long-term cadmium uptake and accumulation.[35] It is well established that kidney damage and impairment of renal function is commonly associated with increased cadmium levels in the diet. Cadmium can also disrupt a number of important human enzyme systems crucial to human health. It induces changes in several gluconeogenic enzymes and depresses the trypsin inhibitor capacity of serum.[36] When the gluconeogenic enzyme systems become dysfunctional, the effects upon the human body may include metabolic changes which have the same physiological impact as drug-induced diabetes. A hypothesis yet to be explored is that the increased incidence of diabetes within the Western world is at least partly explained by increased exposure to cadmium toxicity.

In addition to the foregoing biochemical disruptions, cadmium is known to suppress testicular function and has been linked with an increased incidence of prostate cancer.[37] It is now established that cadmium triggers hormone-like responses in certain mammalian cells[38] which can result in cell proliferation[39] and cancer.

Animal studies have provided alarming insight into the carcinogenic nature of cadmium. Oral administration of cadmium chloride (a soluble cadmium salt) to rats potently induced tumours in the prostate, testes and the systems associated with blood cell formation.[40] A single injection of 7.3 milligrams of cadmium chloride under the skin of rats produced a high incidence of benign tumours in the testes, uncontrolled cell growth in the prostate and malignant tumours at the injection site.[41] Exposure of rats to an aerosol containing 25 micrograms per cubic metre of cadmium chloride (that is only 15 micrograms of cadmium per cubic metre) produced a 50 percent incidence of lung tumours.[42] This finding is particularly worrying given the United States Occupational Safety and Health Administration standard for workplace cadmium fumes of 100 micrograms per cubic metre.[43]

In 1992, researchers from the United States National Institute for Occupational Safety and Health (NIOSH) calculated on the basis of lung cancer death rates of cadmium-exposed workers, the lung cancer risk for workers exposed to the maximum NIOSH standard. They estimated from 5–11 percent of workers exposed to this level of cadmium over 45 years of working life, would die of lung cancer.[44]

Another important and provocative research finding involved experiments in which rats were fed trace amounts of cadmium as part of their normal diet. Accumulations of cadmium in their kidneys to levels less than those commonly found in adult Americans corresponded with increased mortality rates and decreased lifespan in the rat population.

Observed as the manifestations of cadmium toxicity, these findings have been viewed as extremely worrying in regard to human health.[45] Furthermore, the life span of rats and mice fed on cadmium-free diets was decidedly longer than that of rats fed on the cadmium diet. The researchers concluded that cadmium was essential neither for the growth nor the reproduction of these animals. Since that time, other experiments demonstrate that cadmium seems to be without any beneficial physiological role in the maintenance of health.[46] Instead, the human body is equipped with mechanisms designed to minimise the uptake of cadmium and render harmless even the low levels of cadmium which would normally be absorbed from an unpolluted environment.

Cadmium protection mechanisms

Up to 50 percent of cadmium which enters the body via the lungs and about 5 percent of cadmium entering via the gastrointestinal tract is absorbed into the bloodstream.[47] In the bloodstream, cadmium tends to be bound to albumin and other high molecular weight proteins, which are then subsequently taken up mainly by the liver. The human body will endeavour to protect itself from cadmium toxicity. It does so by reacting to the intake of cadmium by inducing the synthesis of special protein metallothionein in the liver. Metallothionein binds the cadmium into a stable biocomplex[48] which prevents cadmium from binding with other essential proteins. In this regard, metallothionein is not toxic and accumulates harmlessly in the kidneys.[49] However, when a certain plateau level of cadmium absorption is reached the ability of the kidneys to maintain the cadmium, bound as safe metallothionein, is exceeded and the level of free cadmium ions in the kidneys increases as does cadmium toxicity. The toxic properties of these free cadmium ions are believed to be responsible for the renal damage associated with long-term cadmium exposure.[50] The plateau level of cadmium tolerance most probably varies with the individual and may also be dependent on age and sex.

Of considerable interest is the fact that in nature cadmium occurs mainly in association with zinc. The main industrial sources of cadmium are in fact zinc ores, and cadmium minerals are rare.[51]

Recent studies have found that zinc appears to protect plants from cadmium. When zinc was applied with fertilisers used on durum wheat crops, the zinc-treated soil decreased the cadmium concentration in the grains, leaf, stem and root of the wheat plants. Foliar applications of zinc had little effect and the researchers concluded that soil-applied zinc may be a potential means of reducing cadmium concentration in durum wheats.[52] It is revealing that in test animals zinc serves to stimulate the production of protective metallothionein and may actually be the primary inducer.[53] Thus it appears that zinc increases the body's capacity to protect against cadmium toxicity. This explains why dietary zinc deficiency increases the level of toxic cadmium retention.[54] Furthermore, interesting research with calves has established that ruminants have a strong protective mechanism which allows for the absorption of zinc but rejects that of cadmium.[55] Other research with dairy cattle suggests that there is a natural mechanism in the animals which prevents increased cadmium uptake in the diet from entering the milk they produce, thus protecting young offspring from increased cadmium levels.[56] This protective mechanism in cattle which minimises cadmium absorption should make us wary of the cadmium levels found in the foods we feed our children. According to a 1987 survey of the cadmium levels in babyfoods sold in the United States, exceedingly high levels were found in a number of vegetable preparations, with a level of 2467 micrograms per kilogram dry weight being reported for carrots.[57]

Conclusion

The full health hazards associated with this level of intake are still insufficiently appreciated. What is clear is that blind technological tampering with nature is producing an array of adverse health effects, many of which have been only partially detected and exposed. The emerging problem of toxic cadmium levels in the food chain is a problem of significant proportions and can no longer be overlooked. As long as we continue to acidify natural phosphate rock with sulfuric acid to produce superphosphate, or acidify soil bound cadmium with acid rain from our industrial processes and power production, we are in effect releasing toxic cadmium into our environment. As we efficiently separate zinc and cadmium in our smelters and electrolytic plants from their insoluble earthy forms, and later release these metals in our effluents and wastes, we are unwittingly poisoning our planet and ourselves.

The issue which confronts us presents itself starkly. Technological intervention, guided by a philosophy of science in which power over nature and mindless exploitation of its resources figures prominently, leads inevitably to the disruption of the subtle ecological weave into which the resources of nature are firmly bound.

What you can do

The problem of cadmium is one to be faced by the community at large. However, as individuals there are some things we can do.

1. We can choose to live away from smelters and steel plants.
2. We can choose not to smoke cigarettes.
3. By contacting our local department of agriculture or national health department we can find out which rural areas have known cadmium build-up in the soil, and which foods have been found to contain elevated cadmium levels.
4. Where possible we can buy produce grown in areas away from industrial and power generation fall-out.
5. Vegetables and meat produced by organic farming methods should also be lower in cadmium.
6. We can lobby for farmers to use low cadmium phosphate fertilisers.
7. We can also call for cadmium to be replaced by less toxic alternatives in commercial products such as nails, plumbing fittings and batteries, and thereby make a contribution to the gradual clean-up of cadmium from our environment.

chapter 11

Pesticides: What you need to know to keep your family safe

Just as chemical fertilisers destroy the microbial ecology of our soil and poison our streams and lakes, so the accompanying pesticides have been changing the insect balance and poisoning our food.

Destroying the balance

It is paradoxical that while we have added chemical fertilisers to our soils in the hope of increasing crop yields (and have thereby unwittingly increased their susceptibility to pests), the pesticides we have used to control these pests have actually reduced the soil's capacity to sustain and generate fertility. In her 1963 classic, *Silent Spring*, Rachel Carson points out that the pesticides benzene hexachloride (BHC), DDT, DDD, aldrin, lindane and heptachlor all prevent nitrogen-fixing bacteria from forming the necessary root modules on leguminous plants while other experiments showed that lindane, heptachlor and BHC reduced nitrification after only two weeks in soil.[1]

Of grave concern also, is the staggering effect of our use of these chemicals to correct the imbalance of pests, which has itself been produced by agricultural practices that have become geared to the needs of the machines that were thought to serve agriculture. Human control of plant pests through their eradication, either by chemical or genetic technology, has become the aim of modern agricultural science. Deeper reflection on the issues reveals that the approach is futile. Such a view fails to recognise the complex, precise and highly integrated systems of relationships among living things which presuppose the fundamental interdependency of all nature.

Carson wrote:

Two critically important facts have been overlooked in designing the modern insect control programmes. The first is that the really effective control of insects is that applied by nature, not by man. Populations are kept in check by something the ecologists call the resistance of the environment and this has been so since the first life was created. The amount of food available, conditions of weather and climate, the presence of competing or predatory species, all are critically important ... The second neglected fact is the truly explosive power of a species to reproduce once the resistance of the environment has been weakened.[2]

These two factors are the main reasons why pesticides have not proved to be the simple panacea they were originally thought to be. Robert Metcalf stated, 'The greatest single factor in preventing insects from overwhelming the rest of the world is the internecine warfare which they carry out among themselves'.[3]

Yet as Carson observes, '... most of the chemicals now used kill all insects, our friends and enemies alike'.[4] These friends include birds[5] and spiders.[6] Carson reminds us that in a biologically sound forest there are 50–150 spiders to the square metre and a single spider destroys on average 2000 insects in a life span of 18 months.

Other beneficial insects such as bees are also killed by sprays. In the United States, the extensive use of pesticides resulted in such large numbers of honeybees being killed by these chemicals that the Bee Indemnity Act of 1970 was passed to compensate apiarists. The Act was repealed in 1980. Honeybees are vital to the production of in excess of US$20 billion worth of fruits, vegetables and forage crops in the United States alone. Honeybees killed by pesticide use are estimated to cause losses of at least US$135 million annually as result of reduced crop yields and honey production.[7]

Having declared nature incompetent to maintain the order in nature we demand, we have applied scientific technology to the pest problem and failed miserably. The cost of failing extends far beyond crop losses to insects.

Health effects

In 1955, Wickenden argued that the increasing incidence of hepatitis in the United States was linked to the increasing use of DDT and other chlorinated pesticides.[8] By the late 1950s, a number of commonly used pesticides, including DDT and arsenates, had been linked with a range of disorders from leukemia to skin cancer.[9] Some, such as DDT and BHC, were found to be monstrously mutagenic.[10] Since that time a vast literature has accumulated detailing the mutagenic, teratogenic and carcinogenic properties of pesticides.[11]

The true cost of controlling pests with chemicals is now being revealed. At the World Food Summit held in November 1996, the World Health Organisation revealed that pesticides cause 220 000 deaths and 3 million severe poisoning cases worldwide each year.[12] There has also been increasing concern among farmers in the United States about their higher than average rates of cancer.[13]

A 1986 study by the United States National Cancer Institute found that Kansas farm workers who were exposed to herbicides for more than 20 days per year, had a six times higher risk of developing non-

Hodgkin's lymphomas than non-farm workers.[14] A number of other studies have found similar cancer risks associated with pesticide exposure.[15] The increasing incidence of cancer and growth defects in children perhaps provides the strongest link between chemical contamination of the food chain and health.

Ten years later research continues to reinforce rather than dissipate the pesticide–health link. In 1989 the Natural Resources Defence Council (NRDC) in the United States, alleged that some 5500–6200 children may develop cancer later in life from exposure to just eight pesticides during their preschool years.[16] That same year the Environmental Protection Agency (EPA) released a preliminary report revealing that from 15 000–50 000 infants and children a day run the risk of getting sick from potatoes on which the insecticide aldicarb has been used.[17]

Evidence that human contamination with pesticides is occurring at an early age was noted ten years earlier by J Harrington:

> The intensive spraying of insecticide over many years in North Western Mississippi has, however, now led to such high concentrations of these pesticides in breast milk that if it was cow's milk it would be banned! In one such study, despite the lack of DDT usage during the study period, over 90 percent of the mothers, 84 percent of the black newborns and 45 percent of the white newborns demonstrated evidence of recent DDT exposure.[18]

Recently the North Carolina State Center for Health and Environmental Studies reported data which suggests that children under 14 have four times the normal risk of tumours of connective tissue if the gardens at their homes are treated with pesticides or herbicides. The researchers also found that in homes where insecticidal strips were used, children were one and a half to three times more likely to develop leukemia.[19] Massive pesticide health effects are emerging in other parts of the world. In the late 1980s, soil from the dry bed of the Aral Sea in the Soviet Union was dispersed up to 5000 kilometres distant by wind. This fine dust contained many chemicals and pesticide residues, particularly DDT, which had been washed into the sea over many years. A subsequent investigation revealed that 70 percent of mothers in the region were anaemic, sterility was widespread and infant mortality was the highest for any region in the Soviet Union. In addition, there was also a massive increase in dental disease and congenital deformities.[20]

In 1990, an abnormally high incidence of deformities was reported in the embryos of certain fish in the North Sea. The study found that 50 percent of all whiting embryos, and between 5 and 20 percent of the embryos of sole, cod, flounder and plaice tested over the previous three years were fatally deformed. While the exact chemical cause was not determined, the researchers suspected that organochlorine compounds such as pesticides were responsible.[21] Another revealing discovery was the finding by Australian scientists of excessive levels of DDT, dieldrin, BHC and eldrin in the blubber of a stillborn, full-term baby killer whale.[22] Low concentrations of these pesticides are believed to now contaminate whales and dolphins in all the world's oceans.[23] Although organochlorines are only present in very low concentrations in the sea, they can accumulate in animal fats up to 50 000 times the concentration in surrounding waters.[24]

The extent of the organochlorine pesticide problem is illustrated by a 1995 study of villages living in two separate cultural and environmental areas in the Canadian Arctic. Up to 50 percent of women who consumed meat and blubber from ringed seal, and blubber from walrus and narwhal had intakes of organochlorines which exceeded tolerable daily intake levels. Caribou, whitefish, trout and duck were also found to be sources of organochlorine contamination in the diets of these women.[25] This level of organochlorine contamination may pose health risks for the children of these women. Another study has found that babies born to mothers who had eaten similarly contaminated fish from Lake Ontario, were less able to cope with stress and more easily frustrated than those who had not consumed fish.[26]

A recent Australian study has shed more light on a pesticide–'less able to cope with stress' link. Patients suffering from chronic fatigue syndrome were found on average to have twice the levels of pesticides in their blood compared with controls[27] (see Figure 11.1). These diverse incidents establish unequivocally that the global pesticide problem is rapidly becoming a global health problem.

Figure 11.1 Blood pesticide levels in people with/without chronic fatigue syndrome

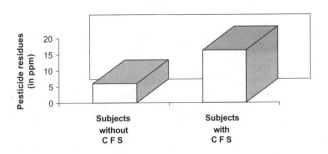

Bioscreen Pty Ltd, University of Newcastle, NSW, 1997

In the early 1990s a new side to the pesticide saga began to emerge. Dr Niels Skakkebaek of the University of Copenhagen sounded the first alarm when after analysing male sperm count data from many studies, he and co-workers concluded that there had been a dramatic decline in sperm counts worldwide over the past 50 years.[28] The researchers have argued that the decline in semen together with an increased incidence of testicular cancer might be linked to oestrogen-like compounds such as the pesticides DDT, DBCP, endosulphane and so on, and other pollutants like PCBs and dioxins.[29] Further evidence for the pesticide link comes from a comparison of the sperm counts of members of the Danish organic farmers association which was more than double the count of a pooled group of blue collars workers in the same community.[30] (See Figure 11.2.)

Figure 11.2 Possible effect of pesticides on human sperm counts

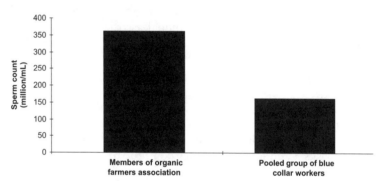

A Abell et al., *The Lancet*, 11 June 1994

In 1995, Lars Mortensen Television produced a documentary on the environmental and health effects of oestrogen-type compounds such as pesticides, for BBC World Wide Television. In Costa Rica, at least 1000 men have been made infertile or impotent or have children with physical and intellectual disabilities, due it is believed, to exposure to the pesticide DBCP, which is now banned. At Grenada University Hospital in Spain, researchers are studying why so many boys have testicle problems. It is believed that there is a link between the massive use of pesticides, and in particular endosulphane, in the plastic houses which are used for the vegetable growing industry of the region. High levels of pesticide residues have been found in both adults and children of the region. Cancer of the testes is increasing world-wide, with a 300 percent increase in Denmark which again appears to be due to exposure of mothers to pesticides during pregnancy. In the Netherlands, a study of fruit growers by Professor Dick Heederik has found that male farm workers exposed to pesticides have slower or less mobile sperm, take longer to make their wives pregnant, and father more girls than boys, compared with the average.

The television documentary also referred to other animal studies which reveal similar findings. During the early 1980s there were large spills of chemicals such as DDT in Florida lakes. The lakes have now been cleaned up but the food eaten by alligators has accumulated DDT. Professor Louis Guillette from Florida University has found that there is a 90 percent decline in the survival of next-generation alligators. Only one out of ten eggs hatch and males have deformed genitals and have lost the urge to mate.

Whether the eostrogen-type pesticides and other industrial wastes such as PCBs and dioxins contribute to the worldwide rise in hormonal associated cancers such as cancers of the breast, uterus, ovaries and prostrate is a moot point. Female hormonal associated cancers have risen by 30–80 percent since the 1940s and cancer of the prostate has tripled over the same 50 year period.[31]

The emerging pesticide problem

Government authorities continue to assure us that the substantial benefits associated with the use of most pesticides in agriculture outweigh their risks[32] and large quantities continue to be used. In 1975

in Australia, the total value of pesticides used was about $30 million. By the early 1990s, sales of insecticides alone were around $50 million, with fungicide sales of $130 million and a massive $400 million spent on herbicides.[33] This latter figure reflects the increase in the use of minimum tillage farming systems to control soil erosion but which rely on agricultural chemicals much more heavily than traditional systems.[34]

Similarly in the United States, huge quantities of agrochemicals continue to be used and pesticide levels in foods continue to be a problem. In 1995, the United States Food and Drug Administration Total Diet Study which surveyed the levels of selected pesticides, found widespread contamination of foods commonly consumed by adults.[35] (See Figure 11.3.)

Figure 11.3 Pesticide contamination of food in the USA

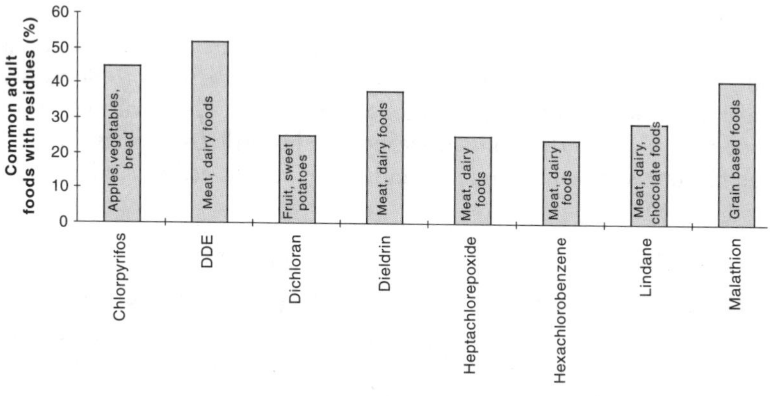

Note: Major contributing food products are listed from the 201 foods surveyed.
EL Gunderson, *Journal of AOAC International*, vol. 78, no. 6, 1995

The Washington DC based non-profit research organisation known as the Environmental Working Group (EWG), found that produce sold in the United States from both domestic and imported sources was almost twice as likely to be contaminated with pesticides above the legal limit than the FDA had reported.[36] The EGW found that residues of 66 pesticides on 42 different fruit and vegetable crops sold in the United States during 1992 and 1993 were at levels higher than legally allowed. Overall an incredible 5.6 percent of all the fruits and vegetables tested were contaminated with illegal levels of pesticides. In 13 percent of violations the pesticides were not even supposed to have been used. An author of the report stated that 'A disturbing number of fruit and vegetable growers routinely break the law'.[37]

Another side to the pesticide picture was revealed at the November 1996 World Food Summit. KL Heong from the International Rice Research Institute (IRRI) in the Philippines told delegates that a massive 80 percent of pesticides sprayed on rice in that country are used for the wrong pests and applied at the wrong time. Rice production would remain the same if they were not used at all.[38] Thai researchers reported similar findings with 75 percent of farmers in Thailand spraying their rice crops at unnecessary times.[39] It was also reported to delegates that an IRRI study conducted early in 1996 revealed that many pesticides widely used in Asia had been banned in Western countries for being too toxic to humans.[40]

Spraying pesticides at the wrong time can do additional harm to the environment. It has been determined that much of the damage to rice crops from the brown plant-hopper which spread through Asia in the 1970s and 1980s occurred because farmers sprayed at the wrong time, killing the hopper's predators instead.[41]

Given our unbridled faith in the ability of chemical interventions and our obsession with chemical fixes to environmental problems, we have long forgotten the observations of Sir Albert Howard, that crops grown on highly fertile soils suffer remarkably little from pests. In turn, soil fertility is improved and maintained by regular feedings of humus.[42]

Conclusion

Modern agriculture has come to depend on pesticides but there are alternative methods of farming which have proven to be successful, such as organic farming[43] and the use of Permaculture methods.[44] The feeding of future generations must be based on a sustainable health-promoting agriculture supported by local communities and not based on economic market forces which fail to take into account the true cost of production including the long-term health and environmental costs of modern pesticide-based agricultural practices.

Although we have explored only a few of the more notorious examples of pesticide pollution, it is clear that the continued widespread use of pesticides constitutes a health problem of considerable magnitude. Appreciation of this fact reveals how making the most of nature's resources depends not upon technological control, but rather upon maximising empathetic balance. Pesticides tend to disrupt the established harmony of the ecosystems, within which nature ordinarily maximises the potential yield of its harvest through the preservation of a delicate equilibrium between predator and prey, climactic conditions and human intervention. The technology needed is that which helps us to recognise and to preserve this equilibrium. We need also to limit our own population growth sufficiently that the food production demands we make of nature are possible to achieve, not just in the short run, but in the long run.

What you can do

We need to encourage organic approaches to farming, and the use of permanent tree crops, while at the same time supporting research directed towards a better understanding of the complex interplay among the various natural mechanisms which serve to reduce the need for pesticides. Part of the task confronting us is also to re-educate ourselves in the aesthetics of food production. If, as the consumer public, we expect every fruit or vegetable we eat to be unblemished and aesthetically perfect, we will inevitably come to rely upon a chemical technology designed not so much to reproduce the fruits of nature as to improve upon them. There was a time when we were content with less. Could it be that learning to live well again has something to do with learning to live with less? Whatever the answer, it is clear that the unbridled use of pesticides will in the end yield less good than was promised and more harm than was feared.

We can make a positive contribution to reducing pesticide levels by:

1. buying as much organically grown produce as we can, thereby offering financial and economic support to this environmentally friendly form of agriculture
2. becoming involved in local community groups such as organic gardening, biodynamic farming and permaculture societies which use and promote environmentally friendly alternatives to pesticides.

PART four

Water technology

chapter 12

Water chlorination and your health: A second opinion

Following a workout in the gym, a training run or cycle, a game of squash or a mountain hike, the thought of a refreshing drink of cool, clear water is often uppermost in our minds. Adequately quenching that thirst is a vital aspect of maintaining fitness, health and even beauty.

But how pure is the water we drink and are the chemicals used to purify it serving, paradoxically, to contaminate it? Unless we are fortunate enough to live in an unpolluted rural environment collecting our own water (or have affixed some type of water purifier to our tap), the peculiar odour and taste of the liquid in our glass serves as a persistent reminder that the water we drink contains chemicals. Generally, the most noticeable of these is chlorine.

We have come to accept the presence of chlorine in our drinking water as one of the necessary, though slightly unpleasant, aspects of maintaining community health. Most of us have learned that a number of contagious diseases such as typhoid and cholera have been virtually eradicated by filtering and chlorinating our municipal water supplies.

The crucial question which many of us never learned to ask, however, is whether chlorinated water represents a health hazard in its own right. In what follows we shall present some of the more neglected research on the subject, aims to allow the reader to make more informed decisions about the merit of treating our drinking water with chlorine.

What is chlorine and how does it work?

The Australian bicentennial year, 1988, marked the centenary of water chlorination. In 1888, a patent on chlorination of water was granted to Dr Albert R Leeds, Professor of Chemistry at Stevens Institute of Technology, Hobokin, New Jersey. Professor Leeds showed that chlorine could be used as the basis of a method of disinfection to control pathogens responsible for waterborne diseases. In the following year, the first chlorination of a public water supply was initiated at Adrian, Michigan, though it was not until 1908 that chlorination was introduced on a large scale at the then huge Boonton Reservoir waterworks in Jersey City, New Jersey. By World War II the practice of chlorination was widely established in the United States.[1]

Water chlorination involves a relatively straightforward chemical process. Chlorine is one of the most reactive elements in nature and is found in a free form only in volcanic gas. Even a small amount of chlorine will dissolve in water, some of it combining with water to form hypochlorous acid and hypochlorite ion. Chlorination of water is achieved by adding chlorine gas directly to the water or by adding the chemicals calcium hypochlorite or sodium hypochlorite. In these latter forms, chlorine is known as 'free available chlorine' and has effective germicidal powers because of its ability to combine with, or oxidise, classes of organic compounds essential to life.

One theory of how chlorine works suggests that there is a physiochemical reaction between chlorine and the structural proteins of the bacterial microbes, thus causing the disintegration of their cell walls. Another popular theory holds that the process of disinfection works by inhibiting a key enzymatic process which oxidises the glucose of the cell; the bacterial cells die because the chlorine destroys the oxidation process upon which the cells depend.

Chlorine was originally added to the water just prior to filtration, though the more common practice now is to apply chlorine both before and after filtration. The idea is that prechlorination serves to reduce the accumulation in filters of biological material such as algae. Postchlorination is alleged to minimise the number and variety of bacteria which would otherwise enter the distribution system. Chlorine also combines with ammonia and organic nitrogen compounds forming chloramine. When ammonia is combined with chlorine, a slower acting disinfectant results, which has been found to be beneficial in the suppression of iron-fixing or slime-forming types of bacterial growths.

Chlorination is not employed as a substitute for other forms of water treatment; on the contrary, the effectiveness of the process to some extent depends upon other treatments such as filtration. Being a very reactive element, chlorine will readily react with many other substances which may be found in water. If some of these substances are present in the water even in minimal quantities, they may create a chlorine demand which significantly reduces the available chlorine for germicidal purposes. Consequently, chlorination is often accompanied by filtration to reduce the presence of substances which may create excessive chlorine demand.

Chlorine is notorious for its pungent and disagreeable odour. The human olfactory sense is capable of detecting only a few parts of chlorine per million in the atmosphere, and a concentration of only 50–60 parts per million (ppm) can cause serious illness 30–60 minutes. Being a toxic substance, it is capable of causing major congestion of lung tissue and even death, if breathed in sufficient quantities.

The odour and taste of chlorinated water may be produced by the presence of excess chlorine or, strangely, it may also occur if insufficient chlorine has been added. In the latter case, the characteristic odours and taste are produced when chlorine reacts with organic matter such as algae in the water. When stronger chlorine levels are used, the organic matter in the water is destroyed completely, and the result is water which is virtually odour free. When water odour is minimal it would thus be misleading to conclude that chlorine levels are low and vice versa.

The potential health risks from chlorinated water

It is amazing to think that a very reactive and poisonous chemical could be deliberately added to public drinking water without an extensive study of the possible harmful health effects being carried out beforehand. Yet with chlorination, this appears to have been the case. In 1951, Dr WJ Llewellyn wrote to the editor of the *Journal of the American Medical Association*:

> What studies have been made to determine the deleterious effects of heavily chlorinated water that is used for drinking purposes? The water supply in our town is chlorinated but not filtered. At times it is possible to smell the chlorine. Could this harm the gastrointestinal or the genitourinary tract?[2]

The editor replied:

> A search of the literature did not reveal any organised investigations on the problem of the effect of heavily chlorinated water on the human body. Allergic manifestations of chlorinated water have been reported. Many cases of asthma have been traced to an allergy to chlorinated water. In all these cases the asthma was relieved or disappeared when the patient drank distilled or unchlorinated water.[3]

In the 1970s popular health writers such as Linda Clark[4] and Richard Passwater[5] warned their readers that water chlorination may be a significant factor in heart disease.

In his 1984 book titled *Coronaries, Cholesterol, Chlorine*, Dr JM Price reports a study where in contrast to the control group, cockerels reared on highly chlorinated water all showed evidence of either atherosclerosis of the aorta or obstruction of the circulatory system. He writes:

> The abdominal aorta (the place where atherosclerosis is known to occur in chickens) of all the cockerels dying after four months were carefully examined. In more than 95 percent of the experimental group grossly visible thick yellow plaques of atherosclerosis protruding into the lumens were discovered![6]

These chickens were noted to have an extremely high apparently spontaneous death rate and common findings on examination of the carcasses were haemorrhage into the lungs and enlarged hearts ... At seven months there were so few experimental chickens remaining alive that the survivors were sacrificed, with identical findings. At the same time one-third of the apparently healthy control group was also sacrificed with not one abnormal aorta found!

Price goes on to describe how he repeated the experiment replacing the distilled water in the diet of half the remaining control group with the heavily chlorinated water. Again, gross atheromas of the aortas were found in the animals fed the chlorinated water.

Subsequent studies by NW Revis and co-workers from the Oak Ridge Research Institute in Tennessee and the United States Environment Protection Agency added more weight to the insidious link between water chlorination and heart disease. They report that they observed 'hypercholesteraemia [excess cholesterol in blood] and cardiac hypertrophy [enlarged heart] in pigeons and rabbits exposed to chlorinated drinking water' and 'significant increases in plasma cholesterol and aortic atherosclerosis in pigeons exposed to three commonly used drinking water disinfectants [chlorine, chlorine dioxide and monochloramine]'.[7]

Water chlorination has also now been linked to an increased risk of cancer.

In Massachusetts, a case-control study of 614 individuals who died primarily of bladder cancer and 1074 who died of other causes, revealed a positive association between bladder cancer incidence and chlorinated drinking water.[8] Another study involving 2805 cases of bladder cancer and 5258 controls from ten areas in the United States revealed that the risk of bladder cancer increased with intake level of beverages made with chlorinated tap-water.[9]

In Colorado, where surface water had been chlorinated for drinking purposes for many years, University of Colorado researchers found

189

that prolonged exposure to chlorinated water significantly increased the risk for bladder cancer. The risk of bladder cancer was 80 percent higher in individuals using chlorinated water for 30 years or more compared with individuals who had not used chlorinated water.[10] In 1993, JK Dunnick and RL Melnick from the National Institute of Environmental Health Sciences in North Carolina assessed the carcinogenic potential of chlorinated water. They found that the chlorination by-products, trihalomethanes, were carcinogenic in the liver, kidney and intestine of rodents.[11] Their results reinforced earlier animal studies which linked water chlorination and cancer.[12] A chlorinated organic compound is now believed to be one of the principal agents responsible for the mutagenic activity of chlorinated tapwater.[13] This powerful mutagen is produced as a result of the reaction of chlorine with humic substances in the water and has the potential to persist in significant quantities throughout the water distribution system.[14] Water chlorination is now known to produce a range of chlorinated organic compounds which are euphemistically termed 'disinfection by-products'.[15] In 1992, RD Morris and co-workers from the Medical College of Wisconsin in Milwaukee, carried out a major review of reported studies on water chlorination and cancer. Using an accurate meta-analysis technique to analyse the data, they found a positive association between the consumption of chlorination by-products in drinking water and bladder and rectal cancer in humans.[16]

Recently, LS Pilotto from the National Centre for Epidemiology and Population Health at the Australian National University in Canberra, has also reported that there are now several studies which suggest a clear link between exposure to chlorinated drinking water and the development of urinary bladder cancer.[17]

Chlorination of water and nutrition

We have seen that chlorine is a very reactive chemical in water, but what are its effects once this chemical enters our body, where it is exposed to a complex and delicately balanced biochemical organism? Chlorine is a powerful oxidising agent that readily destroys, oxidises or combines with organic substances such as certain vitamins, enzymes, unsaturated fatty acids and beneficial bacteria.

Let us first consider its impact upon ascorbic acid (vitamin C). Among other important functions, this vitamin is associated with the body's protective action against pollutants. Vitamin C, however, is destroyed by chlorine. At the same time, if vitamin C is present in sufficient quantities, it can over time, reduce chlorine to harmless chloride ion, provided the chlorine has not combined with some other organic compound in the meantime. Thus fruit juices which contain vitamin C might plausibly offer some protection against the chlorine contained in the water added to reconstitute them.

Another important vitamin affected by chlorine is vitamin E. It is well known that vitamin E is essential for maintaining the integrity of the coronary and reproductive systems. Once again, chlorine destroys vitamin E. It may be that drinking large amounts of chlorinated water may destroy vitamin E in the body. If this hypothesis were true, it would help explain the purported correlations between chlorinated water and heart disease mentioned earlier. Excess saturated fats in the diet have been associated with heart disease for some time now, and most people are aware of the increasing emphasis in health education to encourage people to replace certain types of saturated fats in their diets with unsaturated fats, coupled with a reduction of their overall fat intake. Chlorine reacts readily with unsaturated compounds, and whenever chlorinated water and unsaturated fats are mixed, the resultant organochlorine compounds are likely to have toxic properties. The chlorination of flour produces chlorinated fatty acids and in animals fed these fatty acids significant increases in heart weight have been observed.[18] The implications of these increases for the human heart needs to be seriously investigated.

Chlorinated water and the unknown hazards of the shower

There is another side to the chlorine-water story. When we return from a gym workout or a jogging session, not only are we thirsty, but we usually shower or bathe to wash away waste products and perspiration. We have been taught that cleanliness and health go together, and indeed they do, when chemical-free water is used. When chlorinated water is used, however, bathing may be much less healthy than we ever supposed.

Gases are, as a rule, less soluble in hot water, and when water is heated or boiled dissolved gases are released. Boiling water is a way in which the free chlorine content in water is greatly reduced, the chlorine escaping into the air. When we have a hot shower or run a bath we can sometimes smell the chlorine released as it escapes from the hot water. In a confined shower recess, however, especially one with poor ventilation, the chlorine escapes from the water as we continue our hot shower and steadily increases in concentration in the air we breathe. The olfactory threshold for chlorine is about 3.5 parts per million (ppm) so when we can smell chlorine the concentration is already above this level. The lethal concentration for ten minute exposure is about 600 ppm. We suggest that regularly taking long hot showers with chlorinated water could pose a health risk. Chlorine causes pulmonary oedema, and it would seem likely that regular exposure to chlorine gas even at low levels such as in normal showering may reduce the oxygen transfer capacity of the lungs. This could be a critical factor for athletes and for others prone to heart failure.

Another aspect to be considered is our skin, which is an important protective barrier for our bodies. When we shower with chlorinated water we are essentially exposing our skin to a relatively large volume of a dilute chlorine solution. Some of this chlorine reacts with the oils in the skin to form chlorinated compounds, and it is these compounds which may then be absorbed by the body. Recent research has also shown that the concentration of certain chlorination by-products known as trihalomethanes (THM) is actually about 50 percent higher in hot water than in cold water. The researchers point out that the estimated lifetime cancer risk associated with exposure to THM in water during showering is therefore underestimated by 50 percent if the concentration in cold water is used to estimate the risk.[19] It seems very likely, considering the strong oxidising power of chlorine, that regular exposure to chlorinated water promotes the aging processes of the skin, not unlike extended exposure to sunlight. Moreover, chlorine may actually enhance the aging effects of ultraviolet radiation by reinforcing the process of cell deterioration. In particular some health researchers from

the University of Nijmegan in The Netherlands are now suggesting that water chlorination may be a significant factor in melanoma risk. They point out that the skin pigmentary system appears to be a target organ for chlorine compounds and chlorination by-products.[20] The researchers have also found that people who regularly swim in chlorinated swimming pools have more than twice the incidence of melanoma compared with those who swim in unpolluted waters.[21]

Another skin factor to be considered is the destruction by chlorine of the natural bacteria balance on our skin. Our skin has an ecology all of its own, which needs to be preserved in order to maintain healthy skin and its associated beauty.

Despite the overwhelming evidence of the health dangers of water chlorination, chlorine has remained the disinfectant of choice for most water authorities because it is relatively cheap and is promoted as being highly effective. It is argued that while there are disadvantages associated with the formation of the organochlorine disinfection by-products, the risk due to the presence of these impurities is far outweighed by the reduction in mortality due to waterborne diseases.[22] The 1991 Peruvian cholera epidemic is often cited as a case in point. Peruvian authorities abandoned the practice of disinfecting water because of concerns about the health risks associated with the presence of chlorinated by-products in drinking water. This led to a major cholera epidemic which spread to other countries in South America and caused widespread illness and death.[23]

Let us make clear that we are not advocating that it is unnecessary to ensure that the water we drink is safe. What we are arguing is that water chlorination is not as safe as we are often led to believe. In addition to the subtle health effects discussed earlier, chlorination does not necessarily destroy all viruses, parasites and harmful microorganisms such as cryposporidium and giardia.[24] Both these latter organisms which can be present in water supplies can cause diarrhoea and a flu-like intestinal illness. Both can survive normal chlorination, and this fact was made apparent in the city of Milwaukee in the United States in 1993. In the massive disease outbreak, an estimated 400 000 cases of diarrhoea occurred over a 14-day period in a single city.[25]

There is probably no cheap, foolproof water treatment, which shows the need to consider the total community hygiene problem, particularly the disposal of sewerage and other wastes in our environment.

We believe however that nature provides clues to the solution of this controversial environmental problem. It is commonly known that sunlight and oxygen disinfect water, and the efficiency of sunlight as a drinking water sterilising agent has been promoted by A Acra and others.[26] Other research has shown that sunlight can purify water highly contaminated with faecal bacteria.[27] In another experiment, the organochlorine solvent tricholoroethylene, which pollutes ground water in some parts of the United States, was reduced from a concentration of 1–2 parts per million to less than 50 parts per billion with a five-minute exposure to concentrated sunlight.[28]

Ultraviolet light-based water treatments which add no chemicals to the environment can be extremely effective. When correctly used, UV treatment can kill all bacteria and all forms of virus in drinking water, although bacterial spores and some parasitic protozoa can survive UV treatment.[29]

The educational goal

The complete chlorine–health picture is even larger and more disconcerting than we have intimated in the limited space available. Whenever a toxic substance is added to the environment in large quantities, the overall balance of nature is bound to be changed, and thus the chain of reactions produced are likely to be very disruptive, far-reaching, and often unexpected. One need only mention, for example, the detrimental impact upon the ozone layer caused by water chlorination by-products such as chloroform, which may find their way to the upper atmosphere.[30]

With education we can begin to discern that nature carries within itself an important pattern for the design of our health, and optimum fitness, even for the natural beauty of ourselves and our planet. When all is said and done, it is apparent that nature did not intend that we despoil it and ourselves by contaminating it with the waste products of the technology we have used to dominate it. After 100 years of chlorination it is surely time to express our misgivings about the prospect of continuing until we celebrate its bicentenary.

Since fluoride, chlorine and aluminium are standardly added to most of our municipal water supplies, the problem of finding chemical-free drinking water is difficult. When we appreciate the extent to which drinking water has been degraded, it is evident that we need, as a community, to join together to do something about it. If we cannot avoid the chemical contamination of our municipal drinking water supplies, one of the first things we need to do is to become sufficiently enough informed to lobby influential politicians and to protect against, with the help of the media, the progressive deterioration of the state of our drinking water. In this regard it will probably be necessary to find a spokesperson or spokespersons who can represent accurately and authoritatively the nature of the problems and the options available. In the meantime there are a few things we can do to minimise the harmful effects of the toxic substances contained in water.

Water filters

Water filters are probably the least expensive devices to improve the quality of poor drinking water. Unfortunately, the claims made for water filters by their manufacturers are often grossly exaggerated. A number of filter systems are applauded for their ability to produce pure, crystal-clear and safe water by removing all solids down to the most minute particles, including residues of rust, minerals, toxic metals and organic chemical contaminants. The truth is that the most sophisticated large scale filter systems cannot deliver quite that much.

The best of household filters, however, can be used to reduce turbidity, and to filter to some extent rust particles, dirt and suspended matter. In addition, filters help to ensure cleaner-looking water and to minimise a range of tastes and odours which would otherwise be regarded as objectionable. The idea that tap-water filters eliminate harmful bacteria is somewhat misleading. Because the activated charcoal contained in the filter effectively absorbs chlorine, the reduction in chlorine levels allows certain bacteria to propagate freely. In this regard, filtered water and the filter itself may become a breeding ground for bacteria. The most efficient type of filter seems to be made up of two different filter cartridges, one of which depends upon granulated carbon as a filtering agent. This type of filter has been reported to remove up to 93 percent of trihalomethanes, a group of carcinogenic compounds, one of the best known of which is chloroform.

Water purifiers

Water purifiers can also help to improve water quality by reducing bacterial levels. There are several different types of purifiers. One type utilises ultraviolet light to kill bacteria, but its germicidal effectiveness depends upon factors such as the intensity of ultraviolet light used, coupled with its wave length. This type of purifier becomes less effective when the water temperature drops below 10 degrees Celsius. Its effectiveness is also reduced if the water is turbid or contains organic substances. The presence of colours or iron may further decrease effectiveness, as can the rate of flow of water which passes through the system. Control devices ordinarily restrict the water flow to about 36 litres per minute.

The use of minute amounts of silver ions which are released as the water passes through the filter is another method of purification. As some of the silver escapes into the water which has been purified, you may be imbibing some of the silver. Although there are allegedly no toxic effects which silver is known to have on humans, we recommend caution in the use of this type of purifier.

The 'reverse osmosis' method involves the use of a semi-permeable membrane to eliminate impurities. In addition to being susceptible to clogging, the membrane permits the passage of certain contaminants such as iron and nitrate. Because the membrane traps particles in the process of preventing their passage, some organic chemicals may in fact become concentrated.

The most basic method of water purification is boiling. Boiling does serve to kill virtually all bacteria. It will also remove many organic contaminants, but it may serve to concentrate others. Boil when in doubt, but don't doubt that there are still impurities.

Distilling water

There is some evidence that a household distiller offers the most effective means of purifying water. Distilling involves a process in which water is heated to the point of vaporisation and is then condensed. Although the process eliminates many contaminants, some carcinogens such as benzene may be carried over into the distillate. Despite its benefits as a purifier, the distilling process may not be the best means of purifying water for constant use. The reason for this is that the process of distilling evacuates the water of minerals and upsets the electrolyte balance of the water.

If you persist in drinking distilled water, you may find that the electrolyte equilibrium of your body is disrupted by the deprivation of minerals ordinarily contained in natural water. People who regularly drink distilled water should at least supplement their diets with mineral tablets or some natural mineral source such as dolomite. While supplementation may have some benefits, it would be imprudent to assume that the body's electrolyte balance can easily be restored in this way.

Bottled water: Check the labels

If you live in an area relatively free of air pollution and away from crop spraying, it is possible to collect fresh drinking water during heavy rains. Where it is difficult or impossible to do this, one option is to buy bottled water. There are many types of bottled water commercially available, but not all bottled water is worth buying. Imported bottled water is expensive, especially if you are drinking the recommended eight to ten glasses of water per day, each glass of which is approximately 250 millilitres. For a family of four or five, the weekly drinking water bill quickly becomes prohibitive. Mineral waters contain varying quantities of sodium and some brands are actually sufficiently high in sodium that they can cause water retention and bloating. They can also provide a problem for those people who, for medical reasons, are on low sodium diets. When purchasing bottled water, brands which use glass rather than plastic containers are to be preferred, as glass does not generally transmit odours, or alter the taste of the water. It is also worth checking the bottle cap before purchasing, as seals can sometimes be faulty or otherwise broken in transport or by tampering.

A final word of caution—the word 'spring' is often bandied about in the advertisements for bottled water, or appears on the label in a variety of forms such as 'spring clean', 'spring pure' and 'spring fresh'. Check to determine whether bottled water thus labelled is in fact *natural spring water*. The labels are not simply interchangeable. Remember, when you purchase bottled water—especially if you intend to become a regular customer—you have every right to ask the manufacturers to provide an analysis of the contents of their product. Sad to say, some bottled water has been found to have more impurities than ordinary tap-water.

chapter 13

The fluoridation debate: Can fluoride be damaging to your health?

The controversy surrounding fluoridation raises a number of important socio-ethical issues which cannot be overlooked. One of the most burning questions is whether the fluoridation program represents a milestone in the advancement of community health or the opportunistic outcome of a powerful lobby concerned largely to advance its own vested interests at the expense of the interests of the public. The historical origins of fluoridation are revealing. We shall show in this chapter that the potential and actual health risks associated with fluoridation have not been sufficiently appreciated by those in favour of fluoridation—partly for lack of historical perspective. The introduction of fluorides into drinking water has certainly not received the rigorous scrutiny and testing properly brought to bear on the wide array of available medical drugs, many of which can be bought without prescription. Finally, we shall urge that even if it were determined that the addition of a minimal amount of fluoride to our water supply was both safe and effective in the reduction of caries in children's teeth, the relevant dosage of fluorides could not be satisfactorily restricted to ensure that the harmful effects of fluoride did not outweigh the beneficial effects.

The genesis of fluoridation

Many readers will be surprised to hear that fluorides have been in use for a long time, but not for the prevention of tooth decay. The fluorides we now, in the name of health, add to our drinking water were used as *stomach poison* insecticides and rodenticides for nearly four decades. Fluorides are believed to exert their toxic action on pests by combining with and inhibiting many enzymes that contain elements such as iron, calcium and magnesium. For similar reasons, fluorides are also highly toxic to plants, disrupting the delicate biochemical balance which enables photosynthesis to take place. Nor is there any reason to suspect that humans are immune to the effects of this potent poison. Even a quick perusal of the indexes of most reference manuals on industrial toxicology list a section on the hazards of handling fluoride compounds. In assessing the toxicity levels of fluorides, Sax confirms that doses of 25–50 milligrams must be regarded as 'highly toxic' and can cause severe vomiting, diarrhoea and central nervous system manifestations.[1]

It is crucial to recognise from the outset that fluoride is a highly toxic substance. Appreciation of this simple point makes it easier to understand the natural reluctance on the part of some to accept the compulsory inges-tion of a poison, to obtain partial control of what would generally be regarded as a noncommunicable disease, that is, tooth decay.

The potent toxicity of fluoride and the narrow limits of human tol-erance from 15 parts per million (ppm) in drinking water make the question of optimum concentration of paramount importance.

The fluoridation controversy becomes even more interesting when we realise that industrial fluorine wastes have, since the early 1900s, been one of the main pollutants of our lakes, streams and aquifers, causing untold losses to farmers through poisoning of stock and crops.

Fluorides such as hydrogen fluoride and silicon tetrafluoride are emitted by phosphate fertiliser manufacturing plants (phosphate rock can typically contain 3 percent fluoride). The industrial process of steel production, certain chemical processing and particularly aluminium pro-duction, all release considerable quantities of fluorides into the environ-ment. The fluorides emitted are readily absorbed by vegetation and are known to cause substantial leaf injury. Even at atmospheric concentra-tions as low as 0.1 part per billion (ppb), fluorides significantly reduce both the growth and yield of crops. Livestock have fallen victim to fluo-ride poisoning caused primarily by ingesting contaminated vegetation.[2]

The development of fluoridated drinking water stems from a dis-covery made in 1916 by FS McKay at Colorado Springs. He had con-cluded that the mottling or discolouration of tooth enamel which was referred to locally as 'Colorado brown stain' and in other locations as 'Texas teeth', and which he observed on the teeth of many of his patients, was caused by something in the drinking water. He also reported that people with mottled teeth had very few cavities. It was not until 1931 however, that HV Churchill identified the offending substance as fluoride in high concentrations.[3]

The United States Public Health Service (PHS) subsequently became concerned about the poisoning effects of fluoride and in 1934, HT Dean, a dentist with the PHS published his report *Studies on the Minimal Threshhold of the Dental Sign of Chronic Endemic Fluorosis.*[4] He later reported that although 1 ppm fluoride concentration caused enamel flu-orosis or mottling in a small percentage of children (up to 10 percent), it also served to provide partial protection against dental decay.[5]

Some years later after further studies, the PHS legislated that drink-ing water containing up to 1 ppm fluoride was acceptable, and from 1945 the fluoridation of water for several large cities in North America was begun as a ten-year project, to assess the effect on dental caries of adding soluble sodium fluoride to fluoride-deficient domestic water supplies. In the early 1950s however, many other cities in North

America began fluoridation of the water supply[6] with a number of other countries including Australia following suit.[7]

The effectiveness of water fluoridation as a method of reducing tooth decay in younger members of the population was reported by numerous studies over the next two decades. In 1982, a summary of the results of 95 fluoridation projects in 20 countries initiated between 1945 and 1972 was published. The outcome was that decayed, missing and filled teeth (DMF) indices amongst children had been reduced by amounts ranging from 20–85 percent with a mean of 55.6 percent.[8] Continuing studies in the 1970s and 1980s reported similar findings on the effectiveness of water fluoridation.[9] In Australia, studies carried out in four large cities showed that after ten years of fluoridation the DMF indices of children declined by approximately 50–60 percent. The report of the Australian National Health and Medical Research Council (NHMRC) Working Party on *Fluorides in the Control of Dental Caries*, also concluded that the effect of fluoride on caries is probably greater than indicated in the many reports because the size of the carious lesions has not been taken into account. They go on to offer further evidence by citing three overseas studies in which the number of first permanent molar teeth which had to be extracted was 95, 75 and 85 percent respectively less among children in fluoridated areas compared with children in low-fluoride areas (approximately 0.1 ppm fluoride in the drinking water).[10]

The conclusions of these studies were however refuted by M Diesendorf in 1988[11] who also pointed out that the incidence of dental caries was in any case significantly declining worldwide in non-fluoridated cities as a result of better nutrition and other health care.[12]

In July 1989, Drs M Diesendorf, J Colquhoun and PRN Sutton wrote to the chairman of the National Health and Medical Research Council in Australia, stating that there was little or no benefit from injesting fluoride, yet on the other hand there was evidence of ill-effects. They pointed out that the early experiments on human populations which led to claims that water fluoridation reduced tooth decay by 50–60 percent 'suffered from serious inadequacies of experimental design'. They go on to recommend that the fluoridation of drinking water be terminated.[13] Responding to the submission by Diesendorf, Colquhoun and Sutton, the NHMRC Working Party conceded that:

> In practice, the quality of the early interception trials was generally poor ... Random allocation was usually not possible and the choice of 'control' populations was often inadequate; insufficient baseline data were recorded to allow for adjustment between compared populations; there was a lack of 'blinding' of observers in the recording of dental health outcomes; and detailed and critical statistical analysis were often lacking.[14]

The NHMRC Working Party also argued that 'flaws in study design do not necessarily mean that the results of such studies are "wrong", although their results of such studies may lack accuracy'.[15]

The Working Party then went on to present over 145 pages of evidence supporting the effectiveness of water fluoridation in reducing tooth decay and the need to continue to encourage the fluoridation of reticulated water supplies.[16]

Today, Australia has 'distinguished' itself by promoting the fluoridation program with such vigour that Australia now ranks as the most comprehensively fluoridated country in the world (see Figure 13.1). More than 65 percent of Australians are obliged to drink water to which fluorides have been added. Brisbane is the only State capital city which remains unfluoridated. Australia persists in its policy commitment to artificial fluoridation, despite the fact that 96 percent of the world's population has either discontinued fluoridation programs or never had them. Statistics show that less than 50 percent of the United States is currently fluoridated and less than 10 percent of England. Sweden, Holland and West Germany, have all discontinued fluoridation, to name only a few.[17]

Figure 13.1 Water fluoridation in 1990

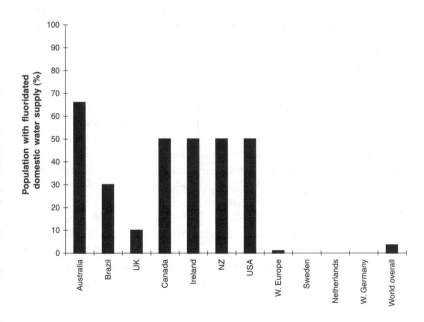

M Diesendorf, *International Clinical Nutrition Review*, April 1990

Can fluoridation be kept at safe levels?

Although 1 ppm is standardly defined as that level of fluoride concentration which provides maximal protection against dental decay, it has become increasingly clear that individual levels of safe fluoride ingestion cannot be adequately controlled. Drinking water dosages of fluoride, for example, will depend partly upon variable factors such as thirst which in turn varies with daily temperature. Liquid intakes also vary according to age, work situation, climate and season, and levels of exercise. Athletes, for instance, tend to consume more water than their non-athletic counterparts. Adjustments to municipal water supplies cannot accommodate satisfactorily the wide array of relevant individual differences of this kind.

In addition, fluorides are ingested in varying quantities from many unsuspected sources. Fluoride toothpastes, gels and even fluoride tablets contribute to dangerous increases in fluoride levels well beyond the recommended 1 ppm contained in drinking water. Even more surprising is the fact that tea leaves contain sufficient fluoride, so that by drinking three to eight cups daily, using fluoridated water, the total fluoride dosage is somewhere between four and six times the safe maximum recommended daily allowance.[18]

A study of a Chinese community where tea was the principal beverage and the water was fluoridated at the optimum level of 1 ppm was reported recently. The mean prevalence rate for dental fluorosis was 80.04 percent for adults and a massive 94.7 percent for children eight to 15 years of age. The mean fluoride intake from water, food and tea was 4.1 milligrams per day on average,[19] which is only 37 percent higher than the Australian average daily intake of fluoride in fluoridated areas of 3.0 milligrams.[20] In addition to endemic fluorides in the natural foods we eat, in many industrial cities we are forced to breathe fluorides from factory emissions.[21] Fluoridated communities also run the risk of accidental fluoride poisoning if the water fluoridation equipment malfunctions. In the United States alone, at least five major outbreaks of fluoride poisoning have resulted in hundreds of people being poisoned and four reported deaths.[22]

In the most recent incident in May 1992, an unreliable control system together with a mechanism that allowed fluoride concentrate to enter a well used as a public water system for a rural village in Alaska, resulted in a drinking water fluoride concentration of 150 milligrams per litre compared with the usual level of 1.1 milligrams per litre. It was estimated that 296 people were poisoned after drinking water and one person died.[23]

Fluoride and disease

Despite health authority endorsement of the effectiveness and safety of fluoridation there is increasing literature which suggests that the panacea has become a poison. One undeniable side effect of fluoridation is dental fluorosis, a condition in which areas of the tooth enamel become softened and porous. This kind of tooth mottling is caused by injesting excess fluoride during the ages when teeth are developing. Numerous studies have reported that in fluoridated communities dental fluorosis affects 25–50 percent of children, some so severely that their teeth are discoloured and pitted.[24] J Colquhoun from the University of Auckland reports that some of the children he examined from Auckland in New Zealand, where they drink fluoridated water, exhibited fluorosis type mottling on all their permanent teeth and appear to have had a high fluoride intake from birth.[25]

In 1977, the first major study linking water fluoridation and cancer were reported by Yiamouyiannis and Burk.[26] Data from 1952–69 for the ten largest fluoridated cities in the United States were compared with statistics from the ten largest cities not fluoridated as of 1969, but with high cancer death rates. They observed:

> The crude cancer death rates of both groups of cities had a strikingly similar trend between 1940 and 1950. Subsequent to fluoridation however, an equally striking divergence could be observed that was maintained through 1969, the last year of study. This increase in crude cancer death rate could be observed in virtually all of the fluoridated cities when compared to the control cities, indicating that the difference in averages was not due to the sharp increase in cancer death rate of only one or two of the fluoridated cities.[27]

Yiamouyiannis and Burk then reported a fluoridation-linked increase in cancer death rates of 50 per 100 000 which means that at least 10 000 people in the United States die every year from fluoride-related cancer. Their review of the literature is also revealing. In their introduction they cite 17 research papers which demonstrate mutagenic affects associated with fluoride. They found studies showing that low levels of fluoride are tumorigenic and that 1 ppm fluoride increases tumour growth rate and interferes with DNA repair in vitro and in vivo.[28] There is now wide consensus within the scientific community that the mutagenic activity of a substance can be regarded as an important indication of its potential cancer-causing activity. Since those provocative studies two decades ago, a vast scientific literature has continued to accumulate which strongly indicates that the practice of fluoridating municipal water supplies is a dangerous one. In 1983 an Australian dental surgeon, G Smith, reported a number of studies which suggest that there is now a serious risk to the public of fluoride overdose. He argues that 'the crucial argument does not concern the fluoride level in a community water supply per se, but rather whether fluoridation increases the risk that certain people develop, even for a short time, levels of fluoride in the blood that can damage human cells and systems'.[29] That same year J Yiamouyiannis published a major review of the accumulating evidence that fluoride ion both accelerates and contributes to the aging process.[30]

In 1985, another Australian scientist M Diesendorf, drew attention to the discovery of a whole new dimension to the health hazards associated with the ingestion of fluorides. Sodium fluoride, for example, had been found to cause unscheduled DNA synthesis and chromosomal aberrations in certain human cells.[31] Both these types of transformations are believed to be responsible for the development of cancer. Other recent studies purport to reveal the actual mechanism by which fluoride can disrupt the DSTA molecule and the active sites of the molecules of many human enzymes which are essential for the maintenance of health.[32] In 1990, the United States National Toxicology Program's findings that there is an equivocal link between fluoride in drinking water and cancer were released. Three out of 80 rats given drinking water at only 20 times the Environmental Protection Agency's 4 ppm standard for fluoride developed osteosarcomas, a rare form of bone cancer. Humans drinking water containing only 4 ppm fluoride ion, have bone fluoride concentrations comparable with the levels of fluoride in the bones of the test animals.[33] In 1993, The Safe Water Foundation (USA) reported that there was now a preponderance of evidence that pointed to fluoridation causing an increase in bone cancer and oral cancer in human populations.[34]

In a New Jersey study, for example, males under 20 years of age were 6.9 times more likely to develop osteosarcoma in fluoridated municipalities, than in non-fluoridated municipalities.[35] Fluoride had been found to inhibit the DNA repair enzyme system, to accelerate tumour growth rate, to inhibit the immune system, to cause genetic damage in a number of different cell lines and to induce tumours. Genetic damage of the type which is also likely to cause cancer had been observed with fluoride exposure as low a 1 ppm in drinking water. The reviewer concluded that water fluoridation would have a generalised effect of increasing cancers overall.[36]

Fluoride is also implicated in the increasing incidence of hip fracture in certain age groups. Of ten studies comparing fracture incidence relative to fluoridation, seven have shown increasing hip fracture rates with increasing fluoridation.[37] In 1990, the *Journal of the American Medical Association* reported that fluoridated water, coupled with some other relevant factors appears to increase the risk of hip fractures in women over 65. This mishap, which is common among osteoporosis sufferers can sometimes be fatal to the women so afflicted.[38] In New Zealand such fractures have increased more than threefold since the 1950s when fluoridation was introduced, and are still increasing.[39] Further evidence comes from a United States study which found that women in an area with 4 ppm fluoride (the claimed 'maximum safe level') had twice the fracture rate of women in an area with 1ppm water fluoride (the claimed 'optimum level').[40]

Another possible effect of excessive fluoride intake is neurotoxicity. Recent studies on rats reviewed in 1996, showed behavioural alterations which increase in severity proportionally to serum fluoride levels. Reviewers were concerned that the serum levels in the rats were the same as that often found in children one hour after topical fluoride treatment.[41] Chinese studies have also indicated fluoride behavioural changes in children.[42]

It is manifestly clear that the time has come for a serious and comprehensive review of the policy which mandates the compulsory fluoridation of our municipal water supplies. Such a review will no doubt require a multi-faceted approach in which reliable research investigations can be integrated with a philosophy of health education to assist their implementation. Through education it may be possible to appreciate that within nature itself there are important patterns of design for an overall program of health. Let us elaborate this point.

Fluoride compounds in nature

Fluorine comprises about 0.08 percent of the earth's crust which makes it the thirteenth most abundant element, roughly as plentiful as nitrogen and several times more plentiful than carbon. However, fluorine only occurs in nature as highly insoluble forms of fluoride. It is a minor constituent of many minerals, including rock phosphate, which typically contains about 3 percent fluoride. The chief ore is fluorspar which is insoluble calcium fluoride. Hydrogen fluoride, noted for its ability to react with silica, may be emitted in volcanic eruptions.[43] However, once in contact with rocks the fluoride is soon bound as a fluorosilicate or similar mineral.

Surface waters around the world are typically low in fluoride concentrations, being typically 0.2 ppm or less.[44] HM Sinclair notes that most of the waters in the United States which naturally contain fluoride at concentrations higher than 0.5 ppm are 'hard'.[45] This means that they are high in calcium and magnesium salts which form insoluble compounds with fluoride, thereby protecting against fluoride toxicity.[46]

The well known Vichy water from France is highly mineralised and contains 6 milligrams per litre of fluoride. A recent 1996 study found that when this water was consumed, 47.4 percent of the fluoride injested was eliminated in the urine within 12 hours. When a weakly mineralised water was adjusted to 6 milligrams per litre of fluoride it was found that only 38.7 percent of the fluoride was eliminated within 12 hours. Thus, a higher level of fluoride was retained by the body in the weakly mineralised water, than in the water that contained a greater fluoride concentration. The researchers reported that they observed no fluorosis in healthy subjects consuming the naturally highly fluoridated Vichy water.[47] By deliberately intervening to make nature's insoluble forms of fluoride soluble, we transform a relatively harmless natural substance into a concentrated and highly toxic substance which can then be indiscriminately dispersed throughout the environment as a poison.

Protection mechanisms

The human organism is equipped with two mechanisms to protect against fluoride toxicity after ingestion. The first of these is to raise the levels of fluoride excreted in urine and the second is deposition of any non-excreted fluorine in the skeleton. The skeleton can absorb only a certain amount of fluoride and once bone saturation is reached, the excess fluoride enters soft tissue and metabolic breakdown within the tissue occurs.[48]

It has long been known that about half the fluoride we ingest is excreted by our kidneys. The rest accumulates in our bones. Children excrete fluoride less efficiently than adults and so retain more in their bones.[49] Indian researchers have reported: 'Fluoride toxicity afflicts children more severely and after a shorter exposure to fluoride than adults, due to the greater and faster accumulation of fluoride in the metabolically more active growing bones of children'.[50]

The subtle constellation of health clues which nature provides in respect of fluorides is further illustrated by the simple but elegant mechanisms of breast-feeding. Breast-fed infants are actually protected from receiving more than extremely low concentrations of fluoride in breast milk by an inbuilt physiological plasma–milk barrier against fluoride.[51] Similarly fluoride concentrations in foetal tissues are relatively low as a result of the placenta being an efficient filter, binding and accumulating fluoride.[52] This is even more remarkable when it is remembered that soft tissue does not normally accumulate fluoride.[53]

Plants have similar mechanisms to limit the amount of fluoride which is absorbed from the soil. As a result plant fluoride levels are relatively low, typically around 1 ppm or less, even when fluoride-containing fertilisers are applied.[54]

Learning from nature

There is evidence suggesting that fluoride is toxic, even at relatively low levels, if the diet is inadequate. Oxford University researchers found dental fluorosis with spinal defects in two villages of Oxfordshire supplied by surface wells with water that varied from 0.3–0.9 ppm fluoride in one case, and 0.3–1.2 in the other. Their conclusion was that fluoride, together with nutritional deficiencies (which was detected in some, but not all, the subjects) may be contributing factors in the development of such defects.[55] Other researchers have also reported cases of skeletal fluorosis in people exposed to water fluoride levels as low as 0.7 ppm.[56]

In discussing these findings, HM Sinclair cites subsequent research which suggested that calcium intake was a mitigating factor in fluoride toxicity. He goes on to warn: 'Far too little attention has so far been paid to the possibility of other factors increasing or decreasing the toxicity of fluoride at 1 ppm.'[57]

Both Yiamouyiannis[58] and Srikantia[59] report an association between symptoms of fluoride toxicity from natural fluoride and poor diet, adding further weight to the argument that an adequate diet contains factors which *protect* against harmful effects of natural fluoride.

Working with nature requires recognising not only the limits set by nature, but also that these limits serve a purpose. The environment which supports the life of our planet ensures fluorine is bound in the form of insoluble fluorides. Other forms of this element are highly toxic. While we still presume that control of their toxicity is easy to secure, we have neglected to recognise that the health systems for the life on this planet are inextricably interwoven. Any manipulation of our environment must therefore be in harmony with the basic patterns which underpin nature. The fact that nature has limited the availability of toxic forms of the element fluorine in the environment appears to be part of the blueprint for health to which our ecological systems are ordered. Deeper reflection upon the patterns of nature reveals that her mechanisms already exist to protect against tooth decay. We now know, for example, that sufficient exposure to sunlight, coupled with a diet rich in whole foods, is in itself a form of protection from tooth decay by increasing the hardness of teeth. For example, at least one study has shown that simply by providing classroom lighting which simulated natural sunlight, it was possible not only to protect against the formation of cavities but in some cases to actually reverse the early signs of tooth decay.[60]

Although refined carbohydrate foods are the most decay-producing of all foods, AK Adatia points out that in 1937, it was found that sugar cane juice and whole wheat, in contrast to refined sugar and white flour, contained factors which reduced the solubility of tooth enamel when incubated with saliva. Factors which reduce the solubility of tooth enamel are also present, for example, in wheat bran, wheat germ, and in the hulls of oats and peanuts. These factors are water soluble and seem to be extracted during mastication.[61] Furthermore, it seems that mastication itself plays a role in incorporating these protective factors into the tooth enamel during the compression processes.[62]

After reviewing dental caries and periodontal disease in the context of refined carbohydrate foods and disease, Adatia concluded that meticulous oral hygiene and cariostatic food additives such as fluorides were of limited benefit.[63] Instead he suggests that:

> Inclusion in the diet of unrefined foods of firm fibrous consistency would encourage vigorous mastication which would aid digestion, induce secretion of saliva of high buffering capacity, promote periodontal health and raise resistance against dental caries. It would seem not unreasonable to suggest that such a scheme of preventive dental medicine based on a natural mechanism and man's scientific dental knowledge could bring the conquest of dental caries and periodontal disease within everyone's reach. It might win general acceptance if it were realised that the dietary factors involved in the pathogenesis of dental caries and periodontal disease are also implicated in the aetiology of many fatal disorders.[64]

This last fact highlights the argument that health is a consequence of the complex interplay of an extensive range of environmental factors, not the least of which is pollution.

Lead exposure and tooth decay

Researchers from the University of Granada in Spain have suggested recently that lead pollution from traffic fumes and tap water may cause tooth decay. They found that in children and adults whose teeth contain high concentrations of lead, there is a significantly higher incidence of caries, dental staining and plaque. Children with ten or more sites of tooth decay were found to have three times as much lead in their blood as children with no decay.[65] The Spanish researchers suggest that the lead from either the mouth or the blood replaces trace metals which are part of the natural constituents of tooth enamel. Once in the tooth, the lead thus weakens enamel, making it more susceptible to attack from bacteria.[66]

Fluoridation was first introduced in Australia in the island state of Tasmania[67] in the early 1950s because of the high levels of tooth decay among children living there. Tasmania, affectionately known locally as the 'Apple Isle', was at that time a major producer of apples, and lead arsenate sprays were widely used to control pests. A 1974 University of Tasmania study, found high blood lead levels not only in inner-city school children but also in school children living in Cygnet, a township in the centre of an apple-growing district.[68] These high lead levels were comparable with the blood lead levels of residents of Los Angeles living near freeways. Given the adverse impact of lead on tooth enamel, the high levels of tooth decay in Tasmania at that time suggest a new and important interpretation.

Furthermore, the role played by lead pollution in tooth decay may prove to be an important factor in assessing the effectiveness of fluoridation, as the decline in tooth decay worldwide, both in non-fluoridated as well as fluoridated areas, as reported by M Diesendorf,[69] may reflect the removal of lead arsenate sprays, lead based paints, and lead fuel additives from the environment. What we thought was a reduction in tooth decay explained by the introduction of fluoridation might better be explained by the reduction in lead pollution.

Whether the fluoridation campaign must be indicted as a public health campaign gone badly wrong is a judgement best reserved for the reader. Whatever the judgement, it is incontestable that the prevention of tooth decay is not the bottom-line of the fluoridation debate, if the panacea has become the poison.

For information on improving the quality of water, see pages 195–97. It is worth noting that a British study found that none of five water filters tested removed fluoride.[70] It is therefore important to check the claims made by the manufacturer of the water purification system used to ensure that it adequately removes fluoride ion.

PART five

Environmental technology

chapter 14

Airconditioning: Finding healthier ways to 'stay cool'

By the turn of the century, over 50 percent of the world population will live in cities. United Nations population statistics now indicate that roughly two billion people (42 percent of the world population) were city dwellers in 1985, as against some 3 billion living in rural areas. The UN experts predict that the world's urban population will have grown to 3.7 billion by the year 2010.[1]

With cities rapidly increasing in size and number, more and more people live in microclimates significantly different from those provided by nature. High-rise apartment blocks, concrete, steel and glass towers housing city offices, giant shopping centres, hospitals, industrial complexes and factories, which constitute the main proportion of the city environment, all isolate their occupants from the circulation of fresh air provided by nature. The barren concrete, brick and stone walls have no natural mechanism to cleanse the air polluted by the wastes of their occupiers. The concrete jungle of the city absorbs the heat from the summer sun, in turn heating the surrounding air to temperatures significantly hotter than the surrounding tree-studded countryside.

Using the technology which produced the structures of our cities we have sought, artificially, to reproduce our own utopian climate. The mechanical means by which we do this is now commonly known as airconditioning and is used almost universally in large buildings. Increasing numbers of suburban homes also proudly display badges of technological affluence in the form of small airconditioning units located in one or more window spaces. Given that we now spend much of our time 'indoors' (see Figure 14.1) one question we should ask is: 'Are there side-effects of the "perfect climate"?'.

Figure 14.1 Time spent indoors vs outdoors by urban dwellers over 11 years of age

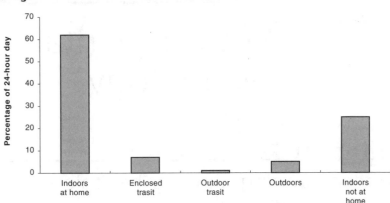

BP Leaderer et al., *Environmental Health Perspectives*, December 1993

Hippocrates, 24 centuries ago, expressed a deep intuition that the air and water around us were primary factors in the maintenance of health.[2] As we shall see in what follows, consideration of the health levels of the inhabitants of airconditioned environments serve as salutary reminders of the truth of his view.

The development of airconditioning

The term 'airconditioning' appears to have been coined by SW Cramer in 1906, who developed a humidity control system for use in textile mills. In 1911, WH Carrier established a scientific basis for airconditioning when he presented his ideas at a conference of the American Society of Mechanical Engineers. Carrier proposed that airconditioning depended on a delicate balance of five interrelated factors: temperature, humidity, particle level (such as dust, pollen, germs), odour level and air movement. His system included fans, water sprays which humidified the air, water screens which cooled the air and a simple system of filtration. The development of modern refrigeration led to the development of summer airconditioning in the early 1920s and the first airconditioning for human comfort was installed in a cinema in 1922. The success of this installation resulted in airconditioning being widely adopted in public buildings in the United States by the end of the decade.

The advent of airconditioning accelerated the development of even larger buildings and significantly influenced their designs, which came to be built around the central air circulation system. Unfortunately, like many other technological advances, airconditioning has created new health problems, as we are thrust into yet another artificial environment. Since the 1960s there has been an increasing literature reporting the adverse effects of airconditioning. These health problems are mainly related to the source of external air, to particles in recirculated air and to contaminated humidifiers.

Circulation of contaminants

Contamination of the air entering intake ducts, for example, has been responsible for a number of outbreaks of legionnaire's disease (described more fully later in this chapter). All too often the warm exhaust from the airconditioning cooling tower is sited near the air intake for the system. Bacteria which have multiplied in the warm environment of the cooling tower (analogous to the warmth at the back of a domestic refrigerator) have subsequently passed into the air-conditioning ducts and spread around the building.[3] Even people passing by the outlet ducts of contaminated buildings have been affected.[4] In one unusual outbreak of legionnaire's disease, golfers were found to be more likely to develop the disease if they played 18 holes rather than nine holes. It was later discovered that the outbreak was caused by an infected evaporative condenser whose exhaust was sited close to the tenth tee.[5]

Recirculating air is also an important vector for disease spread in airconditioned buildings. People often add their own viruses and bacteria to the conditioned air. A normal cough produces about 5000 droplets of liquid while a sneeze may generate a million. Even using a handkerchief does not prevent smaller droplets from contaminating the air. Not surprisingly, bacteria and viruses commonly spread in airconditioned buildings. One of the early studies reporting this phenomenon of airconditioning involved the spread of tuberculosis from one man to 140 of his 350 fellow crew members aboard a United States naval vessel over a period of six months.[6] The subsequent investigation suggested that the airconditioning ducting spread the infection by droplet nuclei. Measles has also been documented as spreading in an epidemic manner among school children in an air-conditioned school.[7] Again the virus was thought to have been distributed through airconditioning ducts as droplet nuclei produced during coughing, sneezing or talking.

Cigarette smoke is another major contaminant circulated by air-conditioning, with serious health consequences. In recent years, there has been a growing awareness of passive smoking. Non-smokers breathing the sidestream smoke of nearby smokers suffer similar health effects as do light smokers. A number of large studies have supported this effect. One study, which involved over one million people, showed that for 30-year-old males, a light smoker (one to nine cigarettes per day) suffered a loss of life expectancy of 4.6 years, while a heavy smoker (over 40 cigarettes per day) suffered a loss of life expectancy of 8.1 years. These results suggest that even a small exposure to smoke may carry a substantial risk.

In another study, involving 2100 office workers, it was concluded that those non-smokers who were continually exposed to cigarette smoke in the work environment suffered about as much damage to the small airways of the lungs as the light smokers. That is, the damage was about half as much as that found in the heavy smokers.[8]

The effect of airconditioning circulating sidestream smoke was investigated by JL Repace and AH Lowrey.[9] Using standard ventilation rates they calculated that a non-smoking office worker in a typical office would inhale the equivalent of between two and three cigarettes per work shift. A subsequent study of 10 000 non-smoking office workers revealed that 52 percent believed that their productivity was impaired by on-the job exposure to tobacco smoke. The Repace and Lowrey study concluded that none of the then existing ventilation standards were adequate and even if the ventilation rate was increased considerably, the situation could not be greatly improved.

The seriousness of the cigarette smoke problem in airconditioned buildings is pointed out by P Carter. He states:

> A standard filter will remove none of the gases and little of the particulate matter in tobacco smoke, because most of the particles are less than a micron in diameter. Therefore, if there are smokers in the airconditioned space, air contamination will build up, undermining the health and, on average, shortening the lives of the occupants whether smokers or non-smokers ... Using a high efficiency filter which costs more and has a higher air resistance, he [an air conditioning designer] can filter out the recirculated particulate matter (not the gases), but he cannot do anything about tobacco smoke on its way between the cigarette and the return air grille. More outside air gives limited improvement for a great increase in operating cost.[10]

Particulate matter in the air is sometimes removed using electrostatic precipitators. These devices also produce an excess of positive air ions in the circulating air as they function. A number of studies have indicated that positive air ions affect serotonin production in the brain and can cause feelings of malaise, headaches and irritability.[11]

Contaminated airconditioning humidifiers are another source of adverse health effects. Humidification involves the addition of water to the circulating air. Despite filtration, particulate matter from external and recirculated air is deposited in water reservoirs, which then become an excellent culture medium for a whole ecosystem of bacteria, fungi, amoebae, worms and mites.

These organisms end up growing in a brown slime, and particles from this sludge are taken into the circulated air and inhaled by the building's occupants. That exposure to humidifier sludge occurs via this mechanism has been confirmed by surveys which show that a substantial proportion of workers in airconditioned buildings (over half in some reports) have blood antibodies to sludge extracts.[12]

This results in more sensitive workers coming down with 'humidifier fever' or 'Monday fever'. This is an influenza-like illness with fever and malaise occurring several hours into the work shift. Sometimes it is accompanied by coughing, chest tightness, breathlessness and weight loss. Asthma has also been shown to be produced in certain workers exposed to contaminated humidifiers.[13]

Another associated adverse effect of airconditioning is byssinosis. This is a condition similar to humidifier fever in that it produces chest tightness characteristically worse on Mondays when staff return to work after spending time in the relatively cleaner air in their homes or the out of doors. It is believed to be caused by the circulation of organic matter such as vegetable dusts and cotton.[14] Mould spores are also spread by airconditioners.[15]

The problem of air contamination in large buildings is widespread and difficult to eliminate. In attempts to reduce some of these adverse effects of airconditioning, we have once again resorted to the use of further technology to try to rectify the shortcomings of the present technology. Thus contaminated humidifiers are treated with special chemicals called biocides, whose toxicology in these situations may not have been studied in detail.[16]

Biocides do not always solve the problem of air contaminations. In February 1995 in Britain, for example, the Inland Revenue decided to demolish a 19-storey office in Bootle, Merseyside, where half of the 2000 staff have suffered frequently from flu-like symptoms for more than five years, even though biocides had been in use.[17]

Out of doors, nature uses the ultra-violet component of sunlight to kill bacteria in the air. Using this principle, Y Goswami and S Block from the University of Florida have found that air-borne microorganisms die when exposed to ultraviolet rays in the presence of a titanium dioxide (a white pigment) filter. They have suggested that it may be possible to develop a system for sterilising the circulating air in buildings using this relatively harmless ultraviolet light process.[18]

More health and environmental effects

Living in an airconditioned environment also affects our natural circadian (daily) temperature rhythms. R Thorne and T Purcell from the University of Sydney found that the body temperature in an airconditioned environment falls significantly below the normal 37 degrees Celsius, and does not follow the daily rhythm. They found that the skin, in contact with the dead air, appears to suffer a kind of sensory deprivation. Because of a lack of skin stimulation from the air, the body slips down into a lethargy in the afternoons. The skin needs constant stimulation from moving air—but not draughts—to maintain the body's normal level of arousal and its normal temperature.[19]

The long-term effects of lowered body temperature include an increased susceptibility to illness, particularly viral infection.

Thus it seems that natural variations in air temperature throughout the day, such as occur in the outdoors, are part of nature's blueprint for our health, which we less than intelligently by pass in the creation of our artificial airconditioned environments.

It should be remembered, however, that airconditioning in buildings, homes and vehicles affects not only the health of their occupants but also the health of the entire planet. Large airconditioning units can leak substantial amounts of refrigeration gases into the atmosphere. These gases are usually chlorofluorocarbons, commonly CFC-12. CFCs are known to destroy the ozone layer which protects life on this planet from harmful short wavelength ultraviolet radiation.[20] The CFCs used for airconditioning have the longest life span in the atmosphere of any of the CFCs used for other purposes, such as in aerosol cans. The life span of CFC-12 in the atmosphere is estimated at 120 years,[21] which means that the environmental cost of our present airconditioning will be with us and our children and our children's children and their children.

Furthermore, to run our electricity powered airconditioning systems, in the majority of instances we burn fossil fuels to generate the electric power. Thus we produce large quantities of carbon dioxide and exacerbate the greenhouse effect in order to aircondition our human-made technological 'utopias'.[22]

It is revealing that nature's own airconditioning systems, the trees, produce life-supporting oxygen as they 'condition' the air. A typical temperate climate forest produces 26 tonnes of oxygen per hectare per year. For tropical rainforests the value is 39 tonnes per hectare per year. Grasslands and meadows produce about 20 percent of this amount.[23] Trees also cleanse the air of dust and chemical pollutants. Research has shown that in tree-filled parks, up to 85 percent of suspended particulates may be filtered out. Trees planted along streets may reduce particulate levels by almost 70 percent. Furthermore, even deciduous trees in winter reduce air particulates at 60 percent of the summer rate.[24]

Trees also exhibit a powerful cooling effect on air temperature. In Frankfurt, West Germany, the green strip 50–100 metres wide on the traffic ring around the city centre has brought about a drop in temperature of up to 3.5 degrees Celsius on hot summer days.[25] Trees also help control the relative humidity in the air.

As an example of how trees condition air, A Bernatzky has calculated that a 100-year-old beech tree, 25 metres high and with a crown volume of 2700 cubic metres, absorbs 2.35 kilograms of carbon dioxide per hour and produces 1.71 kilograms of pure oxygen per hour.[26]

What we need to appreciate is that in addition to their many other health-enhancing functions, trees are part of nature's blueprint for true airconditioning. Our humanmade airconditioning systems do not produce pure air, they only contaminate it. As our politicians glibly announce that we will plant more trees, let us remember it will take 5400 young beech trees with a crown volume of 0.5 cubic metres to replace the oxygen produced from each of the 100-year-old airconditioning trees we have previously cut down.

Indoor air pollution: The tight building syndrome

Today much of our time is spent indoors, whether it be at home, at work or in transit (see Figure 14.1, page 212). The extent to which we have synthesised the environments of our homes and workplaces is in one sense an indicator of technological progress. What we have yet to learn is that the price of progress can be high. Whether we actually improve the quality of our lives through the creation of highly synthesised, artificial environments, is a question we can no longer afford to ignore. The risk of environmentally induced illness (EI) also known as the 'tight building syndrome', is a reminder that something has gone badly wrong.

In the name of progress, timber, instead of being sawn into boards, is now chipped into fragments and reconstituted with plastic glues into boards for furniture and flooring. Plastic polymers replace wool and cotton to form the fabrics of our clothes, floor coverings and curtains. Plastic sheeting replaces leather on our furnishings and shoes. Our homes and offices are insulated against the temperature cycles of nature with plastic foams. These plastics also form the basis of cushions, mattresses and toys. Most of these materials, chemically manufactured, exude trace amounts of unreacted monomers, plasticisers and other chemicals into the air, especially soon after manufacture or when new. The concentration of these substances is commonly less than one part per million and have in the past been regarded as insignificant by the industry and most health boards.

The possibility that chemicals at the levels found in the indoor air environment could provoke illness was recognised by TG Randolph in 1970.[27] Since that time the spectrum of symptoms of the tight building syndrome has grown to include 'attacks of headache, flushing, laryngitis, dizziness, nausea, extreme weakness, joint pains, unwarranted depression, voice impairment, exhaustion, inability to think clearly, arrhythmia or muscle spasms'.[28]

The cases of over 1000 patients with undiagnosed chronic symptoms were examined by S Rogers in the early 1980s. Rogers found that patients were often able to associate the onset of disease with the purchase of new carpeting, furniture, beds, cabinets, renovation of the home or office, moving into a new dwelling or insulating with urea formaldehyde foam insulation (UFFI).[29]

Of the many chemicals released, one of the agents now known to trigger the above symptoms is formaldehyde, a chemical exuded by many glues, resins and UFFI. In the United States, measurements of ambient levels of formaldehyde have shown that new mobile homes, newly constructed homes with particleboard subflooring, offices recently renovated with paneling or prefabricated walls, new clothing, or even new carpeting or new furnishings, and homes with new beds

and cabinetry can accumulate as much formaldehyde, as a result of off-gassing, as a UFFI home.

In the early 1980s, a survey of homes revealed that the average level of formaldehyde found in residences without UFFI was 0.03 ppm and 0.12 ppm for houses with UFFI (both with and without health complaints). Some mobile homes had levels up to 0.4 ppm and the mean level for several hospitals was 0.55 ppm.[30] These levels are particularly significant considering that formaldehyde has been shown to produce cancers in laboratory animals at levels as low as 6 ppm.[31]

As discussed earlier in this chapter, another aspect of the indoor environment–health connection has been the concentration of people in large buildings to meet the demands of limited space and its efficient utilisation, urged upon us by technological progress. The impracticality of incorporating a sufficient number of windows which open, has made it necessary to control the environment of the inner regions of these mega structures with airconditioning. One problem with airconditioning, however, is that it tends to circulate contaminated air and bypasses the sterilising effect of sunlight that would otherwise purify the air. In one instance investigated by the Australian Government Analytical Laboratories (AGAL) in the early 1990s, a 'quick response' crew was suffering from headaches, sinus and respiratory tract problems. Evaluation of the room revealed a high concentration of airborne moulds. In another case, staff at a federal electoral office were complaining of lethargy, headaches, sinus problems and coughing. An investigation by the Occupational Hygiene Section of AGAL found that there were high dust levels, incorrect temperature and humidity control, and high levels of volatile organic compounds (VOCs) in the building.[32]

It is now believed indoor 'air pollution' in modern buildings exposes humans to both a wider range and a higher concentration of VOCs than any other source. Common sources of VOC emissions include synthetic carpets, tiles and flooring, contact adhesives and sealants, paints and varnishes and office furniture. The compound 4-phenylcyclohexane (4-PCH) has been identified as the major source of 'new carpet' odour. Other VOCs emitted from carpets include vinyl acetate and propane-1,2-diol. The health effects of these VOCs when present at low concentrations over long periods, are unknown.[33] A common plasticiser in vinyl flooring is diethylhexylphthalate (DEHP) which can be 30 percent by weight of the vinyl. If the vinyl is overlaying damp concrete the DEHP can hydrolyse to 2-ethyl-hexanol which has a sickly odour. Other VOCs released from flooring materials include toluene and xylene from adhesives, phenol and trimellitic from floor leveling resins. This latter compound is considered to be extremely toxic as it causes immunological sensitisation and severe respiratory irritation.[34]

The gradual breakdown of polystyrene-base adhesives that are used to secure bathroom tiles can result in toxic vapours of the chemical benzaldehyde.

Wall paints are well known for producing high levels of VOCs. Water based acrylate paints release significant levels of butyl methacrylate and contain glycol ethers. These latter compounds are not readily detected by the human olfactory system due to their high odour threshold. However these compounds have been found to persist in the indoor air for up to six months after the application of the paint.[35]

Office and home furniture such as chairs, sofas and mattresses can emit volatile irritants that originate from the sponge rubber padding and upholstery. Polyurethane foam has been known to emit over 70 different chemical compounds after manufacture including aldehydes, chloroform and chlorobenzene.

The most predominate VOCs found in the indoor air of established buildings together with average levels[36] are shown in Figure 14.2. Levels for ethanol (common alcohol) may typically be in the order of 120 micrograms per cubic metre.

Figure 14.2 Typical levels of volatile organic compounds in indoor air of established buildings

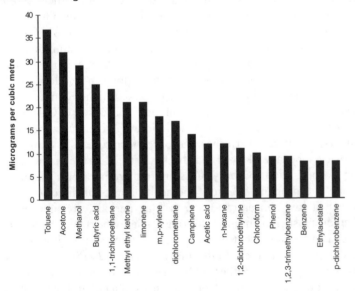

S Brown, *Chemistry in Australia*, January/February 1997

In certain situations, much higher levels of VOCs can be produced. In bathrooms, where chlorinated water is used chloroform levels may be in the range 50–250 micrograms per cubic metre. Fresh magazines may cause room toluene concentrations to rise to 3800 micrograms per cubic metre. Recently dry-cleaned clothing may emit from 100 to 10 000 micrograms of tetrachloroethylene per squaremetre of clothing per hour.[37] Dry building construction products such as plastic, rubber, flooring, carpet and wallpaper can typically emit from 100–10 000 micrograms of VOCs per square metre of surface each hour. Wet products such as polishes, cleaners, disinfectants and deodorises used to clean and maintain walls and floors may emit up to 1 000 000 micrograms of VOCs per square metre of surface each hour.[38]

S Brown from the Commonwealth Scientific Industrial Research Organisation (CSIRO) Division of Building, Construction and Engineering has pooled the measurements of VOCs in buildings from a number of studies. For established dwellings the average total VOC concentration in indoor air was 1130 micrograms per cubic metre, while for buildings less than three months old, the total VOC level was much higher at approximately 4000 micrograms per cubic metre. Brown points out that these values were the geometric means of a large pool of data. For a proportion of buildings the concentrations were substantially greater. For example, the 90 percentile total VOC concentrations were 4600 micrograms per cubic metre in established dwellings, while for new dwellings and offices the total concentrations in the indoor air was a massive 18 000 micrograms per cubic metre.[39] These values are considerably in excess of the maximum indoor air concentrations for total VOCs recommended by the Australian National Health and Medical Research Council (NHMRC) of 500 micrograms per cubic metre.[40]

In poorly air conditioned buildings these VOCs can accumulate or be circulated throughout a number of rooms and offices.

There are thus many aspects of aggravated health conditions and specific health problems arising from airconditioning. A tragic example of one such problem was drawn to public attention by way of a meeting at the Belle Vue Stratford Hotel in Philadelphia in 1976. A convention of the American Legion was in progress when 182 of the war veterans attending contracted what was then an unknown form of pneumonia. Twenty-nine of the legionnaires died. This was the first recorded outbreak of what has since become known as legionnaire's disease. Research undertaken on this condition has revealed that it is mainly associated with poorly designed or improperly maintained airconditioning systems. Another outbreak in the West End of London in 1988 claimed the lives of five people, along with a further 28 suffering pneumonia.[41] Legionnaire's disease, just one of the many diseases induced by so-called technological progress, serves as an incisive example of the truth expressed by Florence Nightingale's celebrated statement, 'There are no specific diseases—there are specific disease conditions'.[42]

What you can do

For good health, there is no substitute for fresh air. Thus, whenever practical, we should have the windows of our homes, schools and workplaces open. However, in these times of widespread air pollution and fixed-window buildings, airconditioning is often essential and unavoidable. In these situations we can improve indoor air quality by introducing as many indoor plants as possible into the workplace, school and home. Special lighting conditions may be necessary for the plants to thrive, but this is a small price to pay for the health-improving qualities of plants, which can absorb particulate matter and harmful ions, while generating healthful oxygen. As a bonus, they provide aesthetic beauty.

When building or renovating our homes and offices, we should avoid using particle-board components. Whole timber is best, even if second hand. Valuing timber and re-using it, as well as recycling solid timber furniture, is also an important step towards saving our remaining forests. Solid timber furniture, compared with veneered particle board furniture, lasts much longer and can be restored a number of times. Similarly, natural covering such as cork and wool should be used as much as practicable instead of vinyls, foams and plastics.

chapter 15

Artificial light: Shedding some light on a dark problem

Throughout the history of the human species, exposure to sunlight has figured as a common experience of everyday life. Today, however, we find in modern urbanised societies that our relationship with the environment has changed markedly. By having progressively synthesised the world in which we live, our contact with the natural environment is rapidly becoming minimal. Many people now make their way to work before the sun has risen and make their way home after the sun has set. For many people the world of artificial light has replaced the world of natural sunlight.

We live, learn and work in buildings where the rooms are artificially coloured with plastic paints and decorated with pictures. Our floors are covered with coloured synthetic, fibres, while artificially coloured or plastic coated furniture together with dyed synthetic fabrics and drapery adorn our rooms. When travelling from one location to another (from house to school or work), a major part of our time is spent in a vehicle usually lined entirely with artificially coloured plastics and fabrics.

The artificial colours with which we have painted our environment have served to ensure that for large periods of time the nature of the light entering our eyes is vastly different from the light reflected from the natural outdoors of meadow, forest and field. In addition, the use of glass, sometimes plastic-coated or tinted, means, that the sunlight entering our building or vehicle is filtered or refined by the removal of certain wavelengths of sunlight by the glass itself. This diminished sunlight is usually supplemented by artificial lighting from incandescent or fluorescent lights.

It is clear then that most of the light striking our skin and entering our eyes comes from artificial light sources, which include VDUs and TV screens, and light reflected from our surroundings. Our modern synthetic surroundings mean that apart from when we step outdoors we are continuously bathing in artificial light. Yet, dare we step outside? Health departments warn us constantly through the media to beware of the skin cancer demon called sunlight. We are urged to protect ourselves from direct sunlight by using sunscreens. We do this willingly, but we fail to recognise that when we coat our skin with sunscreens or tanning oils, and even wear sunglasses, the sunlight reaching the covered areas of skin and our eyes has been modified or refined by the absorption of certain wavelength components. Exposure to natural sunlight has again been reduced or transformed into exposure to artificial light.

This bizarre situation has arisen because little attention has been paid by health educators to the accumulating literature which points to the substitution of artificial light for natural sunlight as an important causative factor in the diseases of civilisation. At the same time, few have appreciated the significance for human health of the unique special distribution of the sunlight that reaches the Earth's surface, and of the variations in intensity and duration of sunlight which are produced by daily and seasonal variations, and which actually regulate our biochemistry and biorhythms.

In this chapter we will present a tapestry of scientific evidence which reveals that daily exposure to natural daylight—far from being a threat to health—is an integral part of nature's blueprint for health. We suggest a far more serious threat to health involves our ever-increasing exposure to artificial light, coupled with exposure to medicines, chemicals and even food additives which serve to increase the photo-sensitivity of the skin.

The melanoma enigma

The conundrum of sunlight and melanoma has challenged health research around the world. There is strong evidence, however, that melanoma once considered a rare disease is yet another disease of civilisation.

Since the 1950s, melanoma mortality rates in the United States have more than doubled in women and tripled in men. There are now about 38 000 new cases of melanoma annually and approximately 7000 deaths.[1] Australian melanoma mortality rates have similarly increased. In New South Wales, for example, mortality rates for melanoma quadrupled in the 33 years from 1954–87.[2] In Queensland, mortality rates doubled in the post-war period.[3] with the incidence of melanoma rising even more steeply. Since 1975 the number of cases of melanoma in Australia has quadrupled with the highest incidence occurring in Queensland.[4]

We are regularly reminded in the media of the progressive depletion of the ozone layer by chlorofluorocarbons and other wastes of our technology, which in turn results in increased levels of ultraviolet light reaching the earth's surface. The popular assumption is that this increased UV flux is responsible for the *increase* in melanoma and we are all earnestly advised to protect ourselves from the sun. Interpretation of the available research, however, suggests that the increase in melanoma which has occurred since the 1950s cannot be adequately explained by environmental exposure to ultraviolet radiation.[5] At the same time, there is consistent evidence of a higher incidence of melanoma in association with a number of environmental factors such as exposure to artificial light, the chronic use of sunscreens that result in artificial light reaching the skin, and environmental pollution among indoor workers and among those of higher social class.[6]

Artificial light and electromagnetic radiation

A recent study of 3527 melanoma cases among white males in Los Angeles County found no significant indoor/outdoor versus outdoor occupation correlation when compared with 53 129 non-melanoma cancer cases. However, the researchers did find that the risk of melanoma was highly correlated with level of education. White males with a PhD were three times more likely to develop melanoma than males in a low-skilled occupation[7] (see Figure 15.1).

Figure 15.1 Level of education and risk of developing melanoma among white males in Los Angeles County

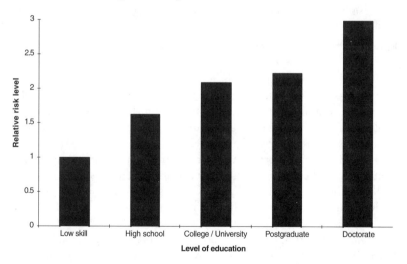

KJ Goodman et al., *Cancer Causes and Control*, September 1995

We suggest that these results may reflect an association with low hours of sleep, or conversely night-time exposure to artificial light and melanoma. RJ Reiter has shown that exposure of humans to artificial light at night rapidly depresses pineal melatonin production and blood melatonin levels[8] (see Figure 15.2). Melatonin is a potent anti-cancer agent and it prevents both the initiation and promotion of cancer[9] (see Chapter 5, pages 79–80). The reduction of melatonin levels at night in people who work long hours or study late into the night may be a significant contributing factor in the aetiology of melanoma.

Figure 15.2 Variation in blood melatonin levels with light and darkness

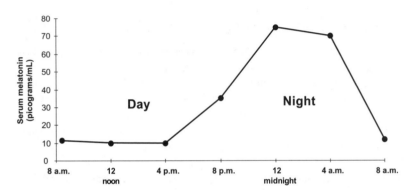

Note: Peak nighttime melatonin can range from 50 to 125 picograms per mL, depending on the individual and the hours of exposure to artificial light.

RJ Reiter, *Reviews of Environmental Health*, vol. 10, no 3-4, 1994

Reiter has found that certain other forms of electromagnetic radiation such as low frequency magnetic fields also reduce night-time melatonin levels and several studies have reported increased risk of melanoma associated with high occupational exposure to electromagnetic fields10 (see Chapter 2, page 45).

JAG Holt has observed another electromagnetic field–melanoma connection. When four melanoma patients were exposed to doses of 434 megahertz (MHz) VHF radio waves, the melanomas grew so rapidly that the average doubling time was reduced from 78 days to just 2.2 days.[11] Holt subsequently wrote to the *Medical Journal of Australia* suggesting that the doubling of the melanoma rate in Queensland after the 1960s could well be related to the introduction of television transmissions and other high powered VHF communication facilities. He pointed out that malignant tissue has a higher dielectric constant and better conductivity than normal tissue and therefore absorbs more power from electromagnetic radiation.[12]

Another part of the melanoma enigma is that the incidence of melanomas is often higher on parts of the body least exposed to sunlight.[13] For example in Scotland, the incidence of melanoma of the foot is five times that of the hands,[14] and the incidence of melanoma on Orkney and Shetland, which are north of Scotland and receive low levels of sunshine, is ten times that on the Mediterranean islands, famous for their sunny climates.[15]

Contrary to the conventional wisdom there is substantial evidence, as we shall see, that adequate sunlight has a therapeutic, if not a preventive role in regard to many of the diseases of modern civilisation. The pattern of evidence also suggests that melanoma is no exception and that regular daily exposure to normal sunlight is an essential preventive measure.[16]

There is also another dimension to the puzzle to be considered. Many of the biological mechanisms associated with our body's hormonal system are regulated not only by the light entering the eyes[17] but also by the light falling on the skin.[18] There also appears to be a delicate balance between the optic pathway and the skin pathway[19] and both these pathways control the mechanisms whereby our skin protects us from harmful effects of sunlight and regulates the essential effects of sunlight.

The human eye is exposed to artificial light from a variety of sources. Indoors, we often work or live under fluorescent lights during the day or night. By wearing sunglasses, the natural light of the sun is converted (as it passes through the glasses) to artificial light with a distribution of wavelengths quite different from sunlight. When we wear a sunscreen, the sunscreen also acts to convert wavelengths of natural light into artificial light. Given the fact that artificial light presents the body with different hormonal stimuli it is unsurprising that the delicate balance of the light-responsive mechanisms in our body will be significantly disrupted as compared to exposures of natural sunlight. The endocrinal sensitivity of our skin is illustrated by the genetic disorders which occur when certain enzymes are absent thus making the skin intolerant of light, with congenital erythropoietic porphyria (excess production of red blood cells in the bone marrow) being an extreme example.[20]

Photosensitising drugs

This brings us to the factors which may increase the sensitivity of our skin to light. By 1964, more than 100 photosensitising agents, including many commonly used drugs, had been documented.[21] It is revealing that photosensitising drugs included hypoglycaemic, anti-hypertensive, tranquilliser and broad spectrum antibiotic drugs, and even sunscreens and tanning lotions.[22] Chemical technology had not only replaced sunlight as a therapy[23] but had actually reduced our tolerance to it. At the same time, reports of photoallergic reactions to artificial sweeteners appeared and in 1967 the now widely used artificial sweetener cyclamate was added to the list.[24] By 1996, the number of known photosensitising drugs had grown to over 1140[25] including oral contraceptives.[26]

Many drugs contribute to our inability to tolerate sunlight, but since light affects the regulation of our hormonal system, drugs or chemicals which affect our hormonal system may have the greatest potential for contributing to the melanoma enigma. For many years, female sex hormones have been suspected as being important in the induction of melanomas. It has been shown that there are oestrogen receptors in melanoma cells and there is evidence to suggest that oral contraceptives may increase the risk of the disease.[27] The hormonal connection may also explain in part why pesticides and other chemical pollutants are possibly involved in melanoma enigma.

Chlorine

We have already discussed in Chapter 11 (page 181) on pesticides, how certain chlorinated pesticides act as synthetic oestrogens and alter the male reproductive balance. They may also interfere with pigment synthesis and contribute to the aetiology of melanoma. Recent research from the University of Nijmegen in the Netherlands has found that people who as children swam in chlorinated swimming pools or in polluted waters have 2.2–2.4 times the risk of developing melanoma, compared with those who did not swim or swam in unpolluted and unchlorinated water[28] (see Figures 15.3 and 15.4).

The fat layer of the skin offers the first protection against ultraviolet light. Early studies showed that ultraviolet radiation increases the cholesterol content of the skin of rats and it was suggested that cholesterol may actually predispose the skin to tumours. However, subsequent more detailed studies showed, surprisingly, that exactly the opposite occurred and where the cholesterol level of the skin was higher, there was more resistance to the effects of ultraviolet radiation.[29] Again, however, there is also an optic light connection. Lipid metabolism and blood cholesterol levels are regulated by light affecting the optic pathway to the pineal system, and different spectral bands appear to affect the eye and skin mechanisms respectively.

Figure 15.3 Risk of melanoma associated with time spent in chlorinated swimming pools as indicated by number of swimming certificates

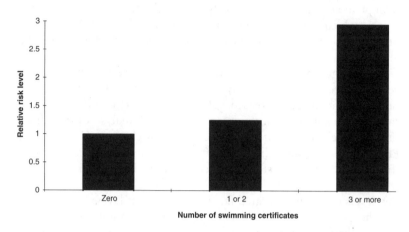

PJ Nelemans et al., *Melanoma Research*, October 1994

Figure 15.4 Risk of melanoma associated with swimming in polluted waters

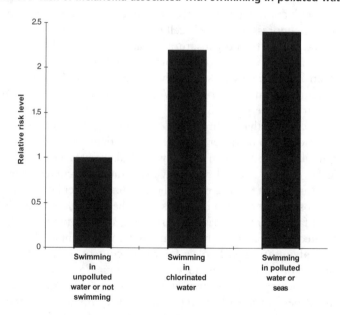

PJ Nelemans et al., *Melanoma Research*, October 1994

Effects of food processing

Using unnatural fat sources (such as polyunsaturated margarines) to manipulate dietary fat is also likely to disturb the balance of the mechanisms which regulate the optimum fat composition of the skin. The possibility that this is a significant factor in the cause of melanoma has also been mooted.[30]

In one study, when melanomas were transplanted into mice, the cancers were accepted more readily and grew more rapidly as the intake of polyunsaturated fats increased.[31] Further support for the polyunsaturated fat–melanoma connection comes from the analysis of the adipose tissue of melanoma patients at the Sydney Melanoma Unit at Royal Price Alfred Hospital. The percentage of polyunsaturated fatty acids in the melanoma patients studied was found to be higher than in the controls and there were significantly more controls than patients who had low levels of polyunsaturated fats in their skin fat layer.[32]

The possible role of dietary fat in the melanoma enigma may involve the function of cancer fighting T-cells. It is recognised that T-cell activity is down-regulated by exposure to UV radiation and that T-cell-medicated immunity is a controlling factor in UV radiation-initiated tumour promotion. Dr V Reeve and co-workers at the University of Sydney, have recently shown that saturated fats protect against the suppression of the T-cell immune system, by UV light. They also found that as the polyunsaturated fat content of the diet increased there was a linear decrease in the protection against immune suppression.[33] In another recent study, the Sydney researchers found that polyunsaturated fat exacerbated UV-induced tumour initiation and led to a remarkable persistence in the UV-induced immunosuppression.[34]

It is fascinating that nature provides for foods grown during the sunny summer months to be richer in saturated fats than their winter counterparts. Similarly, foods grown in sunny tropical climates have a higher saturated fat content and a lower polyunsaturated fat content compared with the same plant when grown in the less sunny cooler latitudes.[35] Even the composition of cow's milk varies with the seasons, with the highest saturated fat content occurring in spring and summer months, while the unsaturated fat content reaches a minimum during summer.[36]

The polyunsaturated linoleic fatty acid content of sunflower seeds can drop from a cold climate value of about 75 percent, down to around 40 percent when grown under hot sunny conditions. Levels as low as 33 percent have been reported for very hot summer temperatures corresponding to high sunshine levels.[37] Tropical oils such as palm oil and coconut oil, have very low levels of polyunsaturated linoleic acid, typically less than 10 percent and around 2.5 percent respectively. These tropical oils are also high in saturated fats, typically more than 90 percent.[38]

Other dietary factors affecting the susceptibility of the skin to cancer may be the processing and refining of foods, especially cereals, which reduce the levels of important vitamins and minerals such as vitamins E, B6 and zinc. Vitamin E is removed when seeds are degermed, such as during the manufacture of peanut butter, corn flakes and the milling of wheat with the associated separation of wheat germ. Zinc is mainly found in the bran of cereals and is mostly removed during the milling process. Vitamin B6 levels are similarly reduced, with white bread containing only 22 percent of the vitamin B6 level of wholemeal bread.[39]

In the early 1950s, HM Sinclair of Oxford University pointed out that a deficiency of B6 in the diet increases photosensitivity to ultraviolet light.[40] Recently, V Reeve and co-workers at the University of Sydney have found that vitamin B6 plays an important role in protecting the immune system against UV suppression.[41] Dietary intakes of vitamins A and E have also been found to be positively associated with a protective effect against melanoma.[42]

The extensive use of refined foods in western countries has lead to zinc deficiency in much of the population. From 54–80 percent of adult Australians have been found to be measurably zinc deficient[43] and in another study 67–85 percent of adults tested had zinc intakes below the recommended daily allowance (RDA).[44] Similarly, widespread zinc deficiencies have been observed in the United States with one recent survey showing that American women of childbearing age had dietary zinc intakes consistently below the RDA.[45]

Epidemiology data suggests that zinc may be protective against melanoma[46] and preliminary research at the University of Sydney has shown that supplementing test animals with zinc protects against UV-induced suppression of the immune system.[47]

These studies suggest that widespread zinc deficiency within the community as a result of dietary intake of refined foods may be a contributing factor in the aetiology of skin cancer in western communities. Other food factors which may help protect against skin cancer include antioxidant polyphenolic compounds such as found in the tannins in teas[48] and garlic.[49] Alcohol, on the other hand may increase the risk of melanoma. Alcohol is known to be a photosensitising agent[50] and for women drinking two or more alcoholic drinks per day the risk of melanoma was 2.5 times higher than for non-drinkers.[51]

More than a decade ago, the editor of *The Lancet*, after reviewing the melanoma enigma, concluded:

What is clearly needed is more information about people in whom melanoma develops: and not so much about their complexion, hair and eye colour, or their skin's reaction to sunlight, which has dominated research for decades, but more about their diet, drug taking, and exposure to environmental factors. Some explanation for the rapid increase in melanoma incidence must exist. The reason may well be simple, but so far overlooked.[52]

What we have argued in this section is that the conventional focus upon sunlight as *the* cause of melanoma has blinded a number of researchers to the increasingly important associations between melanoma and environmental factors such as artificial light, electromagnetic radiation, dietary fats and the photosensitising effects of food processing and some medicinal drugs. Even with conventional interpretation of the data, just more than half of the presented cases of melanoma can be explained solely as 'sun-caused'. The conventional interpretation has little to offer by way of explaining the remainder of these cases, some of which involve melanomas on parts of the body *not* exposed to the sun.

We have argued, on the basis of a multi-disciplinary survey of the literature, that environmental factors contribute to the problem either by (1) altering body chemistry to make it more susceptible to the harmful effects traditionally associated with exposure to the sun, or (2) acting directly as promoters of melanoma.[53] The issue is complex, as many environmental lifestyle factors alter the response of the skin and other bodily mechanisms to sunlight. A case in point is the chronic use of sunscreens. Evidence shows that while they afford a significant degree of photoprotection from ultraviolet light, some sunscreens have proved to be capable of initiating tumours in experimental mice, even without the presence of any ultraviolet radiation source.

Sunscreens

In one study impurities in two common sunscreen ingredients, O-PABA (known as padimate) and 2-ethylhexyl-p-methoxycinnamate (known as cinnamate), were found to induce tumours in 12.5 percent and 4.3 percent of test animals respectively.[54] These impurities have since been removed from the ingredients. In a 1995 study, the chemical psoralen which is used in some sunscreens was suspected of causing cancer in humans. European researchers compared 418 melanoma cases with 438 healthy people. It was found that fair-skinned people who used a psoralen based sunscreen were four times more likely to develop melanoma than fair-skinned people who used psoralen-free lotions.[55] In addition, the researchers found that the users of psoralen-free sunscreens also had a higher incidence of melanoma. After adjusting for age, sex, hair colour, and holiday weeks spent each year in sunny resorts, the melanoma risk was 1.5 higher for regular sunscreen users compared with people who did not use suncreens.[56]

An association between sunscreen use and increased risk of melanoma has been reported in a number of studies since the late 1970s. In 1979, it was reported that men who used sunscreens had 2.8 times the risk of developing melanoma[57] and a large case-controlled study in 1985 found that men who regularly used sunscreens had a 2.2 times higher risk of melanoma than men who used no sunscreens.[58]

A 1990 Swedish study also found an unexpected significant association between frequent sunscreen use and malignant melanoma with the risk factor being 1.8 times higher, for sunscreen users compared with non-users.[59] In a subsequent 1995 study by researchers as the Department of Surgery at the University of Lund in Sweden, regular sunscreen use corresponded to a 1.8 times higher risk of melanoma compared with never using suncreens,[60] and confirmed the earlier findings. The surprised researchers wrote:

> The data of the present study did not offer any explanation for the fact that frequent sunscreens users had an elevated OR (odds ratio or risk) for developing melanoma, although we performed multivariate analysis, taking into account different kinds of possible confounders (constitutional as well as sun exposure variables), the ORs were still significant and essentially unaltered.

The researchers went on to comment that the association was supported by a dose–response relation between the frequency of sunscreen use and malignant melanoma development.[61] (A summary of the reported risks of melanoma associated with sunscreen use is shown in Figure 15.5.)

In animal studies certain sunscreen lotions were found to give no protection against the progression of malignant melanoma and appeared to enhance the cancer.[62] The authors of the study, from the department of Immunology at the University of Texas in the United States, concluded that:

> Protection against sunburn does not necessarily imply protection against other possible UV radiation effects, such as enhanced melanoma growth.[63]

Figure 15.5 Studies reporting risk of melanoma associated with regular sunscreen use

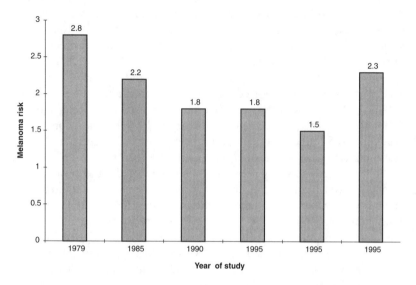

Note: Values greater than 1.0 represent increased risk.
This data is based on studies discussed on pages 232–33.

In 1992, CF Garland and associates from the Department of Community and Family Medicine at the University of California in San Diego, pointed out that worldwide, the countries where chemical suncreens have been recommended and adopted have experienced the greatest rise in cutaneous malignant melanoma. They argue that the rise in melanoma has been unusually steep in Queensland, Australia, where sunscreens were earliest and most strongly promoted by the medical community, and Queensland now has the highest incidence of melanoma in the world. In contrast, the rise in melanoma rates was notably delayed elsewhere in Australia where sunscreens were not promoted until sometime later.[64]

One of the most common sunscreens, para-amino bezoic acid (PABA) was developed in 1922 and commercial products containing sunscreens became available in 1928.[65] It was not until the mid-1950s that sunscreens for cosmetic purposes became popular. As sales increased throughout the 1950s and 1960s so did the melanoma rate. During the 1970s and 1980s, suntan lotions were repositioned as sunscreens with a higher content of PABA or similar sunblock compound. These chemicals reduce the level of UV light reaching the skin. This radiation, which is made up of the shortest wavelengths of the sunlight reaching the surface of the earth, corresponds to the spectral region 286–380 nanometres (nm). The very shortest wavelengths, that is, in the 286–320 nanometre spectral region, are referred to as UV-B, while, the 320–380 nanometre wavelengths are called UV-A. Sunscreen use was recommended for the prevention of skin cancer by health authorities on the basis of the assumption that the spectral component of sunlight which caused sunburn, that is mainly UV-B, was identical to spectral region which initiated and promoted skin cancer including melanoma and basal cellskin cancer. Although suncreens had been found to prevent sunburn, public health authorities who urged routine, liberal use of sunscreens (especially on children) failed to mention that there had never been any laboratory or epidemiological evidence that sunscreens actually prevented melanoma or basal cell skin cancer.[66]

Sunscreens absorb most of the UV-B radiation and varying amounts of UV-A depending on the ingredients. When wearing suncreen we allow the skin to be irradiated with an altered (or refined) UV spectrum which is different from that provided by nature, for long periods of time without the deterrant of sunburn. The long-term effect of refined sunlight exposure is unknown in animals and humans.[67] Many UV absorbers are unstable in protracted sunlight, and the consequences of their breakdown products on the skin are also unknown.[68]

It is known, however, that sunscreens inhibit the skin's production of vitamin D which requires the very wavelengths which sunscreens aim to block, UV-B.[69] Laboratory findings have indicated that the metabolically active form of vitamin D inhibits the growth of human melanoma cells.[70] This suggests that vitamin D deficiency in the skin may have a role in the cause of melanoma.

The use of sunscreens, which appears to offer no protection against melanoma, may in addition actually reduce the effectiveness of one of the body's own mechanisms for protecting against cancer.

There is yet another side to the vitamin D story. W Stumpf's research has shown that the prohormone vitamin D levels are regulated both by sunshine on the skin and by light entering the eyes, again emphasising the interconnection between the optic and skin stimulation by light.[71]

The observation that there appear to be seasonal changes in the skin's response to ultraviolet radiation[72] also suggests a fine-tuned link with the length of daylight as detected by the optic pathway. The use of sunglasses clearly interferes with any such mechanism, and may thus adversely affect the body's capacity to protect against melanoma.

Many people who tried to protect themselves from the sun by using these sunscreens are now counted among the rising toll of victims afflicted with melanoma and other skin cancers.

How artificial light is different

Natural daylight is made up of three components: sunlight coming directly from the sun; skylight which is sunlight reflected off minute particles in the atmosphere; and light reflected off our surroundings such as tree foliage and grass.

The spectral components of natural light

The spectral characteristics of different light sources are often distinguished by referring to their colour temperature.

Colour temperature is the temperature corresponding to the radiant heat energy equivalent to a particular light spectrum or colour. For example, red heat corresponds to a colour temperature of about 1200 kelvin (K—that is, degrees of absolute temperature), while white heat or white light as from an incandescent light bulb results from the filament being at about 3000 kelvin. Midday mid-summer sunlight corresponds to a colour temperature of about 5800 kelvin, with skylight contributing a further 800 kelvin to the colour temperature on a clear day.[73] The colour temperature of daylight not only changes throughout the day, but also throughout the year. These changes in colour temperature associated with the daily cycle and our seasonal changes, in effect mean there are small variations in the distribution of the spectral wavelengths of natural daylight throughout the day and throughout the year.

The light reaching us from the sun approximates very closely the continuous band of radiation emitted by a black body at approximately 5600 kelvin, except that the highly destructive ultraviolet wavelengths below 286 nanometres (nm), ultraviolet type C (W-C), are prevented from reaching the surface of the earth by the protective ozone layer in the upper atmosphere.[74]

Thus the sunlight which reaches the earth's surface consists mainly of ultraviolet light (290–380 nanometres), visible light (380–770 nanometres) and near infrared light (770–1000 nanometres). Radiation of wavelength longer than 1000 nanometres is also present and comprises about 20 percent of the total solar radiation reaching the earth. The middle ultraviolet region or UV-B component of sunlight corresponding to wavelength range 286–320 nanometres, varies mostly with the change in angle of the sun. We shall see later that this particular wavelength region is the source of one of the most important health-enhancing components of sunlight.

Another significant aspect of the light spectrum provided by natural sunlight is its smooth continuous distribution with no sharp peaks of high intensity radiation of a narrow wavelength range. The relative intensities of any of the radiation components which comprise the visible range, for example, do not differ by more than twofold, and the average red to blue to yellow intensities of the midday sun are approximately equal. By contrast, artificial light sources usually contain widely varying intensities in the visible range or display sharp high intensity peaks. We shall see that these variations have an impact upon the health of the human organism which has been sadly neglected.

The intensity of light is measured in lux (lx) where 1 lux is roughly equivalent to the light produced by a bright candle. The natural daylight we experience also has an intensity component which varies with geographical position and time of year, ranging from about 10 000 lux in northern Europe to 80 000 lux on a sunny day near the equator.[75] Levels up to 100 000 lux can be experienced on a bright summer's day in tropical regions.[76]

It can be seen that natural daylight provides a unique rhythm of light intensity and spectra composition to which the human organism is attuned. The extent to which artificial light and the disruption of patterns of exposure to natural daylight results in perturbations of health is becoming more clearly understood.

Before discussing the health aspects of natural light, however, it will be instructive to secure a better understanding of the array of artificial light sources, all of which possess a different spectral composition from natural daylight.

Aspects of artificial light

The earliest forms of artificial light were fire and candle. The light from these sources is vastly deficient in intensity and devoid of the blue, green and ultraviolet wavelengths found in natural daylight. In passing, it is of interest to note that in England from 1696–1851 windows were considered a luxury and window area was actually taxed.[77] The advent of glazed windows introduced another form of artificial light to which we have exposed ourselves. Ordinary window glass transmits UV-A radiation but mostly absorbs the biologically important UV-B rays.[78] Thus the sunlight entering a room or vehicle through glass contains virtually no wavelength component below 320 nanometres, and the same is true of daylight which is filtered through ordinary spectacles, yet another source of artificial light. Even air pollution provides another source of artificial light, this time in the form of smog which again changes the spectral composition of natural daylight by absorbing the sun's rays, especially in the UV-B region. That the smoggy industrial cities of England and California's San Joaquin Valley were noted for the high incidence of rickets last century may, as we shall see, be more closely connected with our exposure to artificial light than has been suspected. Before the clean-up of the smog in London, rickets was even observed in mammals, reptiles and birds in the area.[79]

It was in 1879, however, that Edison invented the incandescent lamp, an event which heralded the process of the gradual replacement of daylight by artificial light which has come to dominate our world today. Commercially oriented, subsequent research on the technology of the incandescent lamp concentrated on producing the brightest light at the lowest cost. The question was virtually never raised whether this different kind of light might have an adverse effect on health. When one considers the radically different nature of artificial light, it becomes manifestly clear that this is a question which should have been raised.

The light produced by the typical modern tungsten filament incandescent bulb, for example, has a colour temperature of about 2850 kelvin. Most of the visible radiation is at the red end of the spectrum with virtually no blue, green or ultraviolet emission, making this form of artificial light vastly different from sunlight.

Moreover, the lightwaves produced by these electricity powered light sources are monotonous. The constant light intensity which they produce represents yet another departure from the character of sunlight. Natural daylight, for instance, varies in intensity from morning to afternoon and exhibits other variations of intensity produced by changes in cloud cover, the orientation of shadows, sunrise and twilight.

Since the 1930s, fluorescent lights have become increasingly popular as indoor light sources. These lights consist of a mercury vapour discharge tube which has been coated internally with a phosphor to convert the ultraviolet radiation produced in the tube into visible light for illumination. Different phosphors have been developed which produce different colour temperatures and a different colour rendering index (CRI). This latter term refers to how closely the colours of illuminated objects compare with their colour in natural daylight which has a CRI of 100.

Fluorescent lights are manufactured with phosphors which produce colour temperatures ranging from 2700 kelvin (that is, 'warm light' similar to incandescent lights) up to 6500 kelvin with CRI close to 100. Fluorescent lights may thus simulate this aspect of natural daylight. All fluorescent lights differ from natural light, however, in that they produce much lower light intensities in the red region and contain sharp intense peaks of light at 406, 436, 546 and 578 nanometres caused by the nature of the discharge spectrum. Again, the extent to which these variations constitute a health risk has been insufficiently appreciated.

The standard 'cool white' lamp which is used extensively in offices and schools was originally developed to maximise the output in the yellow green area of the spectrum, where the eye is most sensitive to brightness. These lights are thus perceived to be bright. However, the CRI value of these lights is low, being in the range of 68–85 and does not approximate natural light. RJ Wurtman points out that 'cool-white fluorescent lamps are notably deficient precisely where the sun's emission is strongest'.[80]

The intensity of light produced by these electricity powered devices is also far different from the level of intensity of natural outdoor light. Incandescent lights matching sunlight in output would produce large and uncomfortable amounts of waste heat. Similarly the energy requirements of equivalent output fluorescent lights would be high, making them uneconomical. Consequently, artificial lighting levels range from about 150 lux, suitable for reading in homes, and up to 700 lux used for office drawing board work.[81] These values are but a fraction of the average outdoor light intensity of 10 000 to 80 000 lux. They are also well below the level of 2500 lux which appears to be the threshold level of light intensity necessary to suppress the symptoms of light deficiency disorders such as seasonal affective disorder (SAD) and to entrain the biological rhythms required for the synthesis of the important hormone, melatonin.[82] Melatonin is responsible for synchronising our sleep–wake cycles[83] and also appears to be a potent anti-cancer factor[84] (see Chapter 5, page 79–80).

The extent of this light deprivation is revealed when, for example, the total amount of visible light to which an individual is exposed who spends 16 hours per day in an artificially lit indoor environment is less than would impinge on that person in 1 hour spent out-of-doors at the latitude of Boston, Massachusetts.[85] Furthermore, it is now known that when used for interior lighting, artificial light of sufficient intensity (say 3000 lux) has the potential to trigger stress reactions in a relatively large number of people.[86]

Space does not permit us to discuss the details of the many other artificial light sources which may be found in the home and workplace. These include quartz halogen reading lamps, sunlamps, 'neon lights' and TV and VDU screens. It is important to note here, however, that halogen lamps emit high levels of ultraviolet light which have been compared to the light intensities associated with midday sun. This being so, it would not be surprising to find that their continued use could either initiate or promote various forms of skin cancer, including melanoma, just as excessive intermittent exposure to sun can. However, there are other completely different ways by which we alter the natural light component of our environment.

Whenever light is reflected off a surface, with the exception of a mirror surface, its spectral composition is changed by absorption of part of the light by the surface. Thus the light reaching our eyes or skin by reflection off a pink wall is quite different in spectral composition to the original incident light. Yet little attention is given to the possible health effects of the dyes and pigments used to artificially colour our environments, though we believe that such effects are not negligible.

Another aspect of the transformation of natural light involves the use of sunscreens and sun-tanning oils. These substances filter out specific wavelength bands of light thus allowing sunlight with missing spectral components to reach our skin. The resultant incomplete form of sunlight is yet another example of artificial light. The implications of these varying forms of artificial light exposures for human health are still unclear, though initial investigations reveal that this is a health area which can no longer be neglected.[87]

Even the very clothes we wear can change the composition of the sunlight reaching the skin. It appears that natural fibres tend to transmit more UV light to the skin than synthetic fibres,[88] and it is revealing that Professor JM Elwood of the University of Otago in New Zealand has noted that in England and Wales the incidence of female lower leg tumours reached a peak with the availability of nylon stockings.[89]

Adverse health effects of artificial light

Apart from depriving us of the benefits associated with exposure to natural daylight, sources of artificial light can also produce adverse health effects. Soon after fluorescent lights began to be installed in factories and offices there were reported cases of skin reactions. In 1941, A James reported two cases of dermatitis from fluorescent light exposure, and in 1949, R Bresler reported a case in which five women received cutaneous burns when working near fluorescent light tubes.[90] Three decades later in New South Wales, Australia, a study of 274 women with malignant melanoma and 549 matched controls, revealed that exposure to fluorescent lights at work doubled the risk of melanoma.[91] Subsequent studies in New York in 1983, and in England in 1986, also found a positive trend in the relative risk for malignant melanoma and total exposure to fluorescent lighting.

Sunlamps used to produce an artificial tan have been similarly implicated with an increased risk of melanoma[92] and P Strickland of the John Hopkins School of Public Health in Baltimore, Maryland, has warned that 'contrary to popular belief an indoor-acquired tan may not shield one from the harmful effects of the sun'.[93] Similar warnings come from the Skin Cancer Foundation (New York) in a pamphlet, endorsed by the American Academy of Paediatrics, which urges adults to ensure that 'the entire family avoids all artificial tanning devices'.[94]

Another potential health risk associated with artificial light sources relates to the production of stress. F Hollwich reminds us of the constant stimulus provided to the unconscious nervous system by monotonous artificial light sources and remarks on '... the increasing insomnia of the city dweller who is deprived of relaxation in the evening with the presently used light sources'.[95]

The increased risk of heart disease in shift workers[96] may serve to support the health connection between artificial light sources and stress. Research studies indicate also that cool white fluorescent lights, for example, are more stress-producing than full spectrum fluorescent lights. The German government has taken these studies seriously enough to restrict the use of cool white fluorescent lamps in public buildings.[97]

Fluorescent lights have also been found to promote hyperactivity, not only in animals but also in school children. In 1974, J Ott reported that male rats, mice or rabbits kept under standard fluorescent lights tended to be highly irritable and aggressive, in some cases cannibalising their young. The control group, on the other hand, was friendly and caring towards its young. Ott also reported a number of interesting experiments which showed increased tumour growth on animals raised under various artificial light sources compared with animals raised with daylight or full spectrum light sources.[98] A subsequent study by M Painter in 1976, showed a 32 percent drop in the hyperactivity displayed by school children when the fluorescent lights were removed from the class.[99]

Children's teeth may also be adversely affected by artificial light. Research conducted in the 1930s showed that the number of cavities in children's teeth varied intensely with the amount of sunlight to which the children were exposed.[100] In 1973, J Ott studied the affect of full spectrum fluorescent lighting versus standard cool fluorescent lighting on first grade children in classrooms, in a school in Sarasota, Florida, and found that the type of lighting affected the susceptibility to tooth decay. Children in rooms with full-spectrum lighting developed one-third the number of cavities in their teeth compared with children in the classrooms with the standard cool white fluorescent lights.[101]

Similar findings were reported in 1991 by J Hargreaves, Professor of Dentistry at the University of Alberta, Canada. In this study, 103 children aged 11–12 years were assigned to two classrooms, one with 'full spectrum' fluorescent lighting which simulated natural sunlight and the other with ordinary fluorescent lighting. After two years, those children who had spent their school days in the 'full sunlight' room had fewer cavities than the children in the other room and the difference was statistically significantly. Hargreaves also noted that where there were early signs of cavities at the beginning of the study, in the full spectrum group those early signs were reversed. The cavities had healed up. In the other group, by contrast any early cavities had developed to the point where they needed filling.[102]

The light reflected off the pigments which we use to artificially colour the walls and floors of our human habitats has the potential to affect our health. Modern technology has enabled us to produce large expanses of colours, such as pinks and oranges and reds, which do not occur in large expanses in nature. These colours are often present in sunsets and sunrises but only for a relatively short period of time.

In 1958, RM Gerard reported his results of studies of the effects of coloured lights viewed on a screen on psychophysiological functions. The autonomic nervous system was significantly less aroused during blue light exposure than with red light exposure. Blue light elicited feelings of being relaxed, less anxious and less hostile, while red light increased feelings of tension and excitement and correlated with increased physiological activation. Subsequent studies confirmed that colour affects the non-visual processes in humans, including changes in emotional state, mood, psychomotor performance, muscular activity, rate of breathing and pulse rate.[103]

One of the most convincing examples of the effect of colour on behaviour is the use of pink colours on the rooms of detention cells in the United States and Canada.[104] This colour has been shown to have a sedative effect on newly arrested prisoners, providing they are in the cells for only a short period of about fifteen minutes. When left in the cells for longer periods of two hours or more the beneficial short-term effects begin to give way to the long-term effects of increased depression, violence and aggression.[105]

A University of Texas study in 1988, found that the light viewed by athletes affected their peak performance and endurance. Briefly viewing red lights increased strength by up to 13 percent. Viewing blue lights seemed to assist in sports requiring a steady level of energy output.[106]

The environmental cost of artificial light

Today we illuminate the interiors of our homes and work places for long hours after the sun has set, as well as the outsides of our buildings and streets. In our cities at night we create a whole new world of colourful lights.

Yet rarely is thought given to the cost to our environment of our lavish artificial lighting. It has been estimated that in the United States, outdoor lighting, including street lights, consumes almost 3 percent of the total electricity production. Poor shielding means that about one-third of this light is actually wasted and goes up into the sky producing night-time light pollution. This humanmade light pollution consumes an estimated 17 billion kilowatt-hours per year with a cost in 1989 of about US$1billion.

T Hunter from the Kitt Peak National Observatory in the United States comments: 'In ecological terms, to produce this light pollution we are using 104 million tonnes of coal, destroying land by strip mining and producing acid rain.'[107]

The contribution to the greenhouse effect also needs to be reckoned. Given that the ecological cost of our total light production is many times the amount cited for outdoor lighting in the United States alone, it is clear that the burden on our environment of artificial light production around the world, is enormous.

Another subtle pattern also emerges in our fabric of health. As we shall see in the next section of this chapter many of the diseases of civilisation, such as diabetes, high blood pressure, high blood cholesterol, heart disease, dental caries, insomnia and sunburn, can be prevented or treated therapeutically with natural sunlight.

Sunlight: One of nature's treatments for disease

The use of sunlight or heliotherapy to treat disease dates back to ancient times and was used by the Greeks and Romans. Herodotus is quoted as saying that sunlight:

> is especially necessary for people who need restoration and the increase of the muscles ... One must further take the precaution that in winter, spring and autumn the sun strikes the patients direct; in summer this method must be avoided by weak people.[108]

In more modern times, in 1815, Cauvin has been quoted as saying:

> Careful observations have shown that sunlight is a curative agent for scrophula, rickets, scurvy, rheumatism, paralysis, swellings, dropsy and muscle weakness.[109]

However, it was not until after 1877 when Downes and Blunt showed that sunlight could kill bacteria that the potential of sunlight to treat disease began to be realised. In 1892, M Ward showed that the most intense antibacterial action was from the ultraviolet component. Soon after, N Finsen, a Danish physician, demonstrated that sunlight concentrated by quartz lenses, which transmit the ultraviolet radiation, effected dramatic cures on the ulcers of sufferers of Lupus vulgaris or tuberculosis of the skin. This disease occurred widely in northern Europe where sunlight levels were low during the winter months. Finsen later developed a carbon arc device to produce UV radiation to treat sufferers and was awarded the Nobel Prize for Medicine in 1903 for his work. In the 1920s and 1930s, sunlight treatment for bone tuberculosis and other forms of non-pulmonary tuberculosis were most dramatically effective treatments. Ultraviolet treatments were also used. An example of the type of results obtained for the treatment of intestinal tuberculosis is quoted by F Daniels as follows:

> ultraviolet treatment—59 surviving, 19 dead; surgical operation—4 surviving, 14 dead; medical treatment—13 surviving, 31 dead. ... While the studies were not done double-blind, the investigators were not fools or charlatans and in the various forms of tuberculosis, they were dealing with chronic, debilitating, often fatal diseases. The rapidity of the response in intestinal tuberculosis was as impressive as the numbers given above.[110]

From the time of Finsen up to the late 1930s there was a progressive development of sunlight and ultraviolet therapy as the only curative treatment for a number of infectious diseases. During this period, sanitariums for the treatment of tuberculosis were set up in countryside and mountains such as in the Swiss Alps, where the air was purer and the full quota of ultraviolet light from the sun could be obtained compared with the reduced levels in the smoggy cities. This situation changed dramatically in 1938, after Domagk received the Nobel Prize in Medicine for his success in treating bacterial infections with sulphanilamide. With this, an era of antibiotic and pharmacological dominance of treating infectious diseases began. Heliotherapy became one of the casualties of the dominance of drug therapy. However, Daniels also astutely observes, 'It is now obvious that chemotherapy does not offer the permanent cure of all infectious diseases'.[111]

The control of blood pressure is a case in point. In 1935, JR Johnson and co-workers showed that sunlight can reduce blood pressure in hypertensive subjects.[112] D Downing comments:

> In 1985 the three biggest selling antihypertensives (drugs to control blood pressure) had world-wide sales of 1.2 billion dollars. Nobody was prescribed sunlight for their blood pressure, yet a single dose of sunlight can lower blood pressure for up to a week at no cost to the taxpayer. The higher your blood pressure, the greater the benefit you can derive from sunlight. No-one has ever disproved the studies on the effect of sunlight that were done before the war. The medical community simply lost interest in them because of the more profitable effects of drugs.[113]

Another sunlight treatment which has succumbed to chemotherapies is the treatment of rickets. In 1919 it was established by Huldschinsky, that rickets results from a deficiency of sunlight. A short time later a nutritive factor, vitamin D, was also found to prevent rickets. Later it was shown that it was the UV-B component of sunlight which reacted with a chemical precursor, 7-dehydrocholesterol, found in living skin cells, synthesising the vitamin D from cholecalciferol or Vitamin D_3. This vitamin could more aptly be termed a prohormone as it is now known that there is a vitamin D endocrine system.[114]

Generally vitamin D is present in unfortified foods in only very small amounts and variable quantities.[115] Consequently adequate exposure to sunlight is essential to obtain this vitamin naturally. However, over the years sunlight treatments were replaced by chemical fortification and today, in the United States for example, the daily requirements of vitamin D are met by the artificial fortification of milk, butter, margarine, cereals and chocolate mixes.

This policy to fortify artificially with vitamin D appears to have been premature, if not misguided. Research by J Haddad and others suggests that sunlight is vastly more important than food as a source of vitamin D, and that large amounts of vitamin D2, as used to fortify food, can be toxic, causing general weakness, kidney damage and elevated levels of cholesterol and calcium in the blood.[116]

Neonatal jaundice is another disorder that can be treated by sunlight. In the late 1950s, it was discovered that sunlight bleached and destroyed bilirubin in solution. High levels of bilirubin, a yellow degradation product of dead red blood cells, impart a yellow colour to the skin. Very high levels which can occur before the baby's liver is fully functioning cause brain damage in the newborn. It was subsequently observed that newborn infants whose cribs were near an open window tended to show less evidence of jaundice than infants receiving less sunlight. Artificial blue light or full spectrum white light is used in hospitals to lower bilirubin levels.[117]

A revealing feature of the sunlight treatments is the significance of the ultraviolet component of the sunlight in most instances. Accordingly, in some countries, such as Germany and Russia, miners are required by law to receive ultraviolet therapy.[118] This highlights the inadequacy of artificial light including common light sources and glass filtered daylight, which is devoid of the ultraviolet component.

Sunlight has also been used to treat cancer and a number of interesting cases have been described by Ott.[119]

In one study in 1959 by Dr JC Wright who was then in charge of cancer research at Bellevue Medical Center in New York, 15 terminal cancer outpatients were given instructions to spend as much time as possible out of doors during the warmer months. They were to avoid artificial light or light received by the eyes through glass including sunglasses or reading glasses. Dr Wright had become intrigued with the idea that light energy entering the eyes might conceivably be a growth regulating factor as far as tumour development was concerned. Dr Wright and her assistants observed that at the end of summer, 14 of the 15 patients showed no further advancement in tumour development

and several showed possible improvement. Subsequently it was discovered that the one patient whose condition had deteriorated had continued to wear prescription glasses which blocked transmission of the ultraviolet portion of natural sunlight from entering the eyes.[120] Unfortunately Dr Wright had not used any patients as controls in her preliminary experiment and largely because of this criticism she was unable to convince university authorities to carry out a larger case-controlled study over a longer time period.

Several animal studies followed, however, which suggested that sunlight did slow the progression of cancer. Mice of a strain C_3 H/H which will spontaneously develop cancer and rats subjected to chemical carcinogens were found to take longer to develop cancers when raised in natural daylight compared to animals raised in artificially lit rooms.[121]

It is now known that bright daylight combined with natural night-time darkness is very effective in resetting circadian rhythm,[122] which, in turn, maximises the night-time production of melatonin.

In contrast, indoor or normal intensity artificial lighting is not very effective in resetting circadian rhythms and exposure to artificial lighting at night time reduces the levels of melatonin produced.[123] Melatonin can have a powerful anti-cancer action [124] and may explain, in part, the apparent protection against cancer produced indirectly by natural daylight. For example, sunlight appears to be preventative against breast cancer. FC Garland and co-workers at the University of California, San Diego, have demonstrated a strong negative correlation between breast cancer mortality rates and sunlight received on a daily basis in 29 United States cities. When the effect of air pollution diminishing sunlight levels is taken into account the breast cancer mortality rates are even more strongly negatively correlated.[125] This latter observation in particular suggests that the UV component of sunlight which is significantly reduced by air pollution, is involved and that the mechanism depends on the light entering the eyes and not just that which falls on the skin, adding further support to Dr Wright's earlier intuition. A similar marked negative correlation between sunlight levels and breast cancer has been reported using global data.[126]

The important role of sunlight in human health

In 1796 in his book *Macrobiotics*, Hufeland wrote:

> Even the human being becomes pale, flabby, and apathetic as a result of being deprived of light, finally losing all vital energy—as many a sad example of persons sequestered in a dark dungeon over a long period of time has demonstrated.[127]

The importance of sunlight to general health has been particularly noticeable in studies of inhabitants of the polar regions and of participants in polar expeditions. These groups frequently showed abnormalities in water balance accompanied by oedema; general loss of energy accompanied by deterioration in behaviour; decreased strength; hypoglycaemia; sinking of the basal metabolic rate; decrease in potency and libido; loss of hair; insomnia; feelings of oppression, depression and irritability. The symptoms were observed to recede after sufficient exposure to sunlight.[128]

In 1953, Halberg showed that within the day–night period of 24 hours a circadian rhythm exists in almost all living creatures. Since that time, F Hollwich, RJ Wurtman and other researchers have shown that the stimulatory and regulatory effect of light in entraining these circadian rhythms takes place mainly *via the eye*. That is, independently of the visual process, light entering the eye passes along a neural pathway from the retina to the pineal gland which in turn stimulates secondary hormonal glands,[129] effectively controlling the larger portion of our biochemistry.

Amazingly, the light entering our eyes thus plays a role in regulating our body rhythms, including body temperature; kidney function (incorporating water balance and electrolyte balance, that is, sodium, potassium, calcium and chloride); blood chemistry (including red blood cell count and white blood cell count); metabolic functions such as lipid metabolism, protein metabolism, carbohydrate balance, blood sugar and insulin; sexual function; and the function of the adrenal, pituitary and thyroid glands.[130] Light entering the eyes also affects the function of the pineal gland and the production of pineal melatonin. This raises serious concerns about the custom of wearing sunglasses. The problem is that sunglasses significantly distort the spectral distribution and intensity of the natural sunlight entering our eyes. This being so, although sunglasses may protect against the harmful effects of excessive light exposure, they have the potential to affect adversely the regulation of the various hormone systems involved in protection against disease.[131] The extent to which light entering our eyes is distorted and in that sense made artificial is illustrated by our measurement of the absorbance spectra of various eyeglasses. The results are shown in Figure 15.6.

Regular sunglasses substantially reduce light intensity in both the visible region and the UV region. Plain glass, on the other hand, only begins to absorb substantial amounts of light below 320 nanometres. It is revealing that modern prescription glasses and contact lenses absorb virtually all the light below 365 nanometres. Since the UV

Figure 15.6 Distortion of light by sunglasses and reading glasses

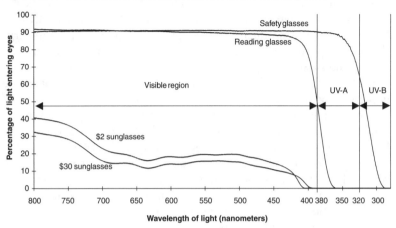

JF Ashton, University of Newcastle, NSW, 1996

wavelengths blocked are precisely those which may be particularly important in respect of the photo-optic stimulation of the pineal gland and the regulation of its hormones, the chronic use of sunglasses, along with some eyeglasses, may inadvertently increase the risk of disease, including skin cancers.

Sunlight also reacts with the largest organ of our body, the skin, which is an important and active element of our immune system. Evidence suggests that the interaction of sunlight with the skin can have a profound impact on other parts of immune system,[132] and produce increased immunologic response.[133] Furthermore, there appear to be seasonal changes in the skin's response to ultraviolet radiation which are independent of the previous exposure of the skin.[134] This has interesting implications for another skin–sunlight interaction which results in the formation of melanin.

When sunlight strikes the skin, particularly the UV-B component of wavelength range 290–310 nanometres, the melanin pigmentary system is activated. Melanocyte cells in the epidermis synthesise specialised pigmented cell-like structures called melanosomes and these particles are transferred to the surface layer of the skin by means of tubular processes.

The melanosomes contain the pigment melanin and skin so pigmented has remarkable resistance to the damaging effects of ultraviolet radiation. It seems that melanin acts (1) by absorbing some of the radiation and dissipating it as heat, (2) by scattering some of the radiation and (3) as a neutral density filter.[135] The production of melanin appears to be regulated in such a way that the UV light entering the skin is limited but not excluded entirely.[136] This means that sufficient radiation still penetrates the skin to activate other biochemical reactions such as the production of vitamin D. It is notable that dark-skinned people living away from the tropics are far more susceptible to rickets and other sunlight deficiency disorders, because the high concentration of pigment blocks too much of the UV component of the limited sunlight they receive.[137]

Thus a regulated amount of UV radiation would normally penetrate the skin and have access to the blood circulated in dermal capillaries,[138] such as in the treatment of jaundice. In the 1930s and 1940s UV irradiation of blood was used successfully to treat patients with bacterial infections.[139] It seems that simple exposure of the skin to sunlight has similar therapeutic potential but without the hazards of the UV-C radiation from the mercury lamp sources.

Professor Wurtman reminds us:

> If people were to expose themselves to sunlight for one or two hours every day, weather permitting, their skin's reaction to the gradual increase in erythemal solar radiation that occurs during late winter and spring would provide them with a protective layer of pigmentation for withstanding ultraviolet radiation of summer intensities.

He observes furthermore, that 'sunburn is largely an affliction of industrial civilisation'.[140] Our lack of regular sun exposure, due to indoor living, industrial smog and other filtration mechanisms, means that we do not build up a protective pigment layer to withstand summer sun exposure, and consequently, we suffer from sunburn.

Sunlight is essential for proper calcium utilisation, which is regulated by vitamin D. Studies have shown that adult males deprived of sunlight for three months during winter develop an impairment in the ability of their intestinal mucosa to absorb calcium. The importance of these findings for the design of lighting in accommodation for the elderly has been pointed out by Wurtman.[141]

The role of sunlight in calcium metabolism has implications for tooth decay, and not surprisingly exposure to sunlight appears to significantly reduce the incidence of dental caries.[142] This finding suggests that water fluoridation may be yet another example of substituting a chemotherapy for natural sunlight.

As referred to earlier, sunlight lowers blood pressure. It also lowers blood cholesterol levels[143] and seems to be generally beneficial to the heart and circulatory system. Further evidence comes from Tapp and Natelson, whose experiments with hamsters suggests that light could be a drug-free agent for slowing the progress of heart failure.[144] A New Zealand study has shown a strong link between low plasma vitamin D levels and ischaemic heart disease. After reviewing similar studies the researchers concluded that: 'Collectively, these studies provide support for the hypothesis that increased exposure to sunlight is protective against CHD [coronary heart disease]'.[145]

Sugar metabolism is also regulated by light.[146] In 1937, L Pincussen showed that daily doses of ultraviolet light very effectively lowered the blood sugar levels of diabetics[147] and more recently the importance of the sunlight prohormone vitamin D has been realised. A deficiency of vitamin D inhibits the pancreatic secretion of insulin, which is essential for regulating blood sugar levels.[148]

The link between carbohydrate metabolism and adequate sunlight exposure has been revealed recently by the research on seasonal affective disorder (SAD), carbohydrate craving obesity (CCO) and premenstrual syndrome (PMS). These disorders can have similar symptoms which include depression, lethargy, an inability to concentrate and episodic bouts of overeating. The symptoms tend to be cyclic, returning late afternoon or evening with CCO, monthly with PMS and yearly, usually autumn and winter with SAD.

Interest in seasonal mood disorders was catalysed by P Muesller in the early 1980s, who proposed that lack of sunlight contributed to the depression in SAD cases. Later studies by M Terman at Columbia University found that by exposing SAD patients to 2500 lux (light of five times the intensity of usual indoor illumination) for two hours in the morning, brought a complete remission from both depression and carbohydrate craving in half the patients with the other half showing definite improvement. It is revealing that researchers find that light administered in the morning is more effective than later in the day.[149] Circadian rhythms can also be effectively reset using light intensities comparable to that found outdoors in the early morning.[150]

Studies by CA Czeisler and associates at the Harvard Medical School in Boston, found that the symptoms of jet lag can be considerably reduced and that travellers can reset their biological clock by exposing the eyes to bright lights. Spending six to eight hours in bright sunshine, upon arriving in a new destination will effectively reset body time with local time within a day or two.[151] Furthermore, the most effective light appears to be light in the blue–green region,[152] remarkably coinciding with the main components of light reflected from the sky and green foliage. These same components, on the other hand, are essentially missing from the output of the common incandescent globe.

The mechanism for these therapies appears to involve, in part, the acute suppression of the production of melatonin by the pineal gland in response to light stimulation. Melatonin induces sleepiness and decreases alertness and the onset of melatonin secretion in the evening is an important promoter of sleep. We are reminded by Professor Hollwich that, 'From morning until evening, daylight yields an uneven gradient taking the form of a curve which is relieved by natural shadows'.[153]

This variation in itself appears to be an important health-giving component of natural daylight.

As research continues, a constellation of intricately connected and interacting biochemical functions regulated by daily and seasonal variations of environmental light is being revealed. In describing his recent findings WJ Stumpf from the University of North Carolina writes:

> Evidence supports the new concept that the skin-derived hormone of sunlight and the pineal hormone(s) of darkness are messengers with comprehensive actions on endocrine, autonomic, sensory, skeletal, and motor functions. In a complementary fashion, both hormone systems appear to correlate biological activities with the daily and seasonal changes of our solar environment.[154]

The importance of sunlight in our environment cannot be overemphasised. Apart from being the ultimate source of our food and, in most instances, our fuel, sunlight cleanses the air we breathe, and prevents the general spread of disease.[155]

Sunlight kills harmful bacteria in water flowing in shallow streams, and helps to break down many pollutants in our environment. Thus sunlight plays a fundamental role in the maintenance, not only of our health, but of the health of the total environment. The time has come to recognise that daylight with its natural variations and rhythms is part of nature's blueprint for the health of this planet, and that it is now time to reinstate unrefined sunlight to its full status of an essential nutrient for human health.

What you can do

1. Arranging your getting up and work start times so that you can spend an hour or so in the sunshine early in the morning before work is ideal. If this early morning experience can be incorporated with exercise, such as walking or gardening in the open air, considerable health benefits will result. Regular cumulative exposure to sunlight each day during the early morning or late afternoon is, we believe, vital for good health and for providing protection against skin cancer.
2. Those of us who are unaccustomed to regular sun should minimise exposure to high intensity summer sun by working or playing in the shade during midday or other high intensity times. Similarly, children should be encouraged to play in the shade and to wear broad-brimmed hats during the midday hours of summer. We do not advocate suntanning/sunbaking binges at any time. Our view is that intermittent high-dose exposure to the sun is dangerous.
3. Clothing of natural fabrics such as wool or cotton should be worn to allow sunlight to reach a larger portion of our skin.
4. Sunglasses are probably best phased-out and replaced with a broad-brimmed hat. The use of sunglasses is best reserved for high glare situations such as snow skiing, desert driving and when travelling on water towards the sun. We believe that it is particularly important not to wear sunglasses during the low intensity periods of sunlight. A delicate balance needs to be preserved between the sunlight entering the eyes and the sunlight falling on the skin. The serious consequences of disrupting this balance are only just beginning to be understood.
5. We recommend caution concerning the regular use of sunscreens. If a sunscreen must be used, inorganic block-outs such as zinc cream are preferred.
6. If fluorescent lighting is installed in your homes, schools or offices, full spectrum tubes approximating natural daylight should be used. Quartz halogen lighting should not be used as the relatively high levels of short wavelength ultraviolet light emitted are damaging to the skin. If these lamps must be used, keep at a considerable distance from them. In the home, subdued lighting in the evenings may help encourage the onset of melatonin production and sleep.

chapter 16

The sounds of technology: Listening for clues to stay healthy

> *There is no quiet place in the white man's cities.*
> *No place to hear the unfurling of leaves in*
> *spring or the rustle of insects' wings. But*
> *perhaps it is because I am a savage and do not*
> *understand. The clatter only seems to insult the*
> *ears. And what is there to life if a man cannot*
> *hear the lonely cry of the whippoolwill or the*
> *arguments of the frogs around a pond at night?*
> *I am a red man and do not understand.*
> *The Indian prefers the soft sound of the wind*
> *darting over the face of a pond, and the smell*
> *of the wind itself, cleansed by a midday rain,*
> *or scented with the pinion pine.*
>
> Chief Seattle, 1854[1]

It was not until some three decades after Chief Seattle's profound statement, 'The clatter only seems to insult the ears', that Western researchers began to systematically study the deafness caused by the sounds of the white man's technology. T Barr reported a study of boilermakers in Scotland, and described their difficulties in hearing in the everyday world. Under the title 'Enquiry into the effects of loud sounds upon the hearing of boilermakers and others who work amid noisy surrounds', Barr's research was published in 1886 in Volume 17 of the *Proceedings of the Philosophical Society of Glasgow*. He demonstrated with sound epidemiological technique that noisy trades were associated with poor hearing.[2]

More than a century after Barr's historic paper, our technology has produced a proliferation of devices which not only produce high intensity sound in the workplace, but also in our homes. Domestic appliances producing high intensity sound include stereo sets, vacuum cleaners, food blenders and numerous power tools. The advent since Barr's time of the automobile, the jet plane and the electric train have meant that environmental noise is no longer confined to the areas near steel foundries and textile mills. Today huge numbers of people live, work, learn or play near noisy highways, airports or rail lines. In the United States, the Environment Protection Agency (EPA) estimated in

1974, that over 70 million Americans live in neighbourhoods with noise levels that interfere with communication and cause annoyance and dissatisfaction.[3]

In an Australian national survey published in 1986, 40 percent of respondents claimed that noise interfered with listening or sleep, with traffic noise being the main source.[4] More recent studies have confirmed the noise problem. In 1993, it was reported that 9 percent of Australians are subject to excessively high levels (that is, greater than 68 decibels) of noise with 39 percent subject to undesirable levels (that is, greater than 58 decibels).[5] A New South Wales Environment Protection Agency study published the same year, found that 33 percent of the population were exposed to road traffic noise, 17 percent were affected by aircraft noise and four percent were affected from nearby industries.[6] Undesirable noise level, as we shall see, can produce harmful health effects.

In the light of fresh research which indicates that exposure to relatively low levels of noise for long periods can affect our health, raising blood pressure, disrupting cognitive development in children, disturbing sleep and prompting psychiatric disorders, the World Health Organisation (WHO), has revised its guidelines on safe noise levels. Due to be published in 1997, the new guidelines reduce the recommended night-time average level of noise suitable for undisturbed sleep from 35–30 decibels. The guidelines also include a peak night-time maximum level of 45 decibels.[7]

Sound and noise

Sound or noise (which is simply unwanted sound) is a mechanical pressure wave produced in air by some object vibrating. How fast the object is vibrating determines the frequency of the sound, which is measured in cycles per second or hertz (Hz). The human ear can detect sounds in the approximate frequency range of 20–20 000 cycles per second. The higher the frequency, the higher the pitch or tone of the sound produced. The loudness of sound is measured using a decibel (dB) scale. This follows a logarithmic function which means, for example, that an increase of 10 decibels makes a sound twice as loud. Examples of noise levels would be about 20 decibels for a whisper, 30 decibels for a quiet country or residential area at night and 40–45 decibels during the day, 60 decibels for normal speech, 80 decibels for a busy office, 80–100 decibels for heavy traffic, trains, power tools and noisy domestic appliances, 120 decibels for a discotheque, and 150–160 decibels for a medium-sized jet engine (see also Figure 16.1).

Figure 16.1 Entertainment noise levels and hearing loss thresholds

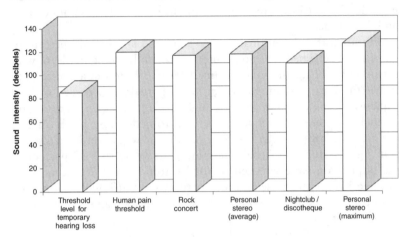

T Patel, *New Scientist*, 27 January & 29 June 1996

It has been found that 80 decibels is about as loud as sound can get without making most people uncomfortable. Sounds rated above 80 decibels produce physiological effects, and long exposures at levels above 100 decibels may produce permanent impairment of hearing. At 140 decibels or more, acute pain is experienced. In 1974, the New York Deafness Research Foundation estimated that one in ten Americans, or some 21 million people in the United States, have impaired hearing.[8] In 1990, T Emmet, Chief Executive of the Australian National Occupational Health and Safety Commission, stated that 'noise-induced hearing loss is the most prevalent industrial disease'.[9]

This is borne out by the fact that Australia has nearly 10 000 workers' compensation cases for industrial deafness every year and in 1996, one in ten Australians suffered from hearing impairment.[10] Noise-induced hearing loss is gradual, non-treatable and permanent.

Nature has provided the ear with two mechanisms to safeguard it from sudden loud noises. The first is an aural reflex (similar to squinting eyes in sudden bright light), which tenses the eardrum and stiffens the lever action of the delicate bones in the ear, thus reducing the amplification of vibrations by up to 20 decibels. The second mechanism is affected by noises exceeding 140 decibels. In this situation the hairs in the organ of Corti, which lies within the cochlea, are caused to rock from side to side rather than backwards and forwards, resulting in a drop in loudness. Gradual hearing loss during a lifetime is attributed to the loss of the sensory hair cells in the organ of Corti. Very loud noises (in excess of 100 decibels) can permanently flatten and damage the sensory hair cells corresponding to their frequency range. Alcohol can impair the reflex action of the autonomic nervous system by between 5–13 decibels. This means that alcohol drinkers are not protected from loud noises until the noise is much louder than usual.[11] Alcohol mixed with noise, such as at a rock concert or discotheque can be a hearing hazard for the unwary.

The effects of noise on individuals

Personal stereo systems pose a significant threat to the hearing of young people. In January 1990, the National Deaf Children's Society in Britain (NDCS) issued a warning that it is 'highly likely' that prolonged exposure to stereos at high volume will damage hearing irreparably.[12] The society measured the level of sound from seven brands of personal stereo and found sound peaks of about 100 decibels in the frequency range 1000 and 4000 cycles per second, where human hearing is most sensitive. Hearing damage produced by these sound peaks would occur slowly and imperceptibly.

In 1996, the French government passed a law limiting the noise levels of personal stereos sold in France to 100 decibels to protect young people's hearing.[13] This action was taken following indications that there was an upsurge in the number of young people suffering partial deafness. In one survey, 20 percent of 400 youths surveyed had some degree of hearing loss.[14] Another study has found that rock concerts are more likely to damage hearing than listening to a personal stereo or going to a nightclub.[15] Evidence of increasing occurrence of impaired hearing in young people adds support to the views of the NDCS. According to officials at the British Broadcasting Commission (BBC), about 3 percent of job applicants are rejected because they suffer from hearing loss. The results of medical tests kept by the BBC show that some young people cannot hear noises of up to 60 decibels at 4000 cycles per second, one of the most sensitive hearing frequencies for humans.[16]

Toys and games can also pose a threat to the hearing ability of children. For example, cap pistols may create a bang up to 138 decibels,[17] and one researcher has suggested that the noise levels of electronic arcade games are such as to pose a potential hearing hazard to children.[18]

A far cry from the peaceful sounds of nature, the noises of our modern technology can induce a number of serious health problems other than hearing loss. For example, accumulating evidence suggests that prolonged exposure to high intensity noise is associated with increased risk of cardiovascular disease, high blood pressure, high cholesterol levels, decreased mental ability and birth defects, such as spina bifida and absence of brain. In humans, auditory stimulation evokes responses by the autonomic and reticular nervous systems and the brain. These responses, in turn, may elicit a spectrum of activity in a number of the body's systems and organs, including changes in cardiovascular volume, heart rate, blood pressure, endocrine function and gastrointestinal motility.[19]

In 1959, G Jansen reported the results of one of the first studies of the physiological effects of noise. He had obtained data from nearly 1400 workers in noisy German factories and found that in industries where the noise levels were 100 phons (equivalent to 100 decibels at a frequency of 1000 hertz) and above, the incidence of heart problems and circulatory disorders was significantly higher than in factories where the sound levels were less than 80 phons.[20] Three years later in 1962, NN Shatalov and co-workers reported that high blood pressure and heart abnormalities were frequent among 300 Russian workers exposed to noise levels of 85–120 decibels in textile and ball-bearing plants.[21] As the harmful effects of noise became more apparent, in December 1971 the Environmental Protection Agency stated in a special report to the President and Congress of the United States that:

> There is some evidence that workers exposed to high levels of noise have a higher incidence of cardiovascular disease, ear-nose-and-throat disorders, and equilibrium disorders than do workers exposed to lower levels of noise.[22]

Within another decade, more than 15 studies had reported that the prevalence of hypertension is increased by at least 60 percent among workers chronically exposed to sound levels over 85 decibels.[23]

In 1995, F Tomei and colleagues in the Department of Occupational Medicine at La Sapienza University in Rome, also found that high blood pressure was common in men exposed to high noise levels (that is, around 90 decibels) at work. They warned that people living or working in noisy environments were at considerable risk of developing cardiovascular disorders. It appears that the risk is increased with the intensity of noise and the period of exposure.[24]

Experiments with animals provide additional information on the noise–heart disease link. When rabbits were exposed to 120 decibels noise levels for ten weeks, their blood cholesterol levels increased significantly and many of the rabbits became ill with atherosclerosis.[25]

Harmful physiological effects of noise are not confined to workers in noisy industries. The effects of high noise levels in the community have also been examined. In a series of studies conducted in the neigh-bourhoods adjacent to an airport in Amsterdam the research showed that residents in areas with high levels of aircraft noise were more likely to be under medical treatment for heart trouble and hyperten-sion. Women especially were more likely to be taking drugs for car-diovascular problems and high blood pressure than an unexposed population. These differences could not be explained by age, gender, smoking habits or obesity. Furthermore, another study showed that increases in the purchase of cardiovascular drugs were positively cor-related with the number of aircraft overflights at night.[26]

One striking aspect of these studies is the evidence that children, as well as adults, show noise-associated cardiovascular effects. In fact, there is reason to suspect that exposure to high intensity noise is a greater threat to children than to adults. In 1968, a German investiga-tion of children in the seventh–tenth grades reported higher systolic and diastolic blood pressure for children from noise-affected schools.[27] Over a decade later, another study of children attending schools under the air corridor of the Los Angeles International Airport, and children of similar socio-economic condition, age and race living and attending schools in quiet Los Angeles neighbourhoods, found identical results. In the noisy schools, subject to one overflight every two-and-a-half minutes, the children had significantly higher systolic and diastolic blood pressures.[28]

In another airport study published in 1995, G Evans and co-work-ers at Cornell University, New York, found that both blood pressure and the level of the hormone adrenaline, which is associated with stress, were considerably higher in children living near Munich's international airport than among children living in quieter areas of the city.[29]

Busy roads can also produce high noise levels. V Regecova and E Kellerova from the Slovak Academy of Sciences in Bratislava have recently reported that children attending kindergarten near busy roads where traffic noise was greater than 60 decibels were found to have high blood pressure.[30]

The Los Angeles Noise Project also revealed another subtle adverse health effect. The mental performance of children in noisy schools was noticeably lower than for quieter schools. Children from the air corri-dor schools were less likely to solve the test puzzle and more likely to take longer to solve it than children from quiet schools.[31] These stud-ies reinforced earlier observations by other researchers. For example, AL Bronzaft and DP McCarthy found that children in classrooms on the side of a school facing train tracks performed worse on a reading achievement test than children in the classrooms on the quiet side of the building.[32] In another study conducted a few years later on children aged from four-and-a-half to six-and-a-half, H Heft found that children from homes described as noisy, when tested in quiet settings, did worse on both a matching and incidental memory task than those from quieter homes.[33]

Similar findings were reported in 1995 by G Evans and colleagues. They found that the reading skills and long-term memory of the children living near the airport were impaired. The researchers concluded that chronic noise exposure in young children may affect more complex, higher order skills such as reading, problem-solving and comprehension of difficult materials.[34]

The researchers did however discover some good news. During the study, the old Munich airport was closed down and it was found that the affected children's memories and reading deficiencies recovered two years after the aircraft noise stopped.[35]

High levels of noise also seriously affect the unborn child. Early studies with animals had suggested that noise may have teratogenic effects. In 1966, W Geber reported a study of pregnant rats exposed to intermittent audiovisual stress in the range of 74–94 decibels. He observed significantly reduced litter size and a significant increase in the number of foetuses which failed to develop and were reabsorbed per litter among the exposed animals. Geber also noticed a significant reduction in mean body weight of foetuses as well as an increased frequency of congenital anomalies, including bone and brain defects, and neural tube defects.[36] Since that time a number of other studies of the effects of noise on test animals have been carried out. The most frequently reported effects include intrauterine growth retardation, foetal mortality, increased litter reabsorptions and teratogenesis.[37] The serious implications that noise also affects the outcome of human pregnancies has been borne out by several epidemiological studies.

In 1978, FN Jones and J Tauscher reported their findings from an assessment of the potential teratogenic effects of airport noise in Los Angeles County. After comparing over 225 000 birth certificates on a race-specific basis they noticed that there was a significantly higher overall rate of malformations in Afro–American babies residing in the high noise (90 decibels) areas than among Afro–Americans from other areas of the county. Among white-skinned Americans, infants from high noise areas displayed a small but nevertheless significant excess of anencephaly (absence of brain) and spina bifida combined, compared with infants from control areas.[38]

In a similar study of nearly 82 500 births in Atlanta, L Edmonds and co-workers classified high noise levels as those above a mere 65 decibels. Initial examination revealed no significant differences in any of the 17 birth defect categories between the high and low noise census tracts. However, upon closer examination of the neural tube defect category, the researchers observed a significant excess of spina bifida with hydrocephalus (water on the brain) in the noisiest census tracts relative to the control areas.[39]

Another study near Amsterdam airport, also using the low 65 decibels cut-off level for noise, revealed a significant association between noise exposure and low birth weight for female babies, after controlling for parent's income and sex of infant. Evidence of a dose–response effect within the noisiest areas was also reported.[40] It is also known that external noise increases the heart rate and activity of the foetus.

A 1995 study of 1500 pregnant nurses by B Luke at the University of Michigan also found that high noise levels stimulated stress hormones in these mothers-to-be, which invariably led to premature contractions.[41] It is interesting that the foetus is protected by the mother's abdomen from high frequency sound more effectively than from low frequency sound. For example, sound at a frequency of 4000 cycles per second is reduced by more than 70 decibels by the mother's abdomen, while sound at a frequency of 50 cycles per second is reduced in intensity by only 20 decibels. Reviewing these observations RE Meyer and co-authors comment:

> These observations suggest that low frequency noise (such as that associated with many industrial environments) may present a potential risk to pregnant women despite the fact that these noises may not be psychologically annoying to the individual. Additional research will be needed to evaluate this possible health risk.[42]

In the United States the Federal Aviation Administration has set a maximum noise level permitted outside homes, from flying aircraft at 65 decibels for a 24-hour average. However the Natural Resources Defence Council has warned, that limit is not only too high but is misleading as well. The limit is based on average noise levels, rather than the loud spikes of noise from takeoffs and landings that are the most troublesome to communities near airports.[43]

The effects of noise on the community

The sounds and inextricably bound images of rock music can have insidious effects. The glorification of violence in popular music may be a contributing factor in the increase of violence in the community. One behavioural study found that violent music videos desensitised viewers to violence immediately after viewing.[44] Another study involving the analysis of 200 concept-type music videos revealed that violence occurred in 57 percent of the samples. Of the videos containing violence, 81 percent also contained sexual references, most of which portrayed women as sex objects[45] (see Figure 16.2). Other researchers analysed a random sample of rock music videos. They found that 53 percent had violent themes. Further examination revealed that 57 percent of the music videos depicting women portrayed them in a condescending manner while only 14 percent depicted women as fully equal to men.[46] When other researchers subsequently studied the influence of rock music videos on the attitudes of men to violence against women, they concluded the report of their disturbing findings from 144 undergraduate men with the comment:

> That 39.6 percent of men in this study indicated some likelihood of committing rape if they could be assured of avoiding punishment is further evidence that attitudes of violence against women remain widespread.[47]

Figure 16.2 Percentage of concept-type rock music videos with violent content

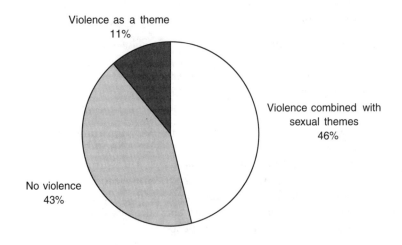

Violence as a theme
11%

Violence combined with
sexual themes
46%

No violence
43%

EF Brown et al., *Journal of the American Medical Association*, vol. 262, no. 12, 1989

On the other hand, music may improve your performance at various tasks. A novel study was carried out by the Centre for the Study of Biobehavioural and Social Aspects of Health, in Buffalo, New York, on a group of 50 male surgeons who regularly listened to music during surgery. Doctors were found to have improved task performance when they were listening to music of their own choice (a range of classical, jazz and Irish folk) rather than music of the experimenter's choice.[48]

There also appears to be health benefits for patients who listen to certain kinds of music before, during and after surgery. In particular an improvement in pain control has been observed for both medical and dental procedures.[49]

As the average Western teenager spends little time listening to the sounds of nature, so too have we given little thought to the cutting down of trees to make way for more factories, freeways and runways. Few of us have realised that we were removing nature's own noise abating devices. Shrubs and trees can help to reduce noise by up to 10–12 decibels in the 1000–11 000 cycles per second frequency range.[50]

The number of truly quiet places in western countries is steadily shrinking.[51] A 1995 survey by the Council for the Protection of England revealed that Britain's area of tranquil land (that is land at least 3 kilometres from large towns and 2 kilometres from major roads has shrunk by 21 percent since the 1960s.[52]

It is becoming more and more difficult to get away from our sounds of technology, though total health and well-being may depend on it. Technological progress which distorts the environment provided by nature is not without a cost. That price must be considered and weighed in terms of human, environmental and social health before the products of our science and technology can be called progress.

What you can do

1. In noisy work environments, which may include using electric saws, grinders, vacuum cleaners and even appliances such as blenders in the home, ear protection should be worn. This may seem extreme, but deafness is irreversible.
2. For reasons mentioned in the chapter, avoid drinking alcohol just before or during periods of exposure to high noise levels, a point which is especially relevant when attending discotheques and listening to loud home stereo systems.
3. Be wary of listening to personal stereo systems or walkmans at high volume settings.
4. Unfortunately it is not always possible to live away from high noise areas such as airports, factories and highways. If you do live in a noisy area, closing doors and windows and using heavy curtains can help to reduce noise levels. In the long term, planting trees and shrubs around your house can also help reduce indoor noise.

PART six

Surviving technology

chapter 17

Finding the way back to good health

If the promises of technological society were being kept, this book would not have been written. The truth is that in addition to the virtues which technological society has bequeathed us, we have also inherited a legacy of toxic time-bombs. Committed to technological solutions to our problems, which involve using more chemicals to make our water cleaner and our foods safer, we simply end up increasing the chemical content of our food and water. In doing so we increase, rather than diminish, the number of toxic time-bombs which threaten us. There is hope, but as a society we need to see what we are doing wrong to know what has to be done to make things right.

It is to be admitted that the mechanical model of the world (together with a reductionist scientific approach) has proved to be remarkable for developing machines of all types—from steam engines to motor cars, microwave ovens, mobile phones, refrigerators, aircon-ditioning, television sets, videos, electric lighting, skyscrapers, free-ways, supertankers, x-ray machines, CAT scanners and pacemakers, to name only a few. There is little doubt that Francis Bacon's vision of Utopia has at least in one respect been achieved. The comforts and facilities of the average, middle class home in Western countries would probably surpass those of the Emperors of Ancient Rome or the palaces of the Kings of Europe in Bacon's time. But while we have reshaped our environment and to some extent subdued the forces of nature, we have also destroyed or polluted many of nature's life-sus-taining and health-giving systems, without which, all the machines we can master have little purpose or value.

For example, using our technological prowess, we have developed massive farm machines such as tractors and harvesters, and chemical fertilisers, but the mechanical model failed to predict that we would lose our topsoil at an increasing rate because of technological interventions. The mechanical model of science enabled the development of steam engines, chainsaws and bulldozers but failed to predict that the massive destruction of forests that followed would affect on the global balance of carbon dioxide or potentially disrupt weather patterns and possibly sea levels.

When the Mechanical Middling Purifier was developed enabling very high quality pure white wheat flour to be produced, it was not anticipated that essential nutrients including the B vitamins, vitamin E and zinc (all essential for human health) would be lost or destroyed, thus predisposing those eating this refined flour to diseases such as bowel cancer, skin cancer and even heart disease.

The chemists who developed the chlorine-based chlorofluorocarbon compounds for refrigerators, air conditioners, aerosol can propellents and expanding disposable polystyrene cups and food trays, did not foresee that these compounds would accumulate in the atmosphere and begin to destroy the earth's protective ozone layer. On 5 March 1996 the thickness of the ozone layer over Britain dropped to a record low of 195 Dobson Units (DUs).[1] The usual reading at that time of year is around 365 Dobson Units. Another reminder of the folly of our throw-away cups and spray cans was given in November 1996 when sunny Perth in Western Australia recorded the highest solar UV radiation levels on record.

Reflecting on relevant interconnections in other than reductionist ways, we now know that the essential fatty acid content of a plant food can depend on the latitude where it was grown and whether artificial fertilisers or organic fertilisers were used during its growth. The fatty acid content of hens' eggs can vary significantly, depending on what the hens are fed; and even the blood cholesterol levels in humans are affected by the diet of animals used for meat and milk. Since the composition of the fat in our diet can affect our susceptibility to heart disease and cancer, how our food is grown has a direct bearing on disease within the community—a point often missed.

There is growing acceptance of the non-reductionistic insight that the levels of natural versus artificial light to which we are exposed each day may affect our mood, our blood pressure and even our craving for sweets and carbohydrate foods, and the rates at which we metabolise these things. Through the convenience of electric lighting we have conquered night and changed our sleeping habits without realising that we have been interfering with one of the body's anti-cancer mechanisms—the night-time production of the hormone, melatonin.

One reason why the reductionist orientation has distracted scientists from the foregoing interconnective subtleties is that the mechanist approach has little place for synergistic events and processes. The reductionist thrust of traditional evolutionism is a case in point. Professor Goodrich[2] from Oxford has recently pointed out, for example, that not all medical ideas based on reductio-evolutionary premises have proved medically productive. Presupposing an evolutionary progression in organic development from simpler forms of organisms, evolutionary science taught that the body contains a number of organs which no longer served any useful *purpose*. More than 150 such organs have been identified, including the appendix, tonsils, thymus, external ear muscle, third molar, coccyx, recurrent laryngeal nerve and knee menisci. Contrary to the conventional wisdom, the appendix is now known to be part of the Gut Associated Lymphoidal Tissue, with subtle but important immunological functions and should not therefore be removed without extremely careful consideration and certainly not routinely, as has all too often been the case.

The synergy to which we are alluding may be defined roughly as that property of a system which results in the whole being greater than the logical sum of the component parts. This observed property of many biological systems stands in stark contrast to the reductionist presumption that the whole is simply the mathematical or quantifiable sum of the component parts.

As intimated above, some of the synergies which can affect our health may be extremely subtle. Consider, for example, the widely used food additive citric acid which is generally considered to be harmless by the majority of food technologists. In itself, citric acid is indeed quite harmless and occurs naturally in oranges and lemons. However, if the citric acid is used in a food mixture which is reconstituted or mixed with tap water containing aluminium salts as a water treatment, the citric acid readily binds with the aluminium ions, forming a new compound which is readily absorbed by the body. In consequence, a much larger proportion of the aluminium contained in tap water is absorbed into the body.

Given that the neurotoxic properties of aluminium have been linked with accelerated aging and possibly Alzheimer's disease, the addition of so-called 'harmless' citric acid to foods which are then in turn mixed with what would be regarded as harmless levels of aluminium in drinking water, is likely to be a far less healthy practice than anticipated.

At this point we are still faced with the problem of choosing the best pathway to health in the maze of information which is now available. How can we know how to make the best choice when we come to the various options? How can we know which choice of diet or lifestyle will have the best outcome?

The model we propose here postulates that all of nature is fundamentally interconnected and the seemingly disparate parts of nature are in fact functionally integrated. By enhancing our understanding of why the systems of nature interface as they do, we become better equipped to make choices which best harmonise with nature.

When an interconnective view of nature is embraced, a whole new interpretive paradigm of nature can emerge which has powerful predictive power. This can be illustrated by the example of the latitudinal variation of the fat content of plants. If we take a handful of sunflower seeds from a single flower head and take half the seeds to northern Queensland to grow them in hot topical conditions and take the remaining seeds to southern Victoria to grow them under cool–temperate conditions, the next generation of seed will have varying fatty acid compositions depending on latitude. Those grown in the hot, sunny climates will have an overall higher level of saturated fatty acid than the seeds grown in the colder climate. Those seeds grown in the colder, less sunny climates will have a higher polyunsaturated oil content than seeds grown in the hot climate.

Because the price paid for sunflower seeds is often on the basis of their polyunsaturated content, scientists are working on genetically engineering sunflower seeds which will produce higher levels of polyunsaturated fatty acids in their oil even when grown in the hot sunny tropics. However, what this research neglects is that seeds native to the tropics such as coconuts and palms tend to be rich in saturated fats. This being so, the conventional advice that encourages people living in hot sunny climates to increase the polyunsaturated fat content of their diet in order to protect against heart disease shows a misunderstanding of the pattern of interconnection between climate and saturated fat content.

Moreover, Dr VE Reeve and colleagues from the University of Sydney, have found that when test animals prone to skin cancer are fed a diet containing mainly saturated fats, the animals have substantially increased immunity to UV-induced skin cancer. When the diet of the test animals is changed over to polyunsaturated fats, skin tumours 'grow like mushrooms' under the UV radiation.

While in nature we find higher levels of saturated fats in natural tropical diets, it is important to observe that the long hours of bright sunshine seem to offer a certain degree of protection against the harmful effects of saturated fats on the heart. In turn, these saturated fats offer increased immunity to skin cancers as a result of living in a sunny climate.

The interconnectedness of nature is thus revealed by patterns within nature which can usefully be exploited to articulate a blueprint for health.

The patterns of purpose in the environment

In the early 1900s, LJ Henderson a Professor of Biological Chemistry at Harvard University, reflected on patterns in the biochemical world and published two works, *The Fitness of the Environment*[3] and *The Order of Nature*,[4] in 1913 and 1917 respectively. Henderson noted the high acidity buffering capacity of phosphoric acid and carbonic acid, which enabled near neutral conditions to be maintained in living organisms. He referred to the anomalous expansion of water near its freezing point, whereby the coldest water floats to the top of a lake and ocean. He observed that water had a number of other environment-regulating properties such as high thermal conductivity and heat of vapourisation, along with a unique ability to dissolve a large range of other substances, all of which were necessary for life as we know it.

Henderson also noted that the elements hydrogen, oxygen and carbon have a peculiar arrangement of properties essential for living organisms. He wrote: 'The unique properties of water, carbonic acid, and the three elements constitute, among the properties of matter, the fittest ensemble of characteristics for durable mechanism'.[5]

Henderson concluded that the biochemical properties of these compounds were so outstanding in their life supporting roles that:

> We were obliged to regard this collocation of properties as in some intelligible sense a preparation for the process of planetary evolution ... Therefore the properties of the elements must for the present be regarded as possessing a teleological character.[6]

In the early 1950s, Fred Hoyle the eminent Cambridge astronomer, noticed that it was necessary for a particular balance between oxygen and carbon to exist in the Universe for it to be able to support life as we know it. On the basis of the observed ideal cosmic-abundance ratios of oxygen 16 to carbon 12 to helium 4, he correctly predicted the necessity for the existence of an excited state for carbon 12. This prediction was confirmed subsequently by experiment.[7] A number of apparent 'coincidences' among the physical constants have been reviewed by Carr and Rees who concluded their report by saying:

> The possibility of life as we know it evolving in the Universe depends on the values of a few basic physical constants—and is in some respects remarkably sensitive to their numerical values ... even if all apparently anthropic coincidences could be explained [by some presently

unformulated unified physical theory], it would still be remarkable that the relationship dictated by physical theory happened also to be those propitious for life.[8]

The precise values of the fundamental constants that control the strength of the physical laws of nature are just as essential for the existence of our world as are the laws of nature themselves. Barrow and Silk write:

> The fact that the laws of nature barely, but only barely, allow stable stars to exist with planetary systems today is not a circumstance subject to evolutionary variation ... The laws of nature allow atoms to exist, stars to manufacture carbon, and molecules to replicate—but only just.[9]

The values of the physical constants which, in effect, tune the general laws of nature determine distribution not only of the chemical elements but also the balance of biodiversity in biological systems and the structure of the galaxies. Thus the universal gravitational constant is just as fundamental a part of nature as the law of gravity. What the gravitational constant does, is set a limiting factor or boundary condition for the law of gravity. Thus the position of the earth's orbit around the sun is determined by the gravitational constant. If this value were different, the earth would have to be closer to or further away from the sun and would be subject to vastly different surface temperatures. Life as we know it could not exist. A multitude of other life-supporting conditions, such as the composition of atmospheric gases would also be affected. Similarly, if the values of the solubility constants for the salts of poisonous ions, such as fluoride, lead, mercury, and aluminium were much larger, life as we know it could not exist on our planet.

It can be seen from these situations that the notion of purpose or a blueprint comes about as a result of the limiting values or boundary conditions imposed by the precise values of fundamental constants. These values determine what we term the 'bounds of purpose'.

Purpose revealed by the limits of chaotic systems

In nature interconnectedness is often masked by chaos. To search for a pattern, in what for all intent and purpose are random interactions, might seem incongruous. However, chaos theory provides new insight into nature's holographic blueprint. Chaos theory embraces our understanding of behaviour which appears random but which is actually determined within the constraints of certain laws. This can be illustrated simply by a computer demonstration. A completely random collection of pixels or dots is generated in different colours (or numbers) on the screen. If the pixels are assigned a simple rule or constraint, for example, either to agglomerate or spread apart according to the colour of each pixel's nearest neighbours, a closely defined spiral picture of colours separated out into definite shapes is formed. The motion of the pixels as they seethe around each other appears completely random or chaotic, although it is actually constrained by our simple rule. Each time a new set of pixels is generated when equilibrium is reached the picture is different, yet it retains the basic spiral form.[10]

If on the other hand, no rule was assigned to the pixels, they would simply continue in random motion. Only on rare occasions would anything like a recognisable pattern be formed momentarily on the screen. The chances of pixels forming a distinct spiral pattern on the screen would be of a similar magnitude to the chances of the same pixels forming the letters C H A O S on the screen. Thus repeated formation of a recognisable pattern within an apparently random system suggests that certain limitations are in effect or that the system is operating to a basic blueprint. Such systems are very difficult to study on the basis of a reductionistic model because of the enormous number of possible interactions to be considered. This problem is manifest even in very simple systems such as the coloured pixels described above. The subtle patterns of these types of systems have only recently been revealed with the advent of high speed computers capable of their articulation.

In most natural systems, an enormous number of interactions are possible, even when relatively limited situations are considered, and only a few variables are involved. One such example is the system studied by WM Schaffer. He spent years with fellow researchers studying the fluctuations of bee populations in the Santa Catalina mountains. He collected data on bee species and flower pollen levels, to find that ants were somehow involved. His reductionist model broke down, as the complexity of the system was impossible to model in this way.[11]

This example also illustrates how species' interactions that are far from obvious, may yet be of crucial importance to a community. This point was also highlighted in a study by RT Paines of an intertidal community comprising 15 species in Mukkaw Bay in the United States, where the top predator was a sea star:

> When Paine excluded the sea star from one area but allowed it to remain in another, the result was a radical change in the species diversity of the first area. Of the original 15 species, only eight remained. One of the two competing species of barnacles was eliminated, because

the sites for its attachment were not cleared of other organisms by the sea star, and the other barnacle was a better competitor for what little space there was. A sponge and its predator also disappeared; apparently the sea star had some sort of indirect influence on them, probably by clearing space for the sponge.[12]

In these examples we see evidence that even chaotic systems must be considered as a whole, and that the properties of the whole are dependent on each individual component. Most population biologists now consider that fluctuating animal populations are examples of chaotic systems, with a variety of limiting systems holding the populations in check.[13]

Snowflakes are often singled out by poet and artist as examples of nature's symmetry and beauty, yet at the same time they are all proverbially 'different'. The six points and symmetry of the snowflake give a hint of an underlying law of growth. As they fall through the turbulent air however each snowflake experiences a different path, a different temperature, along with variant humidity, contours and a different distribution of atmospheric dust particles. Each parameter contributes in such a way that the simple crystal growth laws produce an endless variety of patterns but all to the same blueprint. This variety produces the effect that we so often perceive as beauty. The fine arrangement of dendrites, the intricate patterns on sea shells, the delicate blend of colours in a pansy and the majestic grandeur of the bough arrangement of an old oak tree, are all examples of beauty produced by chaos.

Mechanical systems involving only a few variables can also exhibit chaos. A century ago Henri Poincare proved that the motion of just three bodies under gravity can be extremely complicated or chaotic. What Poincare found was that very slight perturbation of a balanced system, by way of the new imbalance can produce such complicated interactions that the resulting behaviour appears very complex.[14] He later wrote:

> A very small cause which escapes our notice determines a considerable effect that we cannot fail to see, and then we say that the effect is due to chance. If we knew exactly the laws of nature and the situation of the universe at the initial moment, we could predict exactly the situation of that same universe at a succeeding moment. But even if it were the case that the natural laws no longer had any secret for us, we could still only know the initial situation approximately. If that enabled us to predict the succeeding situation with the *same approximation*, that is all we require, and we should say that the phenomenon had been predicted, that is governed by the laws. But it is not always so; it may happen that small differences in the initial conditions produce very great

ones in the final phenomena. A small error in the former will produce an enormous error in the latter. Prediction becomes impossible, and we have the fortuitous phenomenon.[15]

The notion that small variations in conditions may produce quite dramatic consequences is also known as the 'butterfly effect'. This phrase was coined by E Lorenz at the Massachusetts Institute of Technology to explain how small variations in weather data rapidly deteriorate the accuracy of the predictions of computer weather forecasts.

Understood in this light, the many examples of chaos in nature serve to demonstrate the complex interconnections within nature which resist explanation by the reductionistic model. In an early article on chaos Crutchfield and colleagues point out:

> Chaos brings a new challenge to the reductionist view that a system can be understood by breaking it down and studying each piece. This view has been prevalent in science in part because there are so many systems for which the behaviour of the whole is indeed the sum of its parts. Chaos demonstrates, however, that a system can have complicated behaviour that emerges as a consequence of simple, nonlinear interaction of only a few components. The problem is becoming acute in a wide range of scientific disciplines, from describing microscopic physics to modelling macroscopic behaviour of biological organisms ... For example, even with a complete map of the nervous system of a simple organism ... the organism's behaviour cannot be deduced. Similarly, the hope that physics could be complete with an increasingly detailed understanding of fundamental physical forces and constituents is unfounded. The interaction of components on one scale can lead to complex global behaviour on a larger scale that in general cannot be deduced from knowledge of the individual components.[16]

Examples of chaos in nature may be seen also to intimate if not at times reveal the underlying laws or interconnections which limit natural fluctuations to within certain bounds. For example, in the case of Mukkaw Bay, the sea star could be considered as having a purpose in 'limiting' the numbers of certain species in order that other species might survive. The apparently random motion of bacteria is another example. The random nature of their movement is 'limited' by some instinct mechanism for the purpose of finding nutrients.[17]

Polkinghorne, cites the interesting example of how termites begin construction of their well-known mounds:

> The first stage of their construction is a scurrying of termites dropping little lumps of earth at random. Each lump is impregnated with a

hormone which attracts other termites. By chance a few more lumps are deposited in one region than in the rest of the area. This attracts more termites to that region, depositing more lumps of earth, and the positive feedback produced soon snowballs the process until one of the large termite pillars has come into being, all as a result of the initial fluctuation in the random scurrying.[18]

Here again, what appears to be random motion is actually 'limited' by the hormone for the purpose of ensuring that a nest is built. Without the hormone, the chance of the nest being built is very small and survival of the termite colony as we know it, would not continue. With some genera the exact location of the nest, within the constraints of available water and food, is not important, thereby allowing the random scurrying to be performed initially over a large area, but nonetheless limited by ecological factors.

With the South African genus *Odontotermes*, the location is also constrained such that the termites form long low mounds known as 'termite bands' which can be up to a kilometre long. It has recently been discovered that this termite activity is responsible for greatly increasing the fertility of the soil. The behavioural mechanism for producing such large regular patterns is at present unknown. However again, the constraint on the behaviour of termites to produce mounds which effectively regulate the ability of the soil to retain water and thereby enhance the growth of plants is another example of a limit to a random activity which fulfils a purpose.[19]

Another example from Africa illustrates how certain constraints on random behaviour are necessary to conserve nature. In 1990, Hoven reported the observation that giraffe, roaming freely, browsed only on one acacia tree in ten. He subsequently discovered that acacias, when nibbled by animals, begin to produce a lethal tannin and emit ethylene gas into the air which can travel up to 50 metres. The ethylene gas in effect warns other trees of the impending danger, which step up their production of leaf tannin to lethal quantities within just five to ten minutes. Thus the random browsing of herds of animals is regulated for what appears to be the purpose of preventing the bush from being systematically eaten out or destroyed.[20]

The concept of purpose in nature is not new. The Greek philosopher Aristotle described a holistic picture of the Universe. To him, in living systems in particular, the component parts appeared to function in a cooperative way to achieve a final purpose in accordance with a preordained plan or blueprint.[21]

In modern times, the biologist J Monod, Director of the Pasteur Institute has observed:

> One of the fundamental characteristics common to all living beings without exception [is] that of being *objects endowed with a purpose or project*, which at the same time they exhibit in their structure and carry out through their performances ... Rather than reject this idea (as certain biologists have tried to do) it must be recognised as essential to the very definition of living beings.[22]

Monod, a strong reductionist goes on to admit, of the dilemma to science posed by purpose. He writes:

> Objectivity nevertheless obliges us to recognise the teleonomic character of living organisms, to admit that in their structure and performance they act protectively—realise and pursue a purpose. Here therefore, at least in appearance, lies a profound epistemological contradiction.[23]

In the above examples we have shown that while much of nature is chaotic, it is also regulated in such a way that individual systems and species are preserved. While fluctuations in equilibrium occur, these are limited to within bounds that maintain the overall ecology. Over aeons of time, the 10 000-odd species of termites have not destroyed the world's vegetation, nor have giraffes, in more recent times, eaten out the acacia trees in Africa.

The interaction of humans with their environment, especially in recent times, is quite different from all other living things. By applying our intelligence to acquire knowledge about nature we have obtained the power to control it. Termites cannot analyse hormones, resynthesise them in quantities to change the pattern of termite houses to suit their own whims, but we can. Similarly giraffes cannot genetically engineer acacia trees to produce less tannin so that they can eat indiscriminately, but humans can.

As nature has imposed limitations on randomness, we can remove or bypass those limitations. We have unlocked nature's stores of energy and with this energy we use our knowledge to clear forests, pollute seas and transform landscapes. Armed with our technology we now ignore many of nature's limitations. In our rush to discover the 'how' of nature, we have ignored the possibility of a 'why' of nature. Nature seems tuned to maximise our health and the health of our environment, and changing the boundary rules, as we have outlined in the this book, has proved deleterious both to human health and the environment.

The 'bounds of purpose' can often be exceeded in less obvious ways. Removing a single forest species to the extent that its distribution density falls below the natural variation, results in dramatic changes to the remaining ecosystem.

For example, in the northern tablelands of New South Wales, the removal of wattle trees in the vicinity of eucalypt trees has resulted in the decline and death of large numbers of these eucalypts. Investigations revealed that during the spring and summer scarab beetles voraciously defoliated the eucalypts. Normally 80 percent of these and other leaf-eating insects in the forest canopy were consumed by sugar glider possums and birds. During the cooler months the wattles provided food in the form of gums and nectar for these natural predators. When the wattles were removed these predator species all but disappeared and the scarab beetle population exploded.[24]

This example illustrates how human activity that exceeds the limits of natural variation or nature's blueprint can result in unexpected changes in other parts of the natural environment. In much the same way, processing food in such a way as to remove a single component such as dietary fibre can disrupt the balance of biochemical processes in the body and over a period of time result in diseases such as cancer.[25]

Using the human mind: Education for survival

The crucial issue that confronts us is to define the limits of our transformations of nature, while still preserving, if not enhancing, the environmental conditions for personal and planetary health. We have only just begun to recognise ways in which our technological developments can harmonise with the properties of nature's systems which include the experiences of beauty, pleasure and friendship. Strange as it may seem, these experiences also may be part of the 'blueprint' for healthy living. For example, a survey of 7405 people by the Baker Medical Research Institute in Melbourne has revealed that pet owners are less likely to suffer heart attacks than people without animals, even though they drink more alcohol and are fatter.[26] Here we have evidence of animals being more than just another species competing for limited food resources. Animals are part of a deeper network of environmental experiences, involving interdependence, friendship, beauty and pleasure, all of which are in subtle ways essential for human health. Such experiences lie outside the anatomisations of the conventional mechanical paradigm, and are either distorted or lost altogether when they are made to conform to it.

Emeritus Professor Charles Birch from the University of Sydney sees purpose as providing 'a new way of thinking for the new millennium'. He points out that the mechanical model fails to recognise not only purpose but also beauty, feelings and the 'richness of life' which enshrines the potential for expanding purpose in terms of human goals and achievements.[27] Birch attributes the framework of his ideas to the writings of the polymath AN Whitehead, who wrote:

> Science can find no individual enjoyment in nature; science can find no aim in nature; science can find no creativity in nature; it finds mere rules of succession. These negations are true of natural science. They are inherent in its methodology. The reason for this blindness of physical science lies in the fact that such science only deals with half the evidence provided by human experience. It divides the seamless coat—or, to change the metaphor to a happier form, it examines the coat, which is superficial, and neglects the body which is fundamental.[28]

Traditionally in our schools we have separated the arts and the sciences but both are intrinsically interwoven aspects of the 'whole' of reality. Happiness and friendship, health and beauty cannot be dealt with in a complete sense as separate issues. It is time for our education programs to enable the artist to explore science, the scientist to discover art and together to search for the blueprint of nature within the fabric of the 'whole'. The great Albert Einstein expressed this need in these words:

> A human being is part of the whole, called by us 'Universe'; a part limited in time and space. He experiences himself, his thoughts and feelings as something separated from the rest—a kind of optical delusion of his consciousness. This delusion is a kind of prison for us, restricting us to our personal desires and to affection for a few persons nearest us. Our task must be to free ourselves from this prison by widening our circle of compassion to embrace all living creatures and the whole of nature in its beauty. Nobody is able to achieve this completely but the striving for such achievement is, in itself, a part of the liberation and a foundation for inner security.[29]

The survival of our planet and ourselves depends on the realisation that we are part of nature—part of a whole—that what we do affects the whole of nature, not just ourselves. The technological dominance of nature—when seen from this perspective—is little more than a Pyrrhic victory. In the end we simply undermine the very things upon which health depends. Our survival depends on us relinquishing our position as masters of the universe and assuming the more appropriate role as stewards. The task that challenges us now is to recognise that our persistent attempts to subjugate nature are ultimately futile. What we need to do is to reconceptualise our relationship with nature in such a way that we maximise our connective participation with it by becoming the hands through which its purposive unity is achieved. The boundary condition for our survival is thus the *'bounds of planetary purpose'*.

Conclusion

The crisis in health and the environment confronts us starkly. Despite the best efforts of medical and scientific technology, the increasing incidence of chronic diseases such as cancer, along with the accelerating deterioration of our environment, stand as reminders that the Utopia envisioned by our scientist forebears has not been achieved.

A closer examination of this crisis reveals that in many cases a deterioration in human health has been associated with technological progress. The industrial revolution which enabled the processing of nature's resources for profit on an ever increasing scale, resulted in vast changes to the human habitat. This new environment which contains the products of industry such as plastics, pesticides and processed foods, also contains the wastes of industry in the form of carbon dioxide, chlorine compounds, and a vast number of toxic chemicals which are either released into the air or dumped in the waterways and landfills.

While we have come to accept that breathing polluted air, drinking contaminated water and eating altered foods are part of the price of the conveniences of modern living upon which we have come to rely, the inflationary cost in terms of human health is only beginning to be realised. In the diverse range of examples discussed in previous chapters we have endeavoured to show, by way of cumulative argument, that the advancement of public health is not an issue separate from the preservation of ecological integrity. Health and the environment are inextricably interconnected and those things we do which disrupt the harmony of nature inevitably disrupt our own harmony.

In the minds of some, this crisis in health serves simply to reinforce even more strongly the view that nature is a threatening force and that greater technological intervention, in terms of control and resynthesisation of our environment, is required to secure a safe and healthy state of existence. It has been the central contention of this book, however, that this view of nature, far from being the panacea of our ecological crisis, lies at the very root of it. Construing nature as an adversary to be dominated, Western society has evolved a theory of power over nature, which has shaped both the aims and practices of science.

Through the misuse of this power and by blindly implementing changes in nature whose disruptive consequences we have failed to anticipate, we have brought ourselves to the brink of destruction. Motivated by the notion that unlimited economic growth is the mechanism whereby the goal of human happiness can be achieved, we have sought to maximise through the application of technology not only what we produce from the resources of nature but also what we discard back into our environment. Nature is not endlessly exploitable and 'progress' in the guise of unlimited economic growth has, we observed, resulted in the inevitable degradation of the environment and the promulgation of disease.

While the mechanical world view with its reductionist approach has given us a certain amount of control over nature, it has at the same time failed to equip us with the appropriate knowledge to exercise discriminate use of this power.

We have seen for example, that artificial lighting tuned to provide visual efficiency has been accepted as a major advancement of modern technology. Mechanical knowledge has enabled the construction of steam engines, electric generators, power lines and light globes together with the utilisation of these inventions to produce artificial lighting. However this knowledge has not empowered us to see at the same time other consequences of these inventions such as acid rain and the greenhouse effect from the use of fossil fuels, cancer from the exposure to the electromagnetic fields of powerlines and the adverse effects on diseases such as dental caries, carbohydrate craving and heart disease, of the artificial light itself.

We have argued that an interpretation of current knowledge about our environment suggests that nature is ordered to a pattern for the preservation of the integrity of the living ecosystems of our planet—and human health. By way of examples drawn from a range of current environmental health issues we have demonstrated that the harmony of this order within nature has been disrupted by uninformed technological intervention. The natural occurrence and chemistry of the toxic metals, cadmium, lead and aluminium, seems to be such as to minimise their uptake into living organisms. In nature, the bioavailabilty of poisonous fluoride is minimised by the low solubility of its calcium and magnesium salts together with inbuilt mechanisms in plants and animals to limit absorption. In the case of chlorine, the free gaseous form does not occur in the environment to any significant extent. We have argued that the employment of these element in forms and at concentrations other than those dictated by nature has proved to be inimical to health. We have observed in previous chapters that the use of chlorine gas to treat drinking water, the fluoridation of soft water to minimise tooth decay, the inclusion of aluminium compounds in processed foods and water supplies, and the addition of lead to petrol have each resulted in increased human disease and disability.

The constituents of whole foods as they occur in Nature seem tuned to provide not only the nutritional requirements of our bodies but also protection against many forms of illness. We have seen that the processing of these foods destroys this nexus thereby increasing our vulnerability to disease. We have also argued that the progressive synthesisation of our surroundings in urban environments has meant an increased exposure to artificial light and a corresponding diminished exposure to natural daylight. As regular exposure to sunlight plays a crucial role in the maintenance of our biochemical mechanisms it has been argued that reduced exposure in the name of progress has again proved to be more of a health problem than was anticipated. One way of putting the matter is to say that we have sought to transform nature without adequately discerning the extent to which patterns in nature provide clues to our most fruitful scientific interactions with it.

An essential part of the solution to our crisis in health and the environment is to exchange our insatiable desire to subdue and control nature in the name of progress for a new consciousness in which we bond ourselves as the partners and stewards of nature. An education system which empowers us to overcome the limits of nature must, at the same time, provide us with an appreciation of why these limits exist. Education must be seen, not as the vehicle which will bring us to techno-Utopia, but rather as a source of enlightenment out of which emerges a new consciousness of the value of every living thing on our planet.

Nature provides clues for maximising healthy interaction with it, in every area of health and the environment. This is as a way of piecing together our knowledge of synergy in nature. The idea of having technology which recognises the limits of nature provides a basis for discovering how far our transformations of nature may extend without harming our environment and our health. This perspective gives direction for new research in every domain of science.

We propose that the pathway back to health lies in reconceptualising our relationship with nature by virtue of whole new understanding of the relations which exist within it.

ENDNOTES

CHAPTER 1

1. *Nikola Tesla: Lectures, Patents, Articles*, Nikola Tesla Museum, Beograd, 1956, p. LI.
2. B. Bernstein, 'The Unsung Father of the A–Bomb', *Discover*, August 1985, pp. 37–42. See also: G. Farmelo, 'The day Chicago went critical', *New Scientist*, 28 November 1992, pp. 26–9.
3. F. Graham, *Since Silent Spring*, Hamish Hamilton, London, 1970, p. 205.
4. A. Howard, *An Agricultural Testament*, Oxford University Press, New York, 1943, pp. 181–2.
5. R. Dubos, *Mirage of Health*, Harper and Row, New York, 1959, p. 147.
6. J. Bellini, *High Tech Holocaust*, Greenhouse Publications, Richmond, Victoria, 1987, p. 8.
7. Ibid., pp. 11–12.
8. A.C. Giese, *Living with Our Sun's Ultraviolet Rays*, Plenum Press, New York, 1976, p. 20.
9. A. Goudie, *The Human Impact on the Natural Environment*, Basil Blackwell, Oxford, 1986, p. 253.
10. S. Woodin and U. Skiba, 'Liming fails the acid test', *New Scientist*, 10 March 1990, pp. 30–4.
11. M. Bond, 'Dirty ships evade acid rain controls', *New Scientist*, 22 June 1996, p. 4.
12. Ibid.
13. T. Patel, 'Power station failure adds to Delhi's woes', *New Scientist*, 12 October 1996, p. 8.
14. Ibid.
15. J. Heicklen, 'The Formation and Inhibition of Photochemical Smog', *Annals of the New York Academy of Sciences,* vol 502, 1987, p. 145.
16. The Conservation Foundation, *State of the Environment: A View towards the Nineties*, Washington DC, 1987, pp. 78–9. See also: J. Bellini, op. cit., pp. 64–5.
17. 'Europe's trees still dying', *New Scientist*, 2 December 1989, p. 9. See also: L.R. Brown, *State of the World 1990*, Allen and Unwin, Sydney, 1990, p. 107–9.
18. T. Patel, 'Killer smog stalks the boulevards', *New Scientist*, 15 October 1994, p. 8.
19. T. Patel, 'French smog smothers hundreds', *New Scientist*, 17 February 1994, p. 7.
20. C. Read, 'Even low levels of ozone in smog harm the lungs', *New Scientist*, 9 September 1989, p. 14. See also: L.J. Folinsbee, 'Human health effects of air pollution', *Environmental Health Perspectives*, vol. 100, 1992, pp. 45–56.
21. M.G. Mustafa, C.M. Hassett, et al., 'Pulmonary carcinogenic effects of ozone', *Annals of the New York Academy of Sciences*, vol. 534, 1988, pp. 714–21. See also: R. Edwards, 'Ozone alert follows cancer warning', *New Scientist*, 27 May 1995, p. 4.
22. 'Americans overdose on ozone', *New Scientist*, 22 July 1989, p. 5. See also: C. Vaughan, 'Streetwise to the dangers of ozone', *New Scientist*, 26 May 1990, pp. 42–5.

23. R. Edwards, 'Ozone alert follows cancer warning', *New Scientist*, 27 May 1995, p. 4.
24. The Conservation Foundation, op. cit., p. 71.
25. M. Lippmann and R.B. Schlesinger, *Chemical Contamination in the Human Environment*, Oxford University Press, New York, 1979, p. 187.
26. 'Time to get the lead out?', *Choice*, Sydney, September 1980, p. 267.
27. A. Motluk, 'Lead blights the future of Africa's children', *New Scientist*, 23 March 1996, p. 6.
28. D. Bellinger, A. Leviton, C. Waternaux, et al., 'Longitudinal analysis of prenatal and postnatal lead exposure and early cognitive development', *NewEngland Journal of Medicine*, vol. 316, 1987, pp. 1037–43.
29. J.M. Davis, 'Risk assessment of the development neurotoxicity of lead', *Neurotoxicology*, vol. 11, 1990, p. 285–92.
30. J. Schwartz and D Otto, 'Blood lead, hearing threshholds and neuro–behavioural development in children and youth, *Archives of Environmental Health*, vol. 42, 1987, pp. 153–60.
31. R.A. Goyer, 'Lead toxicity: Current concerns', *Environmental Health Perspectives*, vol. 100, 1993, pp. 177–87.
32. Ibid.
33. M. Day, 'Past lead pollution poisons unborn child', *New Scientist*, 22 June 1996, p. 4.
34. C. Joyce, 'Lead Poisoning "lasts beyond childhood"', *New Scientist*, 13 January 1990, p. 4. See also: L.R. Ember, 'Environmental lead: insidious health problem', *Chemical and Engineering News*, 23 June 1980, pp. 28–34; Worldwatch Institute, 'Global choking', *New Scientist*, 27 January l990, p. 9.
35. A. Motluk, op. cit.
36. Ibid.
37. F. Pearce, 'A heavy responsibility', *New Scientist*, 27 July 1996, pp. 12–13.
38. Ibid.
39. J. Donovan, 'No lead is good lead', *Medical Journal of Australia*, vol. 164, 1996, pp. 390–1.
40. J.L. Pirkle, D.J. Brody, E.W. Gunter, et al., 'The decline in blood lead levels in the United States: the National Health and Nutrition Examination Surveys (NHANES), *Journal of the American Medical Association*, vol. 272, 1994, pp. 284–91.
41. D.J. Paustenbach, R.D. Bass and P. Price, 'Benzene toxicity and risk assessment, 1972–1992: Implications for future regulation', *Environmental Health Perspectives Supplements*, 1993, vol. 101 (suppl. 6), pp. 177–200. See also: G.K. Raabe, 'Review of the carcinogenic potential of gasoline', *Environmental Health Perspectives Supplements*, 1993, vol. 101 (suppl. 6), pp. 35–8.
42. W. Brown, 'Dying from too much dust', *New Scientist*, 12 March 1994, pp. 12–13.
43. M. Hamer, 'Cars must go to meet clean air target', *New Scientist*, 18 May 1996, p. 12.
44. W. Brown, op. cit.
45. T. Patel, 'France counts the cost of cheap diesel', *New Scientist*, 22 April 1995, p. 10.
46. R. Edwards, 'Drivers risk cancer in the cab', *New Scientist*, 26 October 1996, p. 8.

47. M. Hamer, 'Urban nightmare drives Americans out of control', *New Scientist*, 14 September 1996, p. 5.
48. J. Gribbin and M. Gribbin, 'The Greenhouse Effect', *New Scientist* (Inside Science), 13 July 1996, pp. 1–4.
49. E.K. Jackson, C.S. Guest and A.J. Woodward, 'Climate, health and medicine in a changing world', *Medical Journal of Australia*, vol. 165, 1996, pp. 597–8.
50. F. Pearce, 'World lays offs on global catastrophe', *New Scientist*, 8 April 1995, pp. 4–5.
51. J. Gribbin and M. Gribbin, op. cit.
52. J. Gribbin and M. Gribbin, op. cit.
53. F. Pearce, 'Fiddling while earth warms', *New Scientist*, 25 March 1995, pp. 14–15.
54. D. Mackenzie, 'Polar meltdown fulfils worst predictions', *New Scientist*, 12 August 1995, p. 4.
55. Ibid.
56. B. Hileman, 'Scientists warn that disease threats increase as earth warms up', *Chemical and Engineering News*, 2 October 1995, pp. 19–20.
57. Ibid.
58. Ibid.
59. P. Aldhous, 'Global heat breeds superaphids', *New Scientist*, 23/30 December 1995, p. 17.
60. J. Burton, 'Global warming', *Chemical and Engineering News*, 4 December 1995, p. 5.
61. F. Pearce, 'Carbon targets up in the air', *New Scientist*, 6 July 1996, p. 9.
62. E.K. Jackson, C.S. Guest and A.J. Woodward, op. cit.
63. J. Burton, op. cit.
64. J. Sarmiento, et al., *Science*, vol. 274, 30 November 1996, p. 1346.
65. The Conservation Foundation, op. cit., pp. 247–8.
66. J. Bellini, op. cit., p. 61.
67. Ibid., pp. 59–97.
68. J. Mullins, 'The case of the missing core', *New Scientist*, 20 April 1996, pp. 41–3.
69. R. Edwards, 'Will it get any worse?', *New Scientist*, 9 December 1995, pp. 14–15.
70. V. Rich, 'Concern grows over health of "Chernobyl children"', *New Scientist*, 21 April 1990, p. 7. See also: S. Borman, 'Chemists reminisce on 50th anniversary of the atomic bomb', *Chemical and Engineering News*, 17 July 1995, pp. 53–63.
71. G. Luning, J. Scheer, et al., 'Early infant mortality in West Germany before and after Chernobyl', *The Lancet*, vol. ii, 1989, pp. 1081–3.
72. R. Edwards, 'Will it get any worse?', op. cit.
73. T. Patel, 'Iodine first aid is not enough', *New Scientist*, 27 April 1996, p. 7.
74. R. Edwards, 'Terrifying outlook for Chernobyl's babies', *New Scientist*, 2 December 1995.
75. T. Patel, 'Iodine first is not enough', op. cit.
76. R. Edwards, 'Will it get any worse', op. cit.
77. D. Trichopoulos, *Nature*, 27 July 1996 (on leukaemia in Greek children).
78. R. Edwards, 'Will it get any worse', op. cit.
79. Ibid.
80. Ibid.

81. Y.E. Dubrova, V.N. Nesterov, N.G. Krouchinsky, et al., 'Human minisatellite mutation rate after the Chernobyl accident', *Nature,* 25 April 1996, pp. 683–6.
82. R. Edwards, 'Radiatioactive forests will bring power to the people', *New Scientist,* 26 October 1996, p. 9.
83. R. Edwards, 'Chernobyl floods put millions at risk', *New Scientist,* 23 March 1996, p. 4.
84. Ibid.
85. T. Patel, 'Clash over radiation safety limits', *New Scientist,* 2 December 1995, p. 10.
86. V. Beral, 'Leukaemia and nuclear installations', *British Medical Journal,* vol. 300, 17 February 1990, pp. 411–12.
87. 'Cancers abound near Soviet Union's nuclear test site', *New Scientist,* 2 September 1989, p. 5.
88. 'Nuclear weapons industry "swept health risks under the carpet"', *New Scientist,* 6 January 1990, p. 8.
89. F. Graham, op. cit., p. 205.
90. M. Lippmann and R.B. Schlesinger, op. cit., p. 77.
91. P. Board, 'Contaminated lands', *New Scientist* (Inside Science), 12 October 1996, pp. 1–4. See also: L.B. Lave and A.C. Upton, *Toxic Chemicals, Health and the Environment,* Johns Hopkins University Press, Baltimore, 1987, p. 81.
92. M. Keynes, 'Chemical industry and the environment: A view from history', *Chemistry in Australia,* June 1996, pp. 265–6.
93. P.L. Layman, 'Global top 50 chemical producers show rise in profits and sales', *Chemical and Engineering News,* 24 July 1995, pp. 23–4.
94. D. Hanson, 'Toxics release inventory: Chemical industry again cuts emissions', *Chemical and Engineering News,* 3 April 1995, pp. 4–5.
95. Ibid.
96. M. Szabo, 'New Zealand's poisoned paradise', *New Scientist,* 31 July 1993, pp. 29–33.
97. Ibid.
98. J.J. Mennear, E.E. McConnell, et al., 'Inhalation toxicology and carcinogenesis studies of methylene chloride (dichloromethane) in F 344/N rats and B6C3F mice', *Annals of the New York Academy of Sciences,* vol. 534, 1988, p. 343.
99. Ibid., p. 348.
100. Ibid., pp. 349–50.
101. G. Clews, 'Chlorine in perspective', *Chemistry in Australia,* January 1996, pp. 45–6.
102. P. Flinn, 'Values of Chlorine to the Australian Economy', *Chlorine in Perspective,* CSIRO Division of Chemicals and Polymers, Melbourne, 1996.
103. D. MacKenzie, 'Clinton backs call to ban chlorine' *New Scientist,* 12 February 1994, p. 10.
104. Ibid.
105. Ibid.
106. Euro Chlor, 'The natural chemistry of chlorine in the environment', *Euro Chlor,* Brussels, 1995.
107. G.W. Gribble, 'The natural production of chlorinated compounds', *Environmental Science and Technology,* vol. 28, no. 7, 1994, pp. 310A–19A.
108. G.W. Gribble, 'Organochlorine compounds', *Chemical and Engineering News,* 13 February 1995, pp. 4–5.
109. D. MacKenzie, op. cit.

110. T.G. Spiro and V.M. Thomas, 'Organochlorine compounds', *Chemical and Engineering News*, 13 February 1995, p. 5.
111. Ibid.
112. K. Kleiner, 'Dioxin authors stand firm against critics', *New Scientist*, 23/30 December 1995, p. 10.
113. E. Longstaff, 'Carcinogenic and mutagenic potential of several fluorocarbons', *Annals of the New York Academy of Sciences*, vol. 534, 1988, p. 297.
114. D. MacKenzie, 'Ozone's future is up in the Air', *New Scientist*, 16 December 1995, pp. 14–15. See also: 'The Greenhouse Effect', *New Scientist*, 15 July 1989, p. 16.
115. F.S. Rowland and M.J. Molina, 'Chlorofluoromethanes in the environment', *Reviews of Geophysics and Space Physics*, 1975, vol. 1, p. l.
116. R.A. Plumb, 'Atmospheric ozone—physics and chemistry', *Transactions of the Menzies Foundation*, 1989, vol. 15, pp. 3–13.
117. See for example National Health and Medical Research Council, *Health Effects of Ozone Layer Depletion*, Australian Government Publishing Service, Canberra, 1989. See also: M.L. Kripke, 'Overview: health consequences of stratospheric ozone depletion', *Transactions of the Menzies Foundation*, 1989, vol. 15, pp. 15–20; G.J. Kelly, 'Effects of W–B radiation on terrestrial planets, ibid., pp. 187–90.
118. W.F. Wood, 'The effects of ultraviolet radiation on the marine biota', *Transactions of the Menzies Foundation*, 1989, vol. 15, pp. 179–85.
119. Ibid.
120. A. Duncan and G. Simpson, 'Compounds containing chlorine: How do we decide what is good for us?' *Chemistry in Australia*, January 1996, pp. 41–2.
121. A. Hinwood, 'The Role of Scientist in the Montreal Protocol', *Chlorine in Perspective*, CSIRO Division of Chemical and Polymers, Melbourne, 1996.
122. D. MacKenzie, 'Ozone's future is up in the air', op. cit.
123. D. MacKenzie, 'Rich and poor split over ozone' *New Scientist*, 9 December 1995, p. 5.
124. P. Zurer, 'Satellite data confirm CFC link to ozone hole', *Chemical and Engineering News*, 2 January 1995, p. 9. See also: A. Tabazadah and R. Turco, *Science*, vol. 260, 1993, p. 1082; and J.R. Clark, 'Hydrogen fluoride sources', *Chemical and Engineering News*, 13 February 1995, p. 5.
125. A. Howard, op. cit., p. ix.
126. E.R. Riegel, *Industrial Chemistry*, Reinhold, New York, 1937, p. 147.
127. Ibid., p. 149.
128. F. Graham, op. cit., p. 223.
129. C. Will, 'Safety in diversity ', *New Scientist*, 23 March 1996, pp. 38–42; see also: E.O. Wilson, 'Threats to biodiversity', *Scientific American*, September 1989, pp. 60–6.
130. R. Edwards, 'Tomorrow's bitter harvest', *New Scientist*, 17 August 1996, p 14.
131. See for example National Research Council, *Alternative Agriculture*, National Academy Press, Washington DC, 1989, pp. 40–8; The Conservation Foundation, op. cit., pp. 143–8, 350–3.
132. L. Wickenden, *Our Daily Poison*, Hillman Books, New York, 1961, p. 36.
133. A. Howard, op. cit., p. 30.
134. Ibid., p. 19.

135. See for example BC Nicholson, *Australian Water Quality Criteria for Organic Compounds*, Australian Water Resources Council Technical Paper no. 82, 1984, pp. 214–15.
136. The Conservation Foundation, op. cit., p. 351.
137. Ibid., p. 146; see also: National Research Council, op. cit., pp. 42–8.
138. The Conservation Foundation, op. cit., p. 350.
139. Ibid., p. 352.
140. J. Reisner, 'Nader's New Beef', *Food Business*, 5 June 1989, p. 30.
141. R. Carson, *Silent Spring*, Hamish Hamilton, London, 1963, pp. 218–22; see also: 'Pesticides—the pests are fighting back', *Ecos*, August 1978, pp. 3–10.
142. S.K. Nigam and A.B. Karnik, et al., 'Experimental and human surveillance on BHC and DDT insecticides commonly used in India', *Annals of the New York Academy of Sciences*, 1988, vol. 534, p. 694.
143. A. Howard, op. cit., pp. 171–80.
144. See for example, R. Taylor and M.J. Nobbs, *Hunza the Himalayan Shangri–La*, Whitehorn Publishing Co., El Monte, California, 1962.
145. E.G. Toomey and PD White, 'A brief survey of the health of aged Hunzas', *American Heart Journal*, 1964, vol. 68, no. 6, pp. 841–2.
146. S. Hedges 'Compost is a natural–born killer', *New Scientist*, 21 September 1996, p. 25.
147. J. Seymour, 'Hungry for a new revolution', *New Scientist*, 30 March 1996, pp. 32–7; see also: R Dubos, op. cit., pp. 151–64.
148. D. Pimentel and C.A. Edwards, 'Pesticides and ecosystems', *Bio Science*, 1982, vol. 32, pp. 595–600; see also: US Congress, Office of Technology Assessment, *Impacts of Technology on US Cropland and Rangeland Productivity*, US Government Printing Office, Washington DC, August 1982, pp. 226–50; N. Senesi and TM Miano (eds), *Humic Substances in the Global Environment and Implications on Human Health*, Elsevier, New York, 1994.
149. J.R. Smith, *Tree Crops: A Permanent Agriculture*, Devin-Adair Co., Old Greenwich, 1977 (first published 1950).
150. R. Taylor, 'What's causing sugar yields to fall', *Rural Research*, vol. 173, Summer 1996/1997, pp. 25–6.
151. I. Anderson, 'Australia's growing disaster', *New Scientist*, 29 July 1995, pp. 12–13; see also: J. Reganold, 'Farming's organic future', *New Scientist*, 10 June 1989, p. 314.
152. State of the Environment Advisory Council, *Australia State of the Environment 1996*, CSIRO Publishing, Collingwood, Australia, 1996, pp. 6–26.
153. The Conservation Foundation, op. cit., pp. 361–6; see also: National Research Council, op. cit., pp. 99–109.
154. The Conservation Foundation, op. cit., p. 351.
155. See for example O. Strebel, W.H.M. Duynisveld and J. Bottcher, 'Nitrate pollution of groundwater in Western Europe', *Agriculture Ecosystems and Environment*, vol. 26, no. 3/4, 1989, pp. 189–214; J.C. Germon (ed.), *Management Systems to Reduce Impact of Nitrates*, Elsevier Applied Science Publishers, Barking, UK, 1989.
156. H.D. Junge and S. Handke, 'Nitrate in vegetables—unavoidable risk? Cultivation of new varieties offers some assistance', *Industrielle Obst–und Gemüseverwertung*, 1987, vol. 71, no. 8, pp. 346–8.

157. G.R. Conway and J.N. Pretty, 'Fertiliser risks in the developing countries', *Nature*, 21 July 1988, pp. 207–8.
158. P.A. Steudler, R.D. Bowden, et al., 'Influence of nitrogen fertilisation on methane uptake in temperate forest soils', *Nature*, September 1989, vol. 341, pp. 314–15.
159. S. Pain, 'Pollutants collect on continental side of North Sea, study finds', *New Scientist*, 11 November 1989, p. 6; see also: F. Pearce, 'Sea life sickened by urban pollution', *New Scientist*, 17 June 1995, p. 4.
160. A. Bernatzky, 'Ecological principles in town planning: the impact of vegetation on the quality of life in the city', in R. Krieps (ed.), *Environment and Health: A Holistic Approach*, Avebury, Aldershot, England, 1989, pp. 136–45.
161. Ibid.

CHAPTER 2

1. US Congress, Office of Technology Assessment, *Biological Effects of Power Frequency Electric and Magnetic Fields—Background Paper*, OTABP–E–53, US Government Printing Office, Washington DC, May 1989, pp. 4–5.
2. R.C. Weast and M.J. Astle (eds), CRC *Handbook of Chemistry and Physics*, 61st edition, CRC Press, Boca Raton, Florida, 1980, pp. F–214–15.
3. J.H. Bernhardt, 'Extremely low frequency (ELF) electric fields', *Non–Ionizing Radiations: Physical Characteristics, Biological Effects and Health Hazard Assessment*, International Radiation Protection Association, Yallambie, Victoria, 1988, p. 236.
4. National Institute of Environmental Health Sciences and US Department of Energy, *Questions and Answers about Electric and Magnetic Fields associated with the Use of Electric Power*, US Government Printing Office, Washington DC, 1995, p. 46.
5. R. Wever, 'ELF-Effects of Human Circadian Rhythms', *VHF Electromagnetic Field Effects*, Plenum Press, New York, 1974, pp. 101–44.
6. R.G. Olsen, 'The magnetic field environment of electric power lines', *Panel Session on Biological Effects of Power Frequency Elecrtic and Magnetic Fields*, IEEE, New York, 1986, p. 8.
7. US Congress, op. cit., p. 15.
8. W.T. Kaune, R.G. Stevens, et al., 'Residential magnetic and electric fields', *Bioelectromagnetics*, vol. 8, 1987, pp. 315–35.
9. National Institute of Environmental Health Sciences, op. cit., pp. 36–7; see also: V. Delpizzo, 'Wire coding as an indicator of residential magnetic field exposure', *Carcinogenic Potential of Extremely Low Frequency Magnetic Fields*, Australian Radiation Laboratory, Yallambie, Victoria, 1989, p. 107.
10. R.J. Caola, D.W. Deno and V.S.W. Dymek, 'Measurements of electric and magnetic fields in and around homes near a 500 kV transmission line', *IEEE Transactions on Power Apparatus and Systems*, 1983, vol. PAS–102, no. 10, p. 3343; see also: N. Wertheimer, D.A. Savitz and E. Leeper, 'Childhood cancer in relation to indicators of magnetic fields from ground current sources', *Bioelectromagnetics*, 1995, vol. 16, no. 2, pp. 86–96.
11. National Institute of Environmental and Health Sciences, op. cit., p. 46.
12. L.J. Dlugosz, T. Byers, et al., 'Ambient 60–Hz magnetic flux density in an urban neighborhood', *Bioelectromagnetics*, 1989, vol. 10, p. 195.
13. Ibid.; see also: V. Delpizzo, op. cit., pp. 104–16.

14. T.E. Aldrich and C.E. Easterly, 'Electromagnetic fields and public health', *Enviromental Health Perspectives*, 1987, vol. .75, pp. 159–63.

15. S.M. Bawin and W.R. Adey, 'Sensitivity of calcium binding in cerebral tissue to weak environmental electric fields oscillating at low frequency', *Proceedings of the National Academy of Sciences USA*, 1976, vol. 73, no. 6, pp. 1999–2003.

16. W.R. Adey, 'Frequency and power windowing in tissue interactions with weak electromagnetic fields', *Proceedings of the IEEE*, 1980, vol. 68, no. 1, pp. 119–25.

17. US Congress, op. cit., p. 28.

18. J.M.R. Delgado, J. Leal, et al., 'Embryological changes induced by weak, extremely low frequency electromagnetic fields', *Journal of Anatomy*, 1982, vol. 134, no. 3, pp. 533–51; see also: A. Ubeda, J. Leal, et al., 'Pulse shape of magnetic field influences chick embryogenesis', *Journal of Anatomy*, 1983, vol. 137, no. 3, pp. 513–36.

19. W.R. Adey, 1980, p. 121; see also: US Congress, op. cit., p. 49.

20. B.W. Wilson, E.K. Chess and L.E. Anderson, '60 Hz electric-field effects on pineal melatonin rhythms: Time course for onset and recovery', *Bioelectromagnetics*, 1986, vol. 7, pp. 239–42; see also: G. Cremer–Bartels, et al., 'Magnetic field of earth as additional zeitgeber for endogenous rhythms? *Naturwissenschaften*, (November) 1984, vol. 71, no. 11, pp. 567–74.

21. See also B. Homberg, 'Magnetic fields and cancer', *Environmental Health Perspectives*, 1995, vol. 103, suppl. no. 2, pp. 63–7; see also: US Congress, op. cit., p. 68.

22. J.G. Llaurado, 'Harmful effects of electromagnetic fields: myth or reality', *International Journal of Bio–Medical Computing*, 1985, vol. 17, pp. 3–4.

23. D.B. Lyle, R.D. Ayotte, et al., 'Suppression of T–lymphocyte cytotoxicity following exposure to 60–Hz sinusoidal electric fields', *Bioelectromagnetics*, 1988, vol. 9, pp. 303–13; see also: US Congress, op. cit., pp. 30–1.

24. N. Wertheimer and E. Leeper, 'Electrical wiring configurations and childhood cancer', *American Journal of Epidemiology*, 1979, vol. l09, no. 3, p. 273.

25. Ibid.

26. L. Tomenius, '50–Hz electromagnetic environment and the incidence of childhood tumours in Stockholm County', *Bioelectromagnetics*, 1986, vol. 7, pp. 191–207.

27. T.S. Tenforde, 'Biological effects of extremely-low-frequency magnetic fields', *Panel Session on Biological Effects of Power Frequency Electric and Magnetic Fields*, IEEE, New York, 1986, pp. 32–4.

28. N. Wertheimer and E. Leeper, 'Adult cancer related to electrical wires near the home', *International Journal of Epidemiology*, 1982, vol. 11, no. 4, p. 354.

29. M.N. Bates, 'Extremely low frequency electromagnetic fields and cancer: The epidemiologic evidence', *Environmental Health Perspectives*, 1991, vol. 95, pp. 147–56.

30. D.L. Davis, D. Hoel, J. Fox and A. Lopez, 'International trends in cancer mortality in France, West Germany, Italy, Japan, England and Wales, and the USA', *Lancet*, 1990, vol. 336, pp. 474–81; see also: D.A. Savitz and D.P. Loomis, 'Magnetic field exposure in relation to leukemia and brain cancer mortality among electric utility workers', *American Journal of Epidemiology*, 1995, vol. 141, pp. 1–12.

31. National Institute of Environmental Health Sciences and US Department of Energy, *Questions and Answers About Electric and Magnetic Fields Associated*

with the Use of Electric Power, US Government Printing Office, Washington DC, 1995, pp. 12–14, 19.

32. N.T. Fear, E. Roman, L.M. Carpenter, et al., 'Cancer in electrical workers: an analysis of cancer registrations in England, 1981–1987', *British Journal of Cancer*, 1996, vol. 73, pp. 935–9.

32a. P. Guenel, J. Nicolau, E. Imbernon, et al., 'Exposure to 50hz electric field and incidence of leukemia, brain tumors and other cancers among French electric utility workers,' *American Journal of Epidemiology*, 1996, vol. 144, pp. 1107–21.

33. D.P. Loomis, D.A. Savitz and C.V. Ananth, 'Breast cancer mortality among female electrical workers in the United States', *Journal of the National Cancer Institute*, 1994, vol. 86, no. 12, pp. 921–5.

34. J. Stather, 'Case unproved', *New Scientist*, 11 November 1995, p. 52.

35. E. Sobel, Z. Davanipur, R. Sulkava, et al., 'Occupations with exposure to electromagnetic fields: a possible risk factor for Alzheimer's disease', *American Journal of Epidemiology*, 1995, vol. 142, pp. 515–24.

36. E. Sobel, M. Dunn, Z. Davanipour, et al., 'Elevated risk of Alzheimer's disease among workers with likely electromagnetic field exposure', *Neurology*, December 1996, vol. 47, pp. 1477–81.

37. M. Reichmanis, F.S. Perry, A.A. Marino and O. Becke, 'Relation between suicide and electromagnetic field of overhead powerlines', *Physiology Chemistry and Physics*, 1979, vol. 11, pp. 395–403.

38. S.F. Perry, M. Reichmanis, A.A. Marino and R.O. Becher, 'Environmental power frequency magnetic fields and suicide', *Health Physics*, 1981, vol. 41, pp. 267–77.

39. S.F. Perry and L. Pearl, 'Power frequency magnetic field and illness in multi-storey blocks' *Public Health*, 1988, vol. 102, pp. 11–18.

40. D.I. Dowson and G.T. Lewith, 'Overhead high voltage cables and recurrent headache and depression', *Practitioner*, 1988, vol. 232, p. 22.

41. C. Poole, R. Kavet, D.P. Funch, et al., 'Depressive symptoms and headaches in relation to proximity of residence to alternating-current transmission line right of way', *American Journal of Epidemiology*, 1993, vol. 137, pp. 318–30.

42. D. Baris, B.G. Armstrong, J. Deadman and G. Therialt, 'A case cohort study of suicide in relation to exposure to electric and magnetic fields among electrical utility workers', *Occupational and Environmental Medicine*, 1996, vol. 53, pp. 17–24.

43. B.W. Wilson, E.K. Chess and L.E. Anderson, '60–Hz electric fields effects on pineal melatonin rhythms: time course for onset and recovery', *Bioelectro-magnetics*, 1986, vol. 7, pp. 239–42.

44. B.W. Wilson, 'Chronic exposure to EMF fields may induce depression', *Bioelectromagnetics*, 1988, vol. 9, pp. 195–205.

45. M. Monk, 'Epidemiolgy of suicide', *Epidemiology Reviews*, 1987, vol. 9, pp. 51–69.

46. D. Concar, 'Happiness is a magnet', *New Scientist*, 5 August 1995, pp. 24–9.

47. National Institute of Environmental Health Science, op. cit, p. 27.

48. Ibid., p. 26.

49. R. Edwards, 'Leak links power lines to cancer', *New Scientist*, 7 October 1995, p. 4; see also: 'Draft NCRP report seeks strong action to curb EMFs', *Microwave News*, 1995, vol. 15, no. 4, pp. 1, 11–15.

50. I. Lowe, 'Power line radiation standards under fire'. *New Scientist*, 18 November 1995, p. 74.
51. J. Firshein, 'US panel says electromagnetic fields not harmful', *Lancet*, vol. 348, 9 November 1996, p. 1305.

CHAPTER 3

1. W. Volkrodt, 'Electromagnetic pollution of the environment', in R. Krieps (ed.), *Environment and Health: A holistic approach*, Avebury, Aldershot, England, 1989, p. 71.
2. P. Brodeur, *The Zapping of America*, WW Norton and Company, New York, 1977, p. 13.
3. K.R. Foster and A.W. Guy, 'The microwave problem', *Scientific American*, September 1986, pp. 28–35.
4. W. Volkrodt, op. cit., p. 74.
5. P.J. Dimbylow and O.P. Gandhi, 'Finite-difference time-domain calculations of SAR in a realistic heterogeneous model of the head for plane-wave exposure from 600 MHz to 3 GHz', *Phys Med Biol*, 1991, vol. 36, pp. 1075–89.
6. V. Anderson and K.H. Joyner,'Specific absorption rate levels measured in a phantom head exposed to radio frequency transmissions from analog hand-held mobile phones', *Bioelectromagnetics*, 1995, vol. 16, pp. 60–9.
7. F. Fraschini, D. Esposti, P. Lissoni, et al., 'The pineal function in cancer patients', in D. Gupta, A. Attanasio, R.J. Reiter (eds), *The pineal gland and cancer*, Brain Research Promotion, Tubingen, West Germany, 1988, pp. 157–66; see also: D. Danforth, L. Tamarkin, B. Chabner, et al., 'Altered diurnal secretory pattern of melatonin in patients with estrogen receptor positive breast cancer', *Clinical Research*, 1982, vol. 30, p. 532A.
8. R.J. Reiter, 'Melatonin suppression by static and extremely low frequency electromagnetic fields. relationship to the reported increased incidence of cancer', *Reviews on Environmental Health*, 1994, vol. 10, nos 3–4, pp. 171–86.
9. I. De Guire, 'Increased malignant melanoma of the skin in workers in a telecommunications industry', *British Journal of Industrial Medicine*, 1988, vol. 45, pp. 824–8.
10. R.J. Wurtman, 'Effects of light on the human body', *Scientific American,* July 1975, pp. 69–77.
11. L. Hardell, B. Holmberg, H. Malker and L.E. Paulsson, 'Exposure to extremely low frequency electromagnetic fields and the risk of malignant diseases: An evaluation of epidemiological and experimental findings', *European Journal of Cancer Prevention*, 1995, vol. 4, suppl. 1, pp. 3–107.
12. D.A. Savitz, 'Overview of occupational exposure to electric and magnetic fields and cancer: advancements in exposure assessment', *Environmental Health Perspectives*, 1995, vol. 103, suppl. 2, pp. 69–74.
13. R.P. Liburdy, T.F. Sloma, R. Sokolic and P. Yaswen, 'ELF magnetic fields, breast cancer and melatonin: 60 Hz fields block melatonin's oncostatic action on ER breast cancer cell proliferation', *Journal of Pineal Research*, 1993, vol. 14, pp. 89–97.
14. F.C. Garland, C.F. Garland, E.D. Gorham and J.F. Young, 'Geographic variation in breast cancer mortality in the United States: A hypothesis involving exposure to solar radiation', *Preventive Medicine*, 1990, vol. 19, pp. 614–22.

15. J.A.G. Holt, 'Changing epidemiology of malignant melanoma in Queensland', *The Medical Journal of Australia*, 1980, vol. 1, no. 12, pp. 619–20.
16. J.A.G. Holt, 'The cause of cancer: Biochemical defects in the cancer cell demonstrated by the effects of electromagnetic radiation, glucose and oxygen', *Medical Hypothesis*, 1979, vol. 5, pp. 109–43.
17. Ibid.
18. 'New focus on broadcast radiation: Is there a leukemia risk?' *Microwave News*, January/February 1997, vol. 17, no. 1, pp. 1, 11–12.
19. Ibid., p. 11.
20. Ibid.
21. S. Milham Jr, 'Increased mortality in amateur raido operators due to lymphatic and hematopoietic malignacies', *American Journal of Epidemiology*, January 1988, vol. 127, pp. 50–4.
22. S. Szmiglielski, 'Cancer morbidity in subjects occupationally exposed to high–frequency (radio frequency and microwave) electromagnetic radiation' *Science of the Total Environment*, 1996, vol. 180, pp. 9–17.
23. 'New focus on broadcast radiation: Is there a leukemia risk?' op. cit., p. 11.
24. H. Dolk, et al., 'Cancer incidence near radio and television transmitters in Great Britain, Part 1. Sutton Coldfield Transmitter, *American Journal of Epidemiology*, January 1997, vol. 145, pp. 1–9.
25. H. Dolk, et al., 'Cancer incidence near radio and television transmitters in Great Britain, Part II. All high-powered transmitters', *American Journal of Epidemiology*, January 1997, vol. 145, pp. 10–17.
26. B. Hocking, I.R. Gordon, H.L. Grain and G.E. Hatfield, 'Cancer incidence and mortality and proximity to TV Towers', *Medical Journal of Australia*, 1996, vol. 165, p. 601–5; see also: R.A. Cartwright, 'Cancer and TV towers: Association but not causation', *Medical Journal of Australia*, 1996, vol. 165, pp. 599–600.
27. R. Bechmann, 'Picoteslas, SQUIDS and "personal magnetism"', *ECOS*, 1989/90, vol. 62, pp. 24–6.
28. G. Cremer–Bartels, K. Krause, G. Mitoskas and D. Brodersen, 'Magnetic field of earth as additional zeitgeber for endogenous rhythms?' *Naturwissenshaften*, 1984, vol. 71, no. 11, pp. 567–74.
29. J. Ashton and R. Laura, 'Mobile phones and UFH TV health hazards of radiation', *Search*, 1996, vol. 27, no. 7, pp. 219–23.
30. 'Higher Leukemia rates among those living near Australian TV towers', *Microwave News*, November/December 1995, p. 16.
31. S.B. Barnett, *CSIRO Report on the Status of Research on the Biological Effects and Safety of Electromagnetic Radiation: Telecommunication Frequencies*, CSIRO, Sydney, 1994, pp. 4–9.
32. A. Cain and W.J. Rissman, 'Mammalian auditory responses to 3GHz microwave pulses', *IEEE Trans. Biomed Eng*, BME–25, 1978, pp. 341–51.
33. R. Edwards, op. cit; see also: L. Tomenius, '50–Hz electromagnetic environment and the incidence of childhood tumours in Stockholm County', *Bioelectromagnetics*, 1986, vol. 7, pp. 191–207.
34. W. Volkrodt, 'Electromagnetic pollution of the environment' in R. Krieps (ed.), *Environment and Health: A Holistic Approach*, Grower Publishing Company, Aldershot, England, 1989, pp. 71–6.

35. M.H. Repacholi, A. Basten, V. Gebski, et al., 'Lymphomas in Eµ-Pim 1, transgenic mice exposed to pulsed 900 MHz electromagnetic fields', *Radiation Research*, vol. 147, no. 5, 1997, pp. 631–40.

36. I.P. Macfarlane, 'Magnitudes of the near e–fields close to hand-held GSM digital mobile telephones', *Australian Telecommunications Research*, 1984, vol. 28, no. 2, pp. 61–71.

37. 'Do cellular phones cause headaches?' *Microwave News*, November/December 1995, pp. 7–12.

38. W. Volkrodt, op. cit.

39. J. Cadzow, 'Upwardly Mobile', *Good Weekend: The Sydney Morning Herald Magazine*, 23 November 1996, pp. 20–6.

CHAPTER 4

1. K.B. Murray, 'Hazard of Microwave Ovens to Transdermal Delivery System', *New England Journal of Medicine*, 15 March 1984, p. 721.

2. B. Holmes, 'This is the age of the microwave', *New Scientist*, 23 September 1995, p. 5.

3. M. Doyle, 'The Microwave Revolution Hits Home', *Food Business*, 24 July 1989, p. 12.

4. P. Brodeur, *The Zapping of America*, W. W. Norton, New York, 1977, p. 317.

5. U. Bronfenbrenner, 'The Origins of Alienation', *Scientific American*, vol. 231, no. 2, 1974, p. 57.

6. B. Messenger, 'Sensationalism Agitates British Microwave Scare', *Food Business*, 20 November 1989, p. 12.

7. B.D. Gessner and M. Beller, 'Protective effect of conventional cooking versus use of microwave ovens in an outbreak of salmonellosis', *American Journal of epidemiology*, vol. 139, no. 9, 1994, pp. 903–9.

8. T. Valentine, 'Is microwave-cooked food safe?', *Natural Health*, December 1995/January 1996, pp. 2–5.

9. D. Jonker and H.P. Til, 'Human diets cooked by microwave or conventionally: comparative sub–chronic (13–wk) toxicity study in rats', *Food and Chemical Toxicology*, vol. 33, no. 4, 1995, pp. 245–56.

10. L. Petrucelli and G.H. Fischer, 'D-aspartate and D-glutamate in microwaved versus conventially heated milk', *Journal of the American College of Nutrition*, vol. 13, no. 2, 1994, pp. 209–10.

11. D.K. Stevenson and J.A. Kerner, 'Effects of microwave radiation on anti-infective factors in milk', *Pediatrics*, vol. 89, no. 4, pt. 1, 1992, pp. 667–9.

12. J.A. Steinke, C. Frick, et al., 'Interaction of Flavour Systems in the Microwave Environment', *Cereal Foods World*, vol. 34, no. 4, 1989, pp. 330–2.

13. M. Sharman, C.A. Honeybone, S.M. Jickells and L. Castle, 'Detection of residues of epoxy adhesive component bisphenol A diglycidyl ether (BADGE) in microwave susceptors and its migration into food', *Food Additives and Contaminants*, vol. 12, no. 6, 1995, pp. 779–87.

14. T.P. McNeal and H.C. Hollifield, 'Determination of volatile chemicals released from microwave-heat-susceptor food packaging', *Journal of AOAC International*, vol. 76, no. 6, 1993, pp. 1268–75.

15. J. Gilbert, L. Castle, S.M. Jickells and M. Sharman, 'Current research on food contact materials undertaken by UK Ministry of Agriculture, Fisheries and Food', *Food Additives and Contaminants*, vol. 11, no. 2, 1994, pp. 231–40.

16. L. Castle, J. Nichol and J. Gilbert, 'Migration of mineral hydrocarbons into foods. 4. Waxed paper for packaging dry goods including bread, confectionery and for domestic use including microwave cooking', *Food Additives and Contaminants*, vol. 11, no. 1, 1994, pp. 79–89.

17. A.B. Badeka and M.G. Kontominas, 'Effect of microwave heating on the migration of dioctyladipate and acetyltributylicitrate plasticizers from food–grade PVC and PVDC/PVC films into olive oil and water', *Zeitschrift fur Lebensmittel–Unterschung und–Forschung*, vol. 202, no. 4, 1996, pp. 313–17.

18. B. Messenger, op. cit.

CHAPTER 5

1. Council on Scientific Affairs, 'Health Effects of Video Display Terminals', *Journal of the American Medical Association*, vol. 257, no. 11, 20 March 1987, p. 1508.

2. F. Hollwich, *The Influence of Occular Light Perception on Metabolism in Man and in Animals*, Springer-Verlag, New York, 1980, Preface.

3. International Non–Ionising Radiation Committee of the International Radiation Protection Association, *Video Display Units–Radiation Protection Guidance*, International Labour Office, Geneva, 1993, pp. 11–12. See also: M.M. Weiss, 'The Video Display Terminals—Is There a Radiation Hazard?', *Journal of Occupational Medicine*, vol. 25, no. 2, 1983, p. 100.

4. C.R. Roy and G. Elliot, 'Radiation emisions from video display terminals and alleged health effects', *Journal of Occupational Health and Safety–Aust./N.Z*, vol. 4, no. 2, 1988, pp. 115–22.

5. National Institute of Environmental Health Sciences and US Department of Energy, *Questions and Answers about Electric and Magnetic Fields Associated with the Use of Electric Power*, US Government Printing Office, Washington DC, 1995, pp. 36–44.

6. Ibid.

7. J.R. Gauger, 'Household Appliance Magnetic Field Survey', *IEEE Transactions on Power Apparatus and Systems*, vol. 104, no. 9, 1985, pp. 2436–44.

8. US Congress, Office of Technology Assessment, *Biological Effects of Power Frequency Electric and Magnetic Fields—Background Paper, OTA–BP–E–53*, US Government Printing Office, Washington DC, May 1989, p. 21.

9. National Insitute of Environmental Health Sciences, op. cit., p. 46.

10. M. Feychting and A. Ahlbom, 'Magnetic fields and cancer in children residing near Swedish high–voltage power lines', *American Journal of Epidemiology*, vol. 138, 1993, pp. 467–81. See also: J. H. Olsen, A. Neilsoen and G. Schulgen, 'Residence near high voltage facilities and the risk of cancer in children', *British Medical Journal Council*, vol. 307, 1993, pp. 891–95; and A. Fajardo-Gutierrez, 'Close residence to high electric voltage lines and its association with children with leukemia', *Boletin Medico del Hospital Infantil de Mexico*, vol. 50, 1993, pp. 32–8.

11. P. Landrigan, *New England Journal of Medicine*, 9 November, 1995. (On incidence of childhood leukemia.)

12. C.R. Roy and G. Elliot, op. cit.

13. B. Hocking, I.R. Gordon, H.L. Grain and G.E. Hatfield, 'Cancer incidence and mortality and proximity to TV towers', *Medical Journal of Australia*, vol. 165, 2/16 December 1996, pp. 601–5.

14. International Non–Ionising Radiation Committee of International Radiation Protection Association, op. cit., p. 21.
15. R. Bechmann, 'Picoteslas, SQUIDS and "personal magnetism"', *Ecos*, vol. 62, 1989/1990, pp. 24–6. See also: ibid., p. 2443.
16. J. Kaiser, 'Please extinguish your laptop', *New Scientist*, 2 December 1995, pp. 35–7. See also: M. K. Goldhaber, et al., p. 704.
17. Ibid. See also: I. A. Marriott, et al., p. 841.
18. Council on Scientific Scientific Affairs, op. cit., pp. 1508–9. See also: M. K. Goldhaber, M. R. Polen and R. A. Hiatt, 'The Risk of Miscarriage and Birth Defects Among Women Who Use Visual Display Terminals During Pregnancy', *American Journal of Industrial Medicine*, vol. 13, 1988, p. 695–6. See also: K. Kurppa, et al., 'Birth Defects and Video Display Terminals', *The Lancet*, 8 December 1984, p. 1339; and W. D. Marbach, et al., 'Are VDTs Health Hazards?', *Newsweek*, 29 October 1984, p. 49.
19. Ibid.
20. J. M. R. Delgado, J. Leal, J. L. Monteagudo and M. G. Gracia, 'Embryological Changes Induced by Weak, Extremely Low Frequency Electromagnetic Fields', *Journal of Anatomy*, vol. 134, no. 3, 1982, pp. 533–51.
21. A. Ubeda, J. Leal, M. A. Trillo, M. A. Jimenez and J. M. R. Delgado, 'Pulse Shape of Magnetic Field Influences Chick Embryogenesis', *Journal of Anatomy*, vol. 137, no. 3, 1983, pp. 513–36.
22. National Institute of Environmental Health Sciences, op. cit., pp. 36–44.
23. M. K. Goldhaber, et al., op. cit., pp. 695–6.
24. I. A. Marriott, and M. A. Stuchley, 'Health aspects of work with visual display terminals', *Journal of Occupational Medicine*, vol. 28, no. 9, 1986, p. 841–6.
25. M. K Goldhaber, et al., op. cit., p. 695.
26. J. Juutilainen, P. Matilainen, S. Saarikoski, et al., 'Early Pregnancy Loss and Exposure to 50-Hz Magnetic fields', *Bioelectromagnetics*, vol. 14, 1993, pp. 229–36. See also: N. Wertheimer and E. Leeper, 'Fetal loss associated with two seasonal sources of electromagnetic field exposure', *American Journal of Epidemiology*, vol. 129, 1989, pp. 220–4.
27. M. L. Lindbohm, M. Heitainen, P. Kyyronen, et al., 'Magnetic fields of video display terminals and spontaneous abortion', *American Journal of Epidemiology*, vol. 136, 1992, pp. 1041–51.
28. E. Roman, V. Beral, M. Pelerin and C. Hermon, 'Spontaneous abortion and work with visual display units, '*British Journal of Industrial Medicine*, vol. 49, no. 7, 1992, pp. 507–12. See also: V. Delpizzo, 'Epidemiological studies of work with video display terminals and adverse pregnancy outcomes (1984–1992)', *American Journal of Industrial Medicine*, vol. 26, no. 4, 1994, pp. 465–80.
29. I. A. Marriott, et al., op. cit., p. 843.
30. Council on Scientific Affairs, op. cit., p. 1509.
31. D. S. Chatterjee and A. J. Crookes, 'VDUs and Relative Annoyance from High Frequency Noise', *Journal of Occupational Medicine*, vol. 26, no. 8, 1984, pp. 552–3.
32. I. A. Marriott, et al., p. 843.
33. Ibid., p. 844. See also: Council on Scientific Affairs, op. cit., p. 1509; A. Michotte and G. Ebinger, 'Television induced seizures in alcoholics', *British Medical Journal*, vol. 289, 3 November 1984, pp. l 191–2.

34. O. Johansson, M. Hilliges, V. Bjornhagen and K. Hall, 'Skin changes in patients suffering from so-called "screen dermatitis": A two case open-field provocation study', *Experimental Dermotology*, vol. 3, 1994, pp. 234–8.
35. Ibid.
36. Council on Scientific Affairs, op. cit., pp. 1509–10.
37. I. A. Marriott, et al., op. cit., p. 843.
38. F. Soyka and A. Edmonds, *The Ion Effect*, Bantam Books, New York, 1981.
39. Editor, 'Uncomfortable Images', *The Lancet*, 12 January 1985, pp. 85–6. See also: J. De Keyser, A. Michotte and G. Ebinger, 'Television induced seizures in alcoholics', *British Medical Journal*, vol. 289, 3 November 1984, pp. 1191–2.
40. C. D. Binnie, C. E. Darby and A. T. Hindley, 'Electroencephalographic Changes in Epileptics while Viewing Television', *British Medical Journal*, 17 November 1973, pp. 378–9.
41. F. Hollwich, op. cit.
42. M. M. Weiss, op. cit., p. 100.
43. R. J. Wurtman, 'The Effects of Light on the Human Body', *Scientific American*, vol. 233, no. 1, 1975, pp. 69–77. See also: R. Horne, op. cit., p. 214.
44. J. R. Salvotore, A. B. Weitberg and S. Mehta, 'Nonionizing Electromagnetic Fields and Cancer: A review', *Oncology*, April 1996, pp. 563–70.
45. R. J. Reiter, 'Melatonin suppression by static and extremely low frequency electromagnetic fields: relationship to the reported increased incidence of cancer', *Reviews on Environmental Health*, vol. 10, nos 3–4, 1994, pp. 171–86. See also: M. Cohen, M. Lippman and B. Chabner, 'Role of pineal gland in aetiology and treatment of breast cancer', *Lancet*, 14 October 1978, pp. 814–16; R. P. Liburdy, T. R. Sloma, R. Sokolic, et al., 'ELF magnetic fields, breast cancer, and melatonin: 60 Hz fields block melatonins oncostatic action on ER and breast cancer cell proliferation', *Journal of Pineal Research*, vol. 14, 1993, pp. 89–97; G. G. Luce, op. cit., p. 244.
46. R. J. Wurtman and J. J. Wurtman, 'Carbohydrates and depression', *Scientific American*, January 1989, pp. 50–7.
47. M. K. Goldhaber, op. cit., p. 705. See also: A. Ericson, et al., op. cit., p. 473.
48. R. J. Wurtman and J. J. Wurtman, op. cit.
49. International Non–Ionizing Radiation Committee of the International Radiation Protection Association, op. cit., p. 13.
50. S. Sidney, 'Television viewing and cardiovascular risk factors in young adults: The CARDIA study', *Annals of Epidemiology*, vol. 6, 1996, p. 154–9.
51. J. Liberman, *Light, Medicine of the Future*, Bear and Company, Santa Fe, New Mexico, 1991. See also: F. Hollwich, op. cit.

CHAPTER 6

1. N. Wertheimer and E. Leeper, 'Electrical Wiring Configuration and Childhood Cancer', *American Journal of Epidemiology*, vol. 109, 1979, pp. 273–84.
2. H. Bacon, 'The Hazards of High-voltage Power Lines', in E. Goldsmith and N. Hildyard, *Green Britain or Industrial Wasteland?*, Basil Blackwell, Oxford, 1986.
3. N. Wertheimer and E. Leeper, op. cit., p. 276. See also: P. Brodeur, *Currents of Death*, Simon and Schuster, New York, 1989, pp. 18–22.
4. P. Brodeur, op. cit., p. 18.
5. N. Wertheimer and E. Leeper, op cit., pp. 282–3. See also: P. Brodeur, op. cit., pp. S–l.

6. N. Wertheimer and E. Leeper, op cit., pp. 282–3.
7. N. Wertheimer and E. Leeper, 'Adult Cancer Related to Electrical Wires Near the Home', *International Journal of Epidemiology*, vol. 11 (4), 1982, pp. 345–55.
8. P. Brodeur, op. cit., p. 21.
9. N. Wertheimer and E. Leeper, 1982, op. cit., pp. 348–50.
10. P. Brodeur, op. cit., p. 23.
11. Ibid., p. 25.
12. D. A. Savitz, et al., 'Case-Control Study of Childhood Cancer and Exposure to 60–Hertz Magnetic Fields', *American Journal of Epidemiology*, 1988, vol. 128, pp. 21–38.
13. R. O. Becker, *Cross Currents*, Tarcher, London, 1990, pp. 267–70.
14. Ibid., p. 269.
15. R. O. Becker, op. cit., p. 269.
16. N. Wertheimer and E. Leeper, 'Possible Effects of Electric Blankets and Heated Waterbeds on Fetal Development', *Bioelectromagnetics*, 1987, vol. 7, pp. 13–22.
17. Ibid. See also: P. Brodeur, op. cit., pp. 147–8.
18. N. Wertheimer and E. Leeper, 1987, op. cit., pp. 18–20.
19. Ibid. See also: G. M. Shaw and L. A. Croen, 'Human Adverse Reproductive Outcomes and Electromagnetic Field Exposures: Review of Epidemiologic Studies', *Environmental Health Perspectives Supplements*, vol. 101, suppl. no. 4, 1993, pp. 107–19.
20. N. Wertheimer and E. Leeper, 'Ceiling Cable Heating in Homes and the Risk of Miscarriage', *American Journal of Epidemiology*, vol. 130, 1989, pp. 18–25.
21. Ibid. See also: P. Brodeur, op. cit., p. 151.
22. Ibid.
23. R. O. Becker, op. cit., p. 270.

CHAPTER 7

1. H. M. Sinclair, 'Food and Health', *British Medical Journal*, 14 December 1957, pp. 1424–5.
2. Ibid., p. 1426.
3. D. O. Rudin, 'Omega–3 Essential Fatty Acids in Medicine', in J. Bland, *1984–85 Yearbook of Nutritional Medicine*, Keats Publishing, New Canaan, Connecticut, 1985, p. 46.
4. Ibid., p. 38.
5. Ibid.
6. L. Gamlin, 'Another man's poison', *New Scientist*, 15 July 1989, p. 33.
7. M. Rogers, 'Protective plant foods: new opportunities for health and nutrition', *CSIRO Human Nutrition 1996 Conference Abstracts*, CSIRO Division of Human Nutrition, Adelaide, 1996.
8. A. E. Bender, *Health or Hoax? The Truth About Health Food and Diets*, Elvendon Press, Goring–on Thames, England, 1985, p. 309.
9. D. K. Thompkinson and B. N. Mathur, 'Comparison of the lipid constituents in human and bovine milks', *Australian Journal of Nutrition and Dietetics*, vol. 46, no. 2, 1989, pp. 40–2.
10. N. F. Sheard and W. A. Walker, 'The Role of Breast Milk in the Development of the Gastrointestinal Tract', *Nutrition Reviews*, vol. 46, no. 1, 1988, pp. 1–8.
11. Ibid., p. 6. See also: L. Gamlin, op. cit., p. 33.

12. A. Berger, 'Milk protein may help beat cancer', *New Scientist*, 26 August 1995, p. 15.
13. A. Coghlan, 'Bottle–fed babies need brain boosting formula', *New Scientist*, 18 May 1996, p. 12.
14. A. L. Drash, M. S. Kramer, J. Swanson and J. N. Udall, 'Infant Feeding Practices and their possible relationship to the etiology of diabetes mellitus', *Pediatrics*, vol. 94, no. 5, 1994, pp. 752–4.
15. See for example: Australian Food Council, *Food Biotechnology in Australia*, Australian Food Council, Barton, ACT, 1996; Monsanto Company, *Biotechnology Solutions for tomorrow's world*, Biotechnology Communications, St Louis, Mo, 1996.
16. B. Hileman, 'Views differ sharply over benefits, risks of agriculture biotechnology', *Chemical and Engineering News*, 21 August 1995, pp. 8–17.
17. International Food Information Service, *Foodinfo*, April 1995, pp. 2–3.
18. Enviro News Service, 'Europeans reject genetically-engineered US soy beans', *Foodmonitor*, vol. 2, no. 3, (November/December) 1996, pp. 29–31.
19. K-H. Engel, G. R. Takeoka, *Genetically Modified Foods, No. 605L*, RACS Symposium Series Teranski, American Chemical Society, Washington DC, 1995.
20. B. Hileman, op. cit.
21. S. Mayer, 'Let's keep the genie in its bottle', *New Scientist*, 30 November 1996, p. 51.
22. D. MacKenzie, 'Europe lets in American supermaize', *New Scientist*, 4 January 1997, p. 8.
23. 'Outlaw corn', *New Scientist*, 14 December 1996, p. 11.
24. D. MacKenzie, op. cit.
25. B. Hileman, op. cit.
26. Ibid.
27. Ibid.
28. Ibid.
29. Centre for Science Communication, *The Biotechnology Revolution*, Sydney, 1995.
30. A. Coghlan, 'Arguing till the cows come home', *New Scientist*, 29 October 1994, pp. 14–15. See also: A. Coghlan, 'Milk hormone data bottled up for years', 22 October 1994, p. 4.
31. K. Kleiner, 'Milk hormone dispute boils over into court', *New Scientist*, 30 April 1994, pp. 6–7.
32. A. Coghlan, op. cit.
33. B. Mepham and P. Schofield, 'Not so safe', *New Scientist*, 10 September 1994, p. 44.
34. S. Epstein, 'BST and cancer', *New Scientist*, 29 October 1994, p. 70.
35. L. Gamlin, op. cit., p. 33.
36. S. I. Barr, 'Nutrition in Food Advertising: Content Analysis of a Canadian Women's Magazine, 1928–1986', *Journal of Nutrition Education*, vol. 21, no. 2, 1989, pp. 64–72.
37. P. Hollingsworth, 'Betcha can't watch just one: Food advertising trends', *Food Technology*, October 1996, pp. 59–62.
38. A. O. Duncan, *Food Processing*, Turner E. Smith and Co., Atlanta, Georgia, 1942, p. 395.

39. R. J. Williams, *Nutrition Against Disease, Environmental Prevention*, Pitman Publishing Co., New York, 1971, p. 200.
40. L. J. Machlin, *Handbook of Vitamins, Nutritional, Biochemical and Clinical Aspects*, Marcel Dekker, New York, 1984, pp. 247–9.
41. R. J. Williams, op. cit., p. 200.
42. H. M. Sinclair, 'Deficiency of Essential Fatty Acids and Atherosclerosis, Etcetera', *The Lancet*, 7 April 1956, pp. 381–3. See also: H. M. Sinclair, 1957, op. cit.; and H. M. Sinclair, 'Supply and Dietary Use of Fats', *The Lancet*, 31 January 1959, pp. 252–4.
43. J. G. Hattersley, 'Vitamin B6: The overlooked key to preventing heart attacks', *Journal of Applied Nutrition*, vol. 47, nos 1 and 2, 1995, pp. 24–31. See also: G. B. Gori, 'Food as a Factor in the Etiology of Certain Human Cancers', *Food Technology*, December 1979, p. 50.
44. R. J. Williams, op. cit., p. 201.
45. Ibid.
46. Ibid., p. 202.
47. D. A. Sampson, Q. B. Wen and K. Lorenz, 'Vitamin B6 and pyridoxine glucoside content of wheat and wheat flours', *Cereal chemistry*, vol. 73, no. 6, 1996, pp. 770–4
48. K. Spasovski, 'Study of macro- and microelements in vertical sections of wheat grain by electron microscopy', *Food Science and Technology Abstracts*, vol. 20, no. 9, 1988, p. 103.
49. H. J. Sanders, 'Nutrition and Health', *Chemical and Engineering News*, 26 March 1979, p. 29.
50. J. L. Beard, 'Are we at risk for heart disease because of normal iron status', *Nutrition Reviews*, vol. 51, no. 4, 1993, pp. 112–15.
51. J. M. McCord, 'Effect of positive iron status at a cellular level', *Nutrition Reviews*, vol. 54, no. 3, 1996, pp. 85–8. See also: S. Chazim, 'Is iron making you sick', *Readers Digest*, August 1996, pp. 115–20; R. C. Atkins, *Dr. Atkins Nutrition Breakthrough, How to Treat Your Medical Condition Without Drugs*, William Morrow and Co., New York, 1981, p. 222.
52. K. Schmidt, 'Too much of a good thing?' *New Scientist*, 2 April 1994, pp. 11–12.
53. Ibid.
54. C. L. Bird, J. S. Witte, M. E. Swendseid, et al., 'Plasma ferritin, iron intake and the risk of colorectal cancer', *American Journal of Epidemiology*, vol. 144, 1996, pp. 34–41.
55. R. J. Williams, op. cit., p. 209.
56. L. Wickenden, *Our Daily Poison*, Hillman Books, New York, 1955, p. 160.
57. See for example G. R. Douglas, E. R. Nestmann and G. Lebel, 'Contribution of Chlorination to the Mutagenic Activity of Drinking Water Extracts in Salmonella and Chinese Hamster Ovary Cells', *Environmental Health Perspectives*, vol. 69, 1986, pp. 81–7.
58. C. I. Wei, H. A. Ghanbari, W. B. Wheeler and J. R. Kirk, 'Fate of Chlorine During Flour Chlorination', *Journal of Food Science*, vol. 49, 1984, pp. 1136–8. See also: M. P. Duviau, H. Yamamoto, P. K. W. Ng and K. Kobrehel, 'Modifications of Wheat Proteins Due to Flour Chlorination', *Cereal Chemistry*, 1996, vol. 73, no. 4, pp. 490–4.
59. *IARC Monographs on the Evaluation of the Carcinogenic Risk of Chemicals to Humans*, vol. 40, 1986, pp. 215–16. See also: Y. Kurokawa, S. Takayama, et

al., 'Longterm in Vivo Carcinogenicity Tests of Potassium Bromate, Sodium Hypochlorite and Sodium Chlorite Conducted in Japan', *Environmental Health Perspectives*, vol. 69, 1986, pp. 221–35.

60. W. Mertz, 'The Essential Trace Elements', *Science*, vol. 213, 18 September 1981, p. 1333.

61. N. I. Sax, *Dangerous Properties of Industrial Materials*, 2nd ed., Reinhold Publishing Corp., New York, 1963, p. 525.

62. 'Fat Facts', *Australian Journal of Nutrition and Dietetics*, vol. 46, no. 2, 1989, p. 35.

63. M. Pyke, *Food Science and Technology*, John Murray, London, 1964, p. 108.

64. A. N. Howard and J. Marks, 'Hypocholesterolaemic Effect of Milk', *The Lancet*, 30 July 1977, pp. 255–6.

65. 'Milk compounds can inhibit bacterial toxins', *Australian Dairy Foods*, vol. 10, no. 3, 1989, p. 29.

66. P. Park, 'Soy cheese for a sunny smile with fewer fillings', *New Scientist*, 6 April 1991, p. 13.

67. G. B. Gori, 'Food as a Factor in the Etiology of Certain Human Cancers', *Food Technology*, December 1979, p. 50.

68. C. Ip, S. P. Briggs, A. D. Haegele, et al., 'The efficacy of conjugated linoleic acid in mammary cancer prevention is independent of the level or type of fat in the diet', *Carcinogenesis*, vol. 17, no. 5, 1996, pp. 1045–50.

69. J. V. Joossens, E. Brems–Heyns, et al., 'The Pattern of Food and Mortality in Belgium', *The Lancet*, 21 May 1977, pp. 1069–71.

70. J. Dobbing, *A Balanced Diet?*, Springer-Verlag, London, 1988, p. 107.

71. M. J. Fregly and M. R. Kare, *The Role of Salt in Cardiovascular Hypertension*, Academic Press, New York, 1982. See also: T. O. Cheng, 'Salt and Blood Pressure', *The Lancet*, 22 July 1989, pp. 214–15.

72. J. Beard, 'Full fat milk may protect common cancers', *New Scientist*, 16 November 1996, p. 8.

73. 'Healthy pinta', *New Scientist*, 18 May 1996, p. 13.

74. U. Erasmus, 'Fats that Heal, Fats that Kill', *Healing*, Gerson Institute, nos 22 and 23, 1988, p. 9.

75. B. S. Reddy and L. A. Cohen, op. cit., p. 92.

76. A. Trichopoulos, K. Katsouyanni, S. Stuver, et al., 'Consumption of olive oil and specific food groups in relation to breast cancer risk in Greece', *Journal of the National Cancer Institute*, vol. 87, no. 2 1995, pp. 110–16.

77. F. Prieto–Ramos, L. Serra–Majem, C. La Vecchia, et al., 'Mortality trends and past and current dietary factors of breast cancer in Spain', *European Journal of Epidemiology*, vol. 12, 1996, pp. 141–8.

78. M. V. Krause and L. K. Mahan, *Food, Nutrition and Diet Therapy, A Textbook of Nutritional Care*, W. B. Saunders Co., London, 1984, p. 43.

79. B. S. Reddy and L. A. Cohen, op. cit., p. 92. See also: Dobbing, op. cit., p. 10; and R. Passwater, *Supernutrition for Healthy Hearts*, Jove Books, New York, 1978, pp. 87–8.

80. G. V. Mann, 'Metabolic consequences of dietary trans fatty acids', *Lancet*, 21 May 1994, pp. 1268–71.

81. W. C. Willett and A. Ascherio, 'Trans fatty acids: are the effects only marginal?' *American Journal of Public Health*, vol. 85, no. 5, 1994, pp. 722–4.

82. V. D'Arrigo, 'Ni in commercial margarines', *Food Science and Technology Abstracts*, vol. 11, no. 10, 1979, p. 160. See also: H. Pihlaja, 'Determination of

traces of metals in Finnish margarines by the flameless atomic absorption spectrophotometry method', *Food Science and Technology Abstracts*, vol. 14, no. 9, 1982, p. 139.

83. T. S. Vodichenska and S. K. Dinoeva, 'Experimental study of the atherogenic effect of nickel on entry to the organism via drinking water', *Food Science and Technology Abstracts*, vol. 20, no. 3, 1988, p. 64.

84. H. M. Sinclair, 1959, op. cit., p. 252.

85. W. E. Stehbens, 'Diet and Atherogenesis', *Nutrition Reviews*, vol. 47, no. 1, 1989, p. 10.

86. Passwater, op. cit., pp. 44–78.

87. A. M. Louheranta, E. K. Porkkals, M. K. Nyyssonen, et al., 'Linoleic acid intake and susceptibility of VLDL and LDL to oxidation in men', *American Journal of Clinical Nutrition*, vol. 63, no. 5, 1996, p. 698–703.

88. J. L. Burton, 'Dietary Fatty Acids and Inflammatory Skin Disease', *The Lancet*, 7 January 1989, pp. 27–31. See also: H. A. Schroeder, 'Losses of vitamins and trace minerals resulting from processing and preservation of foods', *American Journal of Clinical Nutrition*, vol. 24, 1971, p. 562.

89. D. O. Rudin, 1985, op. cit., pp. 37–54. See also: Erasmus, op. cit., p. 14.

90. U. Erasmus, op. cit., p. 14.

91. Ibid., p. 11. See also: B. S. Reddy and L. A. Cohen, op. cit., p. 92.

92. P. A. Bernstein, W. F. Graydon and L. L. Diosady, 'Hydrogenation of Canola Oil Using Chromium Catalysts', *Journal of the American Oil Chemists Society*, vol. 66, no. 5, 1989, p. 680.

93. R. Passwater, op. cit., pp. 83–6.

94. L. H. Kushi, A. R. Folsom, R. J. Prineas, et al., 'Dietary antioxidant vitamins and death from coronary heart disease in postmenopausal women', *New England Journal of Medicine*, vol. 334, no. 18, 1996, pp. 1156–62. See also: 'Serum Vitamin E and Risk of Cancer in Finland', *Nutrition Reviews*, vol. 46, no. 10, 1988, pp. 342–3.

95. D. M. Conning, 'Diet and cancer: Experimental Evidence', *BNF Nutrition Bulletin*, vol. 16, no. 1, 1991, pp. 36–44. See also: 'National Academy of Sciences Report on Diet and Health', *Nutrition Reviews*, vol. 47, no. 5, 1989, p. 145; M. F. Oliver, 'Low Cholesterol and Increased Risk', *The Lancet*, 15 July 1989, p. 163.

96. 'N. R. C.: Low–fat diets for all Americans', *Journal of the American Oil Chemists Society*, vol. 66, no. 5, 1989, p. 636.

97. H. M. Sinclair, 1956, op. cit., p. 382.

98. B. S. Mackie, A. R. Johnson, et al., 'Dietary Polyunsaturated Fats and Malignant Melanoma', *The Medical Journal of Australia*, 23 February 1980, pp. 159–63.

99. M. R. Chedekel and L. Zeise, 'Sunlight, Melanogenesis and Radicals in the Skin', *Lipids*, vol. 23, no. 6, 1988, p. 590.

100. V. E. Reeve, M. Matheson, G. E. Greenoak, et al., 'Effect of dietary lipid on UV light carcinogenesis in the hairless mouse', *Photochemistry and Photobiology*, vol. 38, no. 5, 1988, pp. 689–96.

101. H. S. Black, G. Okotie-Eboh and J. Gerguis, 'Dietary fat modulates in immunoresponsiveness in UV irradiated mice', *Photochemistry and Photobiology*, vol. 62, 1996, pp. 964–9. See also: M. A. Fisher and H. S. Black 'Modification of membrane composition, eicosanoid metabolism, and immunoresponsiveness by dietary omega–3 and omega–6 fatty acid sources,

modulators of ultraviolet carcinogenesis', *Photochemistry and Photobiology*, vol. 54, 1991, pp. 381–7.

102. V. E. Reeve, R. B. Cope, M. Bosnic, et al., 'Protection from ultraviolet radiation pathology by topical and systemic agents', *Environmental Dermatology*, vol. 3, suppl. 1, 1996, pp. 20–5. See also: V. E. Reeve, M. Bosnic and C. Boehm–Wilcox, 'Dependence of photocarcinogenesis and photoimmunosuppression in the hairless mouse on dietary polyunsaturated fat', *Cancer Letters*, 1997 (in press).

103. R. B. Cope, M. Bosnic, C. Beohm-Wilcox, et al., 'Dietary butter protects against ultraviolet radiation-induced suppression of contact hypersensitivity in Skh: HR-1 hairless mice', *Journal of Nutrition*, vol. 126, 1996, pp. 681–92.

104. J. Dobbing, op. cit., p. 101.

105. Ibid.

106. K. Cashel, 'A guide to the fat and cholesterol content of foods', *Journal of Food and Nutrition*, vol. 42, no. 1, 1985, pp. 24, 30. See also: M. B. Katan, P. L. Zock and R. P. Mensink, 'Dietary oils, serum lipoproteins, and coronary heart disease', *American Journal of Clinical Nutrition*, vol. 61, 1995, pp. 1368S–73S.

107. K. Kleiner, 'Friendly fat comes under fire in the .S', *New Scientist*, 3 February 1996, p. 5.

108. M. D. Lemonick, 'Fat–Free Fat?' *Time*, 22 January 1996, pp. 41–6.

109. K. Kleiner, 'Health fears linger over no-calorie fat substitute', *New Scientist*, 25 November 1995, p. 10.

110. Ibid.

111. K. Kleiner, 1996, op. cit.

112. Ibid.

113. J. Giese, 'Fats and fat replacers: Balancing the health benefits', *Food Technology*, September 1996, pp. 76–8.

114. I. Newton and D. Snyder, 'Nutritional aspects of long-chain omega-3 fatty acids and their use in bread enrichment', *Cereal Foods World*, vol. 42, no. 3, 1997, pp. 126–31.

115. Ibid.

116. See also: D. Rudin, C. Felix and C. Schrader, *The Omega-3 Phenomenon, The Nutrition Breakthrough of the '80s*, Sidgwick and Jackson, London, 1988.

117. D. O. Rudin, 1985, p. 44. See also: C. G. Isles, D. J. Hole, et al., 'Plasma cholesterol, coronary heart disease, and cancer in the Renfrew and Paisley survey', *British Medical Journal*, vol. 298, 1989, pp. 920–4.

118. T. Patel, 'How walnuts and purslane can save your heart', *New Scientist*, 18 June 1994, p. 8.

119. Ibid.

120. 'Rice bran offers India an oil source', *Journal of the American Oil Chemists Society*, vol. 66, no. 5, 1989, pp. 620–3.

121. M. Gaynor, 'Multinationals Kick Up Their Heels', *Food Business*, 10 July 1989, p. 35.

122. T. A. Roberts, C. R. Dillon and T. J. Siebenmorgen, 'The Impact of Food Processing on the US Economy', *Food Technology*, October 1996, pp. 65–8.

123. P. M. Desmarchelier, 'Foodborne disease: emerging problems and solutions', *Medical Journal of Australia*, vol. 165, 2/16 December 1996, pp. 668–77.

124. P. Hadfield, 'Japanese swallow Western diseases', *New Scientist*, 2 September 1995, p. 5.

125. Ibid.
126. T. Patel, 'Bhajis not burgers for India', *New Scientist*, 3 August 1996, pp. 14–15.
127. A. Coghlan, 'Junk food ads target the young' *New Scientist*, 30 November 1996, p. 8.
128. Ibid.
129. 'Putting fast food to the test', *Choice*, April 1994, pp. 7–9.
130. B. Rolls, 'Why We Choose What We Eat', *Proceedings, World Sugar Research Organisation Ltd., Scientific Conference*, Sydney Australia, 10–13 August 1986, pp. 68–71
131. K. A. Bettelheim, 'Enerohaemorrhagic Escherichia coli: A new problem, an old group of organisms', *Australian Veterinary Journal*, vol. 73, no. 1, 1996, pp. 20–26.
132. Ibid.
133. Ibid.
134. P. Brown, 'Seabirds dirty diet spreads disease', *New Scientist*, 22 March 1997, p. 11.
135. A. Coghlan, 'Jetsetters send festering faeces round the world', *New Scientist*, 17 May 1997, p. 7.
136. State of Environment Advisory Council, *Australia State of the Environment 1996*, CSIRO Publishing, Collingwood, Victoria, 1996, p. 8–11.
137. P. N. Goldwater and K. A. Bettelheim, 'An outbreak of Hemolytic Uremic Syndrome due to Escherichia coli 0157:H–: Or was it', *Emerging Infectious Diseases*, vol. 2, no. 2, 1996, pp. 152–3. See also: S. K. Crerar, C. B. Dalton, H. M. Longbottom and E. Kraa, 'Foodborne disease: current trends and future surveillance needs in Australia', *Medical Journal of Australia*, vol. 165, 2/16 December 1996, pp. 672–5.
138. F. Pearce, 'Ministers hostile to advice on BSE', *New Scientist*, 30 March 1996, pp. 4–5.
139. S. Pain, 'Ten deaths that may tell a shocking take', *New Scientist*, 30 March 1996, p. 6.
140. F. Pearce, 'BSE may lurk in pigs and chickens', *New Scientist*, 6 April 1996, p. 5.
141. F. Pearce, '...Now fears grow for unborn babies', *New Scientist*, 13 April 1996, p. 5. See also: S. Pain, 'Mad calves fuel BSE fears', *New Scientist*, 10 August 1996, pp. 10–11.
142. M. Denton and R. W. Lacey, 'Intensive Farming and Food Processing : Implications for Polyunsaturated Fats', *Journal of Nutritional Medicine*, vol. 2, 1991, pp. 179–89.
143. M. Noakes, P. J. Nestel and P. M. Clifton, 'Modifying the fatty acid profile of dairy products through feedlot technology lowers plasma cholesterol of humans consuming the products', *American Journal of Clinical Nutrition*, vol. 63, 1996, pp. 42–6.
144. J. Bonner, 'Hooked on drugs', *New Scientist*, 18 January 1997, pp. 24–7
145. Ibid.
146. National Research Council, *Alternative Agriculture*, National Academy Press, Washington DC, 1989, pp. 169–71.
147. J. R. Smith, *Tree Crops, A Permanent Agriculture*, The Devin-Adair Company, Old Greenwich, 1977.

148. B. Mollison, Permaculture, *A Designers Manual*, Tagari Publications, Talgum Australia, 1992.
149. N. Myers (ed.), *The Gaia Atlas of Planet Management*, Pan Books, London, 1985, p. 34.
150. Ibid., p. 43.
151. State of the Environment Advisory Council, op. cit.
152. J. R. Smith, op. cit.
153. National Research Council, op. cit., pp., 40–8.
154. F. Pearce, 'Thirsty meals that suck the world dry' *New Scientist,* 1 February 1997, p. 7.
155. M. L. Dreker, C.V. Maher and P. Kearney, 'The Traditional and Emerging Role of Nuts in Healthful Diets', *Nutrition Reviews*, vol. 54, no. 8, 1996, pp. 241–5. See also: J.F. Ashton and R. S. Laura, *The Life Enhancement Handbook*, Simon and Schuster, Sydney, 1997, pp. 73–4; and A. Tavani and C. La Vecchia, 'Fruits and vegetables consumption and cancer risk in a Mediterranean population', *American Journal of Clinical Nutrition*, vol. 61 (suppl.) 1995, pp. 1374S–75.
156. W. Mertz, op. cit., p. 1333. See also for example: H. H. Sandstead, 'Nutrition and brain function: trace elements', *Nutrition Reviews/Supplement*, May 1986, pp. 37–40.
157. Ibid., p. 1336.
158. Ibid., p. 1337.
159. H. J. Sanders, op. cit., p. 35.
160. Ibid., p. 36.
161. Committee on Nutrition, 'Zinc', *Paediatrics*, vol. 62, no. 3, 1978, p. 408.
162. E. Halmesmaki, O. Elikorkala and G. Alfthan, 'Concentrations of zinc and copper in pregnant problem drinkers and their newborn infants', *British Medical Journal*, vol. 291, 1985, p. 1470.
163. Committee on Nutrition, 'Zinc', *Paediatrics*, vol. 62, no. 3, 1978, p. 408. See also: E. Halmesmaki, O. Elikorkala and G. Alfthan, op. cit., p. 1470.
164. M. J. P. Kruesi, 'Carbohydrate Intake and Children's Behaviour', *Food Technology*, January 1986, p. 150.
165. Ibid.
166. A. E. Harper and D. A. Gans, 'Claims of Antisocial Behaviour from Consumption of Sugar: An Assessment', *Food Technology*, January 1986, pp. 142, 146.
167. H. R. Lieberman and R. J. Wurtman, 'Foods and Food Constituents That Affect the Brain and Human Behaviour', *Food Technology*, January 1986, pp. 139–41. See also: R. B. Kanarek and N. Orthen–Gambrill, 'Complex interactions affecting nutrition-behaviour research', *Nutrition Reviews/ Supplement*, May 1986, pp. 172–5; J. J. Wurtman and M. Danbrot, *Managing Your Mind and Mood Through Food*, Grafton Books, London, 1988.
168. C. E. Leprohon-Greenwood and G. H. Anderson, 'An overview of the mechanisms by which diet affects brain function', *Food Technology*, January 1986, p. 132.
169. Ibid., p. 133.
170. D. P. Burkitt and H. C. Trowell, *Refined Carbohydrate Foods and Disease, Some Implications of Dietary Fibre*, Academic Press, London, 1975.
171. Ibid., pp. 37–8.

172. P. A. Baghurst, K. I. Baghurst and S. J. Record, 'Dietary Fibre, Non-starch polysaccharides and resistant starch—a review', *Food Australia*, vol. 48, no. 3, 1996, pp. S3–S35. See also: D. P. Burkitt, 'Colonic-rectal cancer: fibre and other dietary factors', *American Journal of Clinical Nutrition*, vol. 31, October 1978, pp. 558–64. See also: 'Effects of Short-chain Fatty Acids on a Human Colon Carcinoma Cell Line', *Nutrition Reviews*, vol. 46, no. 1, 1988, pp. 11–12; and H. Trowell, 'Diabetes mellitus and dietary fibre of starchy foods', *American Journal of Clinical Nutrition*, vol. 31, October 1978, pp. 553–7.

173. E. B. Rimm, A. Ascherio, E. Giovannucci, et al., 'Vegetable fruit and cereal fibre intake and risk of coronary heart disease among men', *Journal of the American Medical Association*, vol. 275, 1996, pp. 447–51.

174. T. S. Kahlon, F. I. Chow, B. E. Knuckles and M. M. Chin, 'Cholesterol-lowering effects in hamsters of b–glucan enriched barley fraction, dehulled whole barley, rice bran, and oat bran and their combinations', *Cereal Chemistry*, vol. 70, no. 4, 1993, pp. 435–40.

175. N. Hammond, 'Functional and Nutritional Characteristics of Rice Bran Extracts', *Cereal Foods World*, vol. 39, no. 10, October 1994, pp. 752–4.

176. D. Kritchevsky, 'The rationale behind the use of diet and dietary ingredients to lower blood cholesterol levels', *Trends in Food Science and Technology*, vol. 5, July 1994, pp. 219–24.

177. A. Ramsey, 'Eat to your heart's content', *Ecos*, vol. 90, Summer 1996/1997, pp. 8–11.

178. P. A. Baghurst, K. I. Baghurst and S. J. Record, op. cit.

179. Ibid.

180. Ibid. See also: D. Kritchevsky and C. Bonfield (eds), *Dietary Fibre in Health and Disease*, Egan Press, St Paul, USA, 1995.

181. J. Brand Miller and S. Colagiuri, *The G.I. factor*, Hodder and Stoughton, Sydney, 1996.

182. P. Zimmet and M. Cohen, 'The diabetes epidemic in Australia: prevalence, patterns and the public health', *Medical Journal of Australia*, vol. 163, 7 August 1995, pp. 116–17.

183. J. Brand Miller, 'International tables of glycemic index', *American Journal of Clinical Nutrition*, vol. 62, no. 4, 1995, pp. 871S–93S. See also: I. Bjorck and N-G. Asp, 'Controlling the nutritional properties of starch in foods—a challenge to the food industry', *Trends in Food Science and Technology*, vol. 5, July 1994, pp. 213–18.

184. P. Burkitt and H. C. Trowell, op. cit., p. 266.

185. Ibid., p. 268.

186. C. W. Binns, 'Editorial: Into our third century', *Australian Journal of Nutrition and Dietetics*, vol. 42, no. 2, 1989, p. 27.

187. G. B. Gori, op. cit., p. 49.

188. D. P. Burkitt and H. C. Trowell, op. cit., p. 61.

189. M. Saxon, 'What are we eating?', *Scientific Australian*, March 1978, p. 14.

190. See for example: R. L. Smith, P. Newberne, T. B. Adams, et al., 'GRAS flavouring Substances 17', *Food Technology*, October 1996, pp. 72–81. See also: R. Buist, *Food Chemical Sensitivity*, Harper and Row, Sydney, 1986.

191. J. D. Dziezak, 'Sweeteners 1. Consumption Trends', *Food Technology*, January 1986, p. 112.

192. Ibid.

193. S. S. Schiffman, 'Natural and artificial sweeteners', in M. L. Wahlquist, R. W. F. King, et al., *Food and Health, Issues and Directions*, John Libbey, London, 1987, pp. 46–7.
194. J. D. Dziezak, 'Sweeteners 3. Alternative to Cane and Beet Sugar', *Food Technology*, January 1986, p. 122.
195. A. Thomson, 'Nutra Sweet—A Sweet Taste of Success', *Food Marketing and Technology*, April 1989, p. 12.
196. Ibid.
197. W. C. Monte, 'Aspartame: methanol and the public health', *Journal of Applied Nutrition*, vol. 36, no. 1, 1984, pp. 42–54.
198. S. J. Baum, *Introduction to Organic and Biological Chemistry*, 4th ed., Macmillan Publishing Company, New York, 1987, p. 243.
199. Ibid.
200. T. J. Maher and R. J. Wurtman, 'Possible Neurologic Effects of Aspartame, a Widely Used Food Additive', *Environmental Health Perspectives*, vol. 75, 1987, p. 54.
201. Ibid.
202. H. M. Sinclair, 1957, op. cit., p. 1426.
203. T. J. Maher and R. J. Wurtman, op. cit., p. 53.
204. Ibid., p. 54.
205. Ibid., p. 55.
206. Ibid., p. 56.
207. J. W. Olney, N. B. Farber, E. Spitznagel and L. N. Robins, 'Increasing brain tumour rates: is there a link to aspartame?', *Journal of Neuropathology and Experimental Neurology*, vol. 55, no. 11, November 1996, pp. 1115–23.
208. A. A. Barnett, 'Editor suffers "blitz" by aspartame maker', *Lancet*, vol. 348, 23 November 1996, p. 1435.
209. V. A. Koivista, S. L. Karonen and E. A. Nikkila, 'Carbohydrate ingestion before exercise: comparison of glucose, fructose, and sweet placebo', *Journal of Applied Physiology*, vol. 51, 1981, pp. 783–7.
210. L. Levine, W.J. Evans, et al., 'Fructose and glucose ingestion and muscle glycogen use during submaximal exercise', *Journal of Applied Physiology*, vol. 55, 1983, pp. 1767–70.
211. F. W. Schlutz and E. M. Knott, 'The Use of Honey as a Carbohydrate in Infant Feeding', *Journal of Paediatrics*, vol. 13, October 1938, p. 465.
212. Ibid., p. 472.
213. See for example: L. Van Horn, L. A. Emidy, et al., 'Serum Lipid Response to a Fat-Modified Oatmeal Enhanced Diet', *Preventive Medicine*, vol. 17, 1988, pp. 377–86.
214. B. P. Kinosian and J. M. Eisenberg, 'Cutting into cholesterol: Cost effective alternative for treating hypercholestermia', *Journal of the American Medical Association*, vol. 259, no. 15, 1988, pp. 2249–54.
215. F. H. Lowe, 'Quaker Oats Dishes Up a Winner', *Food Business*, 6 February 1989, p. 32.
216. M. E. Lieb, 'A brand new bran comes to town', *Food Business*, 19 June 1989, pp. 33–5.
217. B. F. Haumann, 'Rice bran linked to lower cholesterol', *Journal of the American Oil Chemists Society*, vol. 66, no. 5, 1989, pp. 615–18.

218. T. S. Kahlon and F. I. Chow, 'Hypocholesterolemic Effects of Oat, Rice and Barley Dietary Fibres and Fractions', *Cereal Foods World*, vol. 42 no. 2, 1997, pp. 86–92.

219. A. P. de Groot, R. Luyken and N. A. Pikaar, 'Cholesterol–lowering effect of rolled oats', *The Lancet*, ii, 1963, pp. 303–4.

220. See for example: L. Van Horn, et al., 'Fish oil and the development of atherosclerosis', *Nutrition Reviews*, vol. 45, no. 3, 1987, pp. 90–1.

221. D. P. Burkitt and H. C. Trowell, op. cit., p. 206.

222. D. Oakenfull, 'Dietary fibre, serum cholesterol and heart disease', *CSIRO Food Research Quarterly*, vol. 36, 1978, pp. 66–70.

223. J. Carper, *The Food Pharmacy, Dramatic New Evidence That Food Is Your Best Medicine*, Simon and Schuster, New York, 1989, p. 324.

224. 'Cholesterol–lowering effects of baked beans in pigs and man', *Food Technology in Australia*, vol. 49, no. 8, 1988, p. 326.

225. T. F. Schweizer and C. A. Edwards (eds), *Dietary Fibre—A Component of Food*, Springer–Verlag, New York, 1992, pp. 295–332.

226. E. M. Kurowska, J. Jordan, J. D. Spence, et al., 'Role of the main components of whole soybean products, soy protein and soy oil, in reducing hypercholesterolemia', *Second International Symposium on the Role of Soy in Preventing and Treating Chronic Disease—Abstracts*, Protein Technologies International, St Louis, Missouri, 1996, p. 22. See also: J. W. Anderson, B. M. Johnstone and M. E. Cook–Newell, 'Meta–analysis of soy protein intake on serum lipids', *New England Journal of Medicine*, vol. 333, 1995, pp. 276–82.

227. M. L. Dreher, C.V. Maher and P. Kearney, op. cit.

228. P. Bobek, E. Ginter and L. Ozdin, 'Oyster mushroom (*Pleurotus ostreatus*) accelerates the plasma clearance of low-density and high-density lipoproteins in rats', *Nutrition Reviews*, vol. 13, 1993, pp. 885–90.

229. M. L. Bierenbaum, R. Reichstein, T.R. Watkins, 'Reducing atherogenic risk in hyperlipemic humans with flax seed supplementation: a preliminary report', *Journal of the American College of Nutrition*, vol. 12, no. 5, 1993, pp. 501–4. See also: S. C. Cunnane, S. Ganguli, C. Menard, et al., 'High a-linolenic acid flax seed (*Lineium usitatissimum*): some nutritional properties in humans', *British Journal of Nutrition*, vol. 69, 1993, pp. 443–53.

230. R. Uauy-Dagach and A. Valenzuela, 'Marine oils: The health benefits of n-3 fatty acids', *Nutrition Reviews*, vol. 54, no 11, 1996, pp. S102–S108.

231. P. J. Nestel, 'Fish oil fatty acids—the answer to heart disease?', *CSIRO Food Research Quarterly*, vol. 47, 1987, pp. 9–12. See also: 'Fish oils as hypotriglyceridemic agents', *Nutrition Reviews*, vol. 46, no. 5, 1988, pp. 198–200.

232. I. S. Newton, 'Food enrichment with long-chain n-3 PVFA', *Inform*, vol. 7, no. 2, (February) 1996, pp. 169–77.

233. Ibid.

234. M. J. Gonzalez, 'Review article: Fish oil, lipid peroxidation and mammary tumour growth', *Journal of the American College of Nutrition*, vol. 14, 1995, pp. 325–35.

235. E. Lund and K. H. Bonaa, 'Reduced breast cancer mortality among fishermen's wives in Norway', *Cancer Causes and Control*, vol. 4, 1993, pp. 283–7.

236. L. Cobiac, P. M. Clifton, M. Abbey, et al., 'Lipid, lipoprotein, and hemostatic effects of fish Vs fish-oil n-3 fatty acids in mildy hyperlipidemic males', *American Journal of Clinical Nutrition*, vol. 53, 1991, pp. 1210–16.

237. R. J. Williams, op. cit., p. 208.
238. J. Edington, M. Geekie, et al., 'Effect of dietary cholesterol on plasma cholesterol concentration in subjects following reduced fat, high fibre diet', *British Medical Journal*, vol. 294, 1987, p. 333.
239. W. E. Stehbens, op. cit., p. 7. See also: Passwater, op. cit., pp. 56–8, 77.
240. B. S. Reddy and L. A. Cohen, op. cit., p. 4.
241. S. Menon, R. Y. Kendal, H. A. Dewar and D. J. Newell, 'Effect of Onions on Blood Fibrinolytic Activity', *British Medical Journal*, vol. 3, 1968, pp. 351–2. See also: A. Bordia and H. C. Bansal, 'Essential oil of garlic in prevention of atherosclerosis', *The Lancet*, 29 December 1973, pp. 1491–2; J. Emsley, 'Onions run rings around chemists', *New Scientist*, 30 September 1989, p. 16.
242. R. C. Jain, C. R. Vyas and O. P. Mahatma, 'Hypoglyeaemic action of onion and garlic', *The Lancet*, 29 December 1973, p. 1491.
243. J. Carper, op. cit.
244. H. M. Sinclair, 1957, op. cit., p. 1425. See also: N. Lyon, 'New Start for omega-3 eggs', *Australian Farm Journal*, May 1996, p. 22.
245. M. Denton and R. W. Lacey, 'Intensive farming and food processing: implications for polyunsaturated fats', *Journal of Nutritional Medicine*, vol. 2, 1991, pp. 179–89.
246. M. Noakes, P. J. Nestel and P. M. Clifton, 'Modifying the fatty acid profile of dairy products through feedlot technology lowers plasma cholesterol of humans consuming the products', *American Journal of Clinical Nutrition*, vol. 63, 1996, pp. 42–6.
247. A. Guest, *Gardening without Digging*, R. Wigfield and Co., Doncaster, 1949, p. 14.
248. Ibid., pp. 14–15.
249. J. Reganold, 'Farming's Organic Future', *New Scientist*, 10 June 1989, p. 31.
250. Ibid., p. 33.
251. National Research Council, op. cit., p. 119.
252. Ibid.
253. J. Reganold, op. cit., p. 34.
254. State of the Environment Advisory Council, op. cit., p. 6–28.
255. J. Reganold, op. cit.

CHAPTER 8

1. J. N. Hathcock, *Nutritional Toxicology*, vol. 1, Academic Press, New York, 1982, p. 196.
2. Committee on Food Protection, Food and Nutrition Board, National Research Council, *Toxicants Occurring Naturally in Foods*, 2nd ed., National Academy of Sciences, Washington DC, 1973, p. 45.
3. Committee on Food Protection, Food and Nutrition Board, National Research Council, op. cit., p. 46. See also: A. C. Alfrey, 'Aluminium Intoxication', *New England Journal of Medicine*, vol. 310, no. 17, 1984, pp. 1113–14.
4. J. N. Hathcock, op. cit., p. 196.
5. A. Prescott, 'What's the harm in aluminium', *New Scientist*, 21 January 1989, p. 60. See also: D. Harris, 'Safe Aluminium', *New Scientist*, 18 March 1989, p. 76.
6. R. B. Martin, 'The chemistry of aluminum as related to biology and medicine', *Clinical Chemistry*, vol. 32, no. 10, 1986, p. 1797.

7. A. Lione, 'Aluminium Toxicology and the Aluminium-Containing Medications', *Pharmacology and Therapeutics*, vol. 29, 1985, p. 277.

8. J. A. T. Pennington, 'Aluminium content of foods and diet', *Food additives and Contaminants*, vol. 5, no. 2, 1987, pp. 161–232.

9. M. McGraw, N. Bishop, et al., 'Aluminium Content of Milk Formulae and Intravenous Fluids Used in Infants', *The Lancet*, 18 January 1986, p. 157.

10. A. Lione, op. cit., pp. 274–6.

11. E. E. Kellog, *Science in the Kitchen*, Modern Medical Publishing Co., Chicago, Illinois, 1893, pp. 115–16.

12. 'Some Kitchen Experiments with Aluminium', *The Lancet*, 4 January 1913, pp. 54–55.

13. A. Lione, op. cit., p. 258.

14. J. Kloss, *Back to Eden*, Message Press, Coalmont, Tennessee, 1972, p. 89–95.

15. J. G. White, *Abundant Health*, Northwestern Publishing Association, Sacramento, California, USA, 1951, pp. 428–50.

16. G. A. Trapp and J. B. Cannon, 'Aluminium Pots as a Source of Dietary Aluminum', *New England Journal of Medicine*, vol. 304, no. 3, 1981, p. 172. See also: J. G. White, op. cit.

17. J. R. J. Sorenson, I .R. Campbell, L. B. Tepper and R. D. Lingg, 'Aluminium in the Environment and Human Health', *Environmental Health Perspectives*, vol. 8, 1974, pp. 3–95.

18. J. D. Kirschmann, *Nutrition Almanac*, McGraw-Hill Book Co., New York, 1975, p. 61.

19. S. E. Levick, 'Dementia From Aluminium Pots?', *New England Journal of Medicine*, vol. 303, no. 3, 1980, p. 164.

20. J. H. Koning, 'Aluminium pots as a source of dietary aluminium', *New England Journal of Medicine*, vol. 304, no. 3, 1981, p. 172.

21. P.O. Ganot, 'Metabolism and Possible health Effects of Aluminium', *Environmental Health Perspectives*, vol. 65, 1986, pp. 363–441.

22. D. R. Crapper, S. S. Krishnan and A. J. Dalton, 'Brain aluminium in Alzheimer's disease and experimental neurofibrillary degeneration', *Science*, vol. 180, 1973, pp. 511–13.

23. A. C. Alfrey, G. R. Le Gendre and W. D. Kaehny, 'The dialysis encephalopathy syndrome: possible aluminium intoxication', *New England Journal of Medicine*, vol. 294, 1976, pp. 184–8.

24. A. C. Alfrey, 'Aluminium toxicity in patients with chronic renal failure', *Theraputic Drug Monitoring*, vol. 15, no. 6, 1993, pp. 593–7.

25. E. Storey and C. L. Masters, 'Amyloid, aluminium and the aetiology of Alzheimer's disease', *Medical Journal of Australia*, vol. 163, 1995, pp. 256–9. See also: C. Exley, 'Amyloid, aluminium and the aetiology of Alzheimer's disease', *Medical Journal of Australia*, vol. 164, 1996, pp. 252–3.

26. R. Doll, 'Review: Alzheimer's Disease and Environmental Aluminium', *Age and Ageing*, vol. 22, 1993, pp. 138–53.

27. Ibid, p. 151.

28. C. Exley and J. D. Birchall, 'Aluminium and Alzheimer's disease', *Age and Aging*, vol. 22, 1993, pp. 391–2.

29. C. Exley, 1996, op. cit. See also: J. G. Joshi, M. Dhar, M. Clauberg and V. Chauthaiwale, 'Iron and aluminium Homeostasis in Neural Disorders', *Environmental Health Perspectives*, vol. 102, suppl. 3, 1994, pp. 207–13.

30. M. J. Rowan, 'Recent research on the causes of Alzheimer's disease', *Proceedings of the Nutrition Society*, *vol. 52*, 1993, pp. 255–62. See also: S. Hill, 'Alzheimer's stepchild', *Discover*, September 1992, pp. 85–94.

31. J. Savory, J. R. Nicholson and M. R. Wills, 'Is aluminium leaching enhanced by fluoride?', *Nature*, vol. 327, 14 May 1987, p. 107.

32. K. Tennakone and S. Wickramanayake, 'Aluminium leaching from cooking utensils', *Nature*, vol. 325, 15 January 1987, p. 202.

33. 'Alzheimer's—the saucepan link', *Chemistry in Britain*, April 1987, p. 307.

34. D. R. Williams and R.J. Duffield, 'Aluminium in food and the environment', *Chemistry in Britain*, vol. 24, no. 8, p. 811.

35. J. M. Duggan, J. E. Dickeson, P. F. Tynan, et al., 'Aluminium beverage cans as a dietary source of aluminium', *Medical Journal of Australia*, vol. 156, 1992, pp. 604–5.

36. J. H. Koning, op. cit., p. 172.

37. D. M. Sullivan, D. F. Kehoe and R. L. Smith, 'Measurement of Trace Levels of Total Aluminium in Foods by Atomic Absorption Spectrophotometry', *Journal of the Association of Official Analytical Chemists*, vol. 70, no. 1, 1987, p. 120.

38. J. A. T. Pennington, op. cit.

39. J. Walton, C. Tuniz, E. D. Fink, et al., 'Uptake of Trace Amounts of Aluminium into the Brain From Drinking Water', *Neuro Toxicology*, vol. 16, no. 1, 1995, pp. 187–90. See also: ibid., pp. 410–15.

40. C. N. Martyn, D. J. P. Barker, C. Osmond, et al., 'Geographical relation between Alzheimer's disease and aluminium in drinking water', *The Lancet*, 14 January, 1989, pp. 59–62.

41. D. R. C. McLachland, C. Bergeron, J. E. Smith, et al., 'Risk for neuro-pathologically confirmed Alzheimer's disease and residual aluminium in municipal drinking water employing weighted residential histories', *Neurology*, vol. 48, no. 1, 1996, pp. 1–4.

42. N. C. Bowdler, D. S. Beasley, E. C. Fritze, A. M. Goulette, et al., 'Behavioral Effects of Aluminium Ingestion on Animal and Human Subjects', *Pharmacology Biochemistry and Behaviour*, vol. 10, 1979, p. 511. See also: W. D. Kaehny, A. P. Hegg and A. C. Alfrey, 'Gastrointestinal Absorption of Aluminium from Aluminium-containing Antacids', *New England Journal of Medicine*, vol. 296, no. 24, 1977, p. 1390; L. W. Fleming, A. Prescott, et al., 'Bioavailability of Aluminium', *The Lancet*, 25 February 1989, p. 433 .

43. J. R. Walton, G. Hams and D. Wilcox, *Bioavailability of Aluminium from Drinking Water: Co-exposure with Foods and Beverages*, Urban Water Research Association, Research Report no. 83, July 1994.

44. P. O. Ganrot, op. cit., p. 395. See also: A. Lione, op. cit., pp. 261–4.

45. Ibid., pp. 410–15.

46. N. Kharasch, *Trace Metals in Health and Disease*, Raven Press Books, New York, 1979, p. 131–2.

47. S. J. Karlik, G. L. Eichhorn, et al., 'Interaction of Aluminium Species with Deoxyribonucleic Acid', *Biochemistry*, vol. 19, 1980, p. 5997.

48. N. C. Bowdlertal., op. cit., p. 511.

49. P. O. Ganrot, op. cit., p. 418.

50. J. Bjorksten, 'The role of aluminium and age dependent decline', *Environmental Health Perspectives*, vol. 81, 1989, pp. 241–2.

51. J. Bjorksten, *Longevity*, Bjorksten Research Foundation, Madison, Wisconsin, 1981, pp. 214–22

52. P. Zatta, R. Giordano, et al., 'A neutral lipophilic compound of aluminium (iii) as a cause of myocardial infarct in the rabbit', *Toxicology Letters*, vol. 39, 1987, p. 188.
53. L. Dayton, 'Brittle bones—the legacy of aluminium pots?', *New Scientist*, 6 November 1993, p. 20.
54. A. Lione, op. cit., p. 276.
55. P. Brown, 'Dirty Secrets', *New Scientist*, 2 November 1996, pp. 26–9.
56. A. Lione, 'More aluminium in infants', *New England Journal of Medicine*, vol. 314, no. 14, 1986, pp. 923.
57. K. Kaaber, A. O. Neilsen and N. K. Veien, 'Vaccination granulomas and aluminium allergy: course and prognostic factors', *Contact Dermatitis*, vol. 26, no. 5, 1992, pp. 304–6.
58. A. Lione, op. cit., p. 276.
59. C. P. Howson and H. V. Fineberg, 'The ricochet of magic bullets: Summary of the Institute of Medicine Report, adverse effects of pertussis and rubella vaccines', *Pediatrics*, vol. 89, no. 2, 1992, pp. 318–23.
60. A. B. Graves, E. White, T. D. Koepsell, et al., 'The association between aluminium-containing products and Alzheimer's disease', *Journal of Clinical Epidemiology*, vol. 43, 1990, pp. 35–44.
61. M. Tesarz and A. Wennberf, 'Effects on the nervous system among welders exposed to aluminium and manganese', *Occupational and Environmental Medicine*, vol. 53, no. 1, 1996, pp. 32–40.
62. R. K. Iler, 'Silica chemistry in nature and industry', *Chemistry in Australia*, October 1986, p. 359.
63. S. Sivasubramaniam and O. Talibudeen, 'Effect of aluminium on growth of tea (*Camellia sinensis*) and its uptake of potassium and phosphorus', *Journal of Food Science and Agriculture*, vol. 22, July 1971, p. 325.
64. D. M. Sullivan, et al., op. cit., p. 119.
65. C. N. Martyn, D. J. P. Barker, C. Osmond, E. C. Harris, et al., 'Geographical relation between Alzheimer's disease and aluminium in drinking water', *The Lancet*, 14 January 1989, p. 59.
66. J. A. T. Pennington and S. A. Schoen, 'Estimates of dietary exposure to aluminium', *Food Additives and Contaminants*, vol. 12, no. 1, 1995, pp. 119–28.
67. *McGraw–Hill Encyclopedia of Science and Technology*, McGraw–Hill Book Co., New York, 1960, vol. 12, p. 108.
68. L. E. Roth, J. R. Dunlap and G. Stacey, 'Localizations of aluminium in soybean bacteroids and seeds', *Applied and Environmental Microbiology*, vol. 53, no. 10, 1987, p. 2551.
69. A. Haug, 'Molecular aspects of aluminium toxicity', *Critical Reviews of Plant Science*, vol. 1, 1984, pp. 342–75. See also: M. Freundlich, et al., op. cit., pp. 527–9.
70. R. B. Martin, op. cit., p. 1803.
71. S. Sivasubramaniam, et al., op. cit., p. 325.
72. R. Bechmann, 'Solving and acid problem', *Rural Research*, vol. 173, Summer 1996/1997, pp. 27–30.
73. R. B. Martin, op. cit., p. 1803.
74. S. Sivasubramaniam, et al., op. cit., pp. 325, 329.
75. A. Lione, op. cit., p. 258.
76. L. W. Fleming, et al., op. cit. See also: A. C. Alfrey, op. cit., p. 1114.

77. N. Bishop, M. McGraw and N. Ward, 'Aluminium in infant formulas', *The Lancet*, 4 March 1989, p. 490.
78. M. Freundlich, G. Zilleruelo, et al., 'Infant formula as a cause of aluminium toxicity in neonatal uraemia', *The Lancet*, 7 September 1985, p. 528. See also: N. Bishop, et al., op. cit.
79. N. Bishop, et al., op. cit.
80. M. Freundlich, et al., op. cit., pp. 527–9.
81. R. B. Martin, op. cit., p. 1804.
82. J. D. Birchall and J. S. Chappell, 'The chemistry of aluminium and silicon in relation to Alzheimer's disease', *Clinical Chemistry*, vol. 34, no. 2, 1988, p. 267. See also: J. Walton and D. Bryson–Taylor, 'Aluminium: a neurochemical cause of Alzheimer's disease?' *Chemistry in Australia*, August 1995, pp. 37–8.
83. 'Fluorine and climate change', *Chemistry in Australia*, April 1995, p. 5.
84. 'Alzheimer's—the saucepan link', op. cit. See also: W. E. Sharpe and D. R. DeWalle, 'The effects of acid precipitation runoff episodes on reservoir and tapwater quality in an Appalachian Mountain water supply', *Environmental Health Perspectives*, vol. 89, 1990, pp. 153–8.
85. State of the Environment Advisory Council, *Australia State of the Environment 1996*, CSIRO Publishing, Collingwood, Victoria, 1996, p. 8–40.

CHAPTER 9

1. M. Byrne, 'Irradiation update', *Food Engineering International*, December, 1996, pp. 37–44.
2. 'Food irradiation', *Food Safety and Hygiene*, May, 1996, p. 4.
3. *Facts about Food Irradiation*, Australian Food Council, Barton, ACT, 1995, p. 33.
4. Ibid, p. 4. See also: P. Loaharanu, 'Status and prospects of food irradiation', *Food Technology*, May 1994, pp. 124–31.
5. Ibid.
6. A. Lehmacher, J. Bockemuhl and S. Aleksic, 'Nationwide outbreak of human salmonellosis in Germany due to contaminated paprika and paprika-powdered potato chips', *Epidemiology of Infections*, vol. 115, no. 5, 1995, pp. 501–11.
7. *Facts about Food Irradiation*, op. cit., p. 4.
8. P. M. Desmarchelier, 'Foodborne disease: emerging problems and solutions', *Medical Journal of Australia*, vol. 165, 2/16 December 1996, pp. 668–71.
9. E. D. Todd, 'Worldwide surveillance of foodborne disease: the need to improve', *Journal of food Protection*, vol. 59, 1996, pp. 82–92.
10. J. Doull, C. D. Klaassen and M. O. Amdur, *Casarett and Doull's Toxicology*, 2nd ed., Macmillan Publishing Co., New York, 1980, p. 505.
11. P. A. Wills, 'Radiation treatment of food', *Nuclear Spectrum*, vol. 2, no. 2, September 1986, p. 5.
12. Ibid. See also: G. Vinning, 'Food irradiation: global aspects', *Food Technology in Australia*, vol. 40, no. 8, August 1988, p. 306.
13. Ibid. See also: J. H. Hayner and B. E. Proctor, 'Investigations relating to the possible use of atomic fission products for food sterilization', *Food Technology*, January 1953, pp. 6–7.
14. H. J. Muller, 'Artificial transmutation of the gene', *Science*, vol. 66, no. 1699, 22 July 1927, p. 84.
15. W. S. Stone, O. Wyss and F. Haas, 'The production of mutations in *Staphylococcus aureus* by irradiation of the substrate', *Proceedings of the*

National Academy of Sciences, vol. 33, 1947, p. 66. See also: W. S. Stone, F. Haas, J. Bennett Clark and O. Wyss, 'The Role of Mutation and of Selection in the Frequency of Mutants Among Microorganisms Grown on Irradiated Substrate', *Proceedings of theNational Academy of Sciences*, vol. 34, 1948, p. 149.

16. S. Swaminathan, S. Nirula, et al., 'Mutations: Incidence in *Drosophila melanogaster* Reared on Irradiated Medium', *Science*, vol. 141, 16 August 1963, pp. 637–8.

17. N. W. Desrosier, *The Technology of Food Preservation*, Avi Publishing Co., Westport, Connecticut, 1963, p. 355.

18. B. J. Shay, A. F. Egan and P. A. Wills, 'The use of irradiation for extending the storage life of fresh and processed meats', *Food Technology in Australia*, vol. 40, no. 8, 1988, p. 310.

19. C. Bhaskaram and G. Sadasivan, 'Effects of feeding irradiated wheat to malnourished children', *American Journal of Clinical Nutrition*, vol. 28, February 1975, p. 134.

20. Facts about Food Irradiation, op. cit., p. 14. See also: D. W. Thayer, 'Wholesomeness of Irradiated Foods', *Food Technology*, May 1994, pp. 132–5.

21. J. Barna, 'Compilation of Bioassay Data on the Wholesomeness of Irradiated Food Items', *Acta Alimentana*, vol. 8, no. 3, 1979, p. 205.

22. P. M. Cullis, G. D. D. Jones, et al., 'Electron transfer from protein to DNA in irradiated chromatin', *Nature*, vol. 330, no. 24, 31 December 1987, p. 773.

23. P. M. Cohen, 'Can DNA in food find its way into cells?', *New Scientist*, 4 January 1997, p. 14.

24. Facts about Food Irradiation, op. cit., p. 9. See also: J. F. Diehl, *Safety of Irradiated foods*, 2nd ed., Marcel Dekker, Inc., New York, 1995.

25. L. A. Mac Arthur and B. L. D'Appolonia, 'Gamma radiation of wheat I. Effects on dough and baking properties', *Cereal Chemistry*, vol. 60, 1983, p. 465.

26. S. R. Rao, Ro. C. Hoseney, K. F. Finney and M. D. Shogren, 'Effect of gamma-irradaition of wheat on breadmaking properties, *Cereal Chemistry*, vol. 52, 1975, pp. 506–12.

27. O. Paredes-Lopez and M. M. Covarrubias-Alvarez, 'Influence of gamma radiation on the rheological and functional properties of bread wheats', *Journal of Food Technology*, vol. 19, 1984, pp. 225–31.

28. H. Koksel, S. Celik and T. Tuncer, 'Effects of gamma irradiation on duram wheats and spaghetti quality', *Cereal Chemistry*, vol. 73, no. 4, 1996, pp. 506–9.

29. Ibid.

30. J. Brand Miller and S. Colagiuri, *The G.I. factor*, Hodder and Stroughton, Rydalmere, NSW, 1996.

31. Facts about Food Irradiation, op. cit., p. 12.

32. Supplement, 'Antioxidant vitamins and b–carotene in disease prevention', *American Journal of Clinical Nutrition*, vol. 62, no. 6S, December 1995, pp. 1299S–1535S.

33. Rice-Evans and N. J. Miller, 'Antioxidants—the case for fruit and vegetables in the diet', *British Food Journal*, vol. 97, no. 9, 1995, pp. 35–40.

34. M. Bitomsky, 'Tomato sauce may protect prostates', *Medical observer*, 29 March 1996, p. 15.

35. N. Ramarathnam, T. Osawa, H. Ochi and S. Kawakishi, 'The contribution of plant food antioxidants to human health', *Trends in Food Science and Technology*, vol. 6, March 1995, pp. 75–82.

CHAPTER 10

1. W. Bown, 'Trees Trigger Heavy–Metal "Time Bomb"', *New Scientist*, 12 January 1991, p. 10.
2. Ibid.
3. Ibid.
4. P. Boffetta, 'Carcinogenicity of trace elements with reference to evaluations made by the International Agency for Research on Cancer', *Scandinavian Journal of Work, Environment and Health*, vol. 19, suppl. no. 1, 1993, pp. 67–70.
5. M. Fleischer, A. F. Sarofim, et al., 'Environmental Impact of Cadmium: A Review by the Panel on Hazardous Trace Substances', *Environmental Health Perspectives*, no. 7, May 1974, pp. 254–5 and 262–4.
6. Ibid., p. 254.
7. B. L. Carson, H. V. Ellis III and J. L. McCann, Toxicology and Biological *Monitoring of Metals in Humans*, Lewis Publishers, Chelsea, Michigan, 1986, p. 52. See also: M. Fleischer, et al., op. cit., p. 263.
8. G. P. Marletta, et al., 'Cadmium in Roadside Grapes', *Food Science and Technology Abstracts*, (FSTA), vol. 19, no. 6, 1987, p. 70. See also: L. G. Favretto, et al., 'Cadmium and Lead Pollution in Grapes and Wine', *FSTA*, vol. 19, no. 3, 1987, p. 40.
9. B. L. Carson, H. V. Ellis III, and J. L. McCann, op. cit., p. 51.
10. 'Guide to Low Cadmium Phosphate Supplements', *Farm*, August 1990, p. 16.
11. K. C. Jones and A. E. Johnston, 'Cadmium in Cereal Grain and Herbage From Long-Term Experimental Plots at Rothamsted, UK', *FSTA*, vol. 21, no. 10, 1989, p. 139.
12. R. Srikanth, D. Ramana and R. Vasant, 'Role of rice and cereal products in dietary cadmium and lead intake among different socio-economic groups in South India', *Food Additives and Contaminants*, vol. 12, no. 5, 1995, pp. 695–701.
13. V. G. Mineev, et al., 'Cadmium in Soil and Potatoes', *FSTA*, vol. 20, no. 4, 1988, p. 69.
14. D. Andrey, et al., 'Monitoring Programme For Heavy Metals in Food II. Lead, Cadmium, Zinc and Copper in Swiss Potatoes', *FSTA*, vol. 22, no. 10, 1990, p. 124.
15. L. Yin–Ming, R. L. Chaney, A. A. Schneiter, et al., 'Genotype variation in kernel cadmium concentration in sunflower germplasm under varying soil conditions', *Crop Science*, vol. 35, no. 1, 1995, pp. 137–41.
16. C. Mena, C. Cabrera, M. L. Lorenzo and M. C. Lopez, 'Cadmium levels in wine, beer and other alcoholic beverages: possible sources of contamination', *FSTA*, vol. 28, no. 6, 1996, p. 107.
17. J. Simpson and B. Curnow, 'Cadmium Accumulations in Australian Agriculture', *Proceedings no. 2*, Bureau of Rural Resources, Canberra, 1988.
18. State of the Environment Advisory Council, *Australia State of the Environment 1996*, CSIRO Publishing, Collingwood, Victoria, 1996, p. 6–49.
19. Ibid.
20. 'Guide to low cadmium phosphate supplements', op. cit.

21. S. Armstrong, 'Marooned in a Mountain of Manure', *New Scientist*, 26 November 1988, p. 54.
22. State of the Environment Advisory Council, op. cit.
23. C. Veron, 'Mechanisms of Contamination of the Food Chain by Cadmium, Principally the air-soil-plant-animal-human trophic chain', *FSTA*, vol. 23, no. 6, 1991, p. 55.
24. H. C. Harrison and J. E. Staub, 'Effects of Sludge, Bed, and Genotype on Cucumber Growth and Elemental Concentrations in Fruit and Peel', *FSTA*, vol. 19, no. 7, 1987, p. 100.
25. A. Coghlan, 'Lamb's liver with cadmium garnish', *New Scientist*, 22 March 1997, p. 4.
26. R. M. Harrison and M. B. Chirgawi, 'The Assessment of Air and Soil as Contributors of Some Trace Metal to Vegetable Plants II. Translocation of Atmospheric and Laboratory Generated Cadmium Aerosols To and Within Vegetable Plants', *FSTA*, vol. 22, no. 10, 1990, p. 103. See also: B. L. Carson, et al., p. 55.
27. K. Stein and F. Umland, 'Trace Analysis of Lead, Cadmium and Manganese in Honey and Sugar', *FSTA*, vol. 19, no. 8, 1987, p. 77.
28. G. P. Lewis, et al., 'Contribution of Cigarette Smoking to Cadmium Accumulation in Man', *The Lancet*, (I), 1972, p. 291.
29. C-G. Elinder, L. Friberg, B. Lind and M. Jawaid, 'Lead and cadmium levels in blood samples from the general population of Sweden', *Environmental Research*, vol. 33, 1983, pp. 233–53.
30. M. Fleischer, et al., op. cit., p. 288.
31. B. L. Carson, et al., op. cit., p. 52.
32. M. Fleischer, et al., pp. 291–2.
33. K Nogawa, 'Biologic Indicators of Cadmium Nephrotoxicity in Persons with Low–Level Cadmium Exposure', *Environmental Health Perspectives*, vol. 54, 1984, pp. 163–9.
34. M. Kaneta, H. Hikichi, et al., 'Chemical Form of Cadmium (and Other Heavy Metals) in Rice and Wheat Plants', *Environmental Health Perspectives*, vol. 65, 1986, p. 33. See also: T. Watanabe, et al., 'Cadmium and Lead Contents in Rice Available in Various Areas of Asia', *FSTA*, vol. 21, no. l0, 1989, p. 145.
35. G. F. Nordberg, 'Chelating Agents and Cadmium Toxicity: Problems and Prospects', *Environmental Health Perspectives*, vol. 54, 1984, pp. 213–18.
36. B. L. Carson, et al., op. cit., p. 54.
37. M. Fleischer, et al., op. cit., pp. 292–3. See also: B. L. Carson, et al., op. cit., p. 56.
38. J. B. Smith, L. Smith, V. Pijuan, et al., 'Transmembrane signals and protooncogene induction evoked by carcinogenic metals and prevented by zinc', *Environmental Health Perspectives*, vol. 102, suppl. 3, 1994, pp. 181–9.
39. S. Hechtenberg and D. Beyersmann, 'Interference of cadmium with ATP-stimulated nuclear calcium uptake', *Environmental Health Perspectives*, vol. 102, suppl. 3, 1994, pp. 265–7.
40. M. P. Waalkes and S. Rehm, 'Carcinogenicity of oral cadmium in the male Wistar (WF/NCr) rat: effect of chronic dietary zinc deficiency', *Fundamental and Applied Toxicology*, vol. 19, 1992, pp. 512–20.
41. M. P. Waalkes and S. Rehm, C. W. Riggs, et al., 'Cadmium carcinogenesis in male Wistar [Crl'(WI)BR] rats: dose-response analysis of tumour induction in

the prostate and testes and at the injection site', *Cancer Research*, vol. 48, 1988, pp. 4656–5663.

42. S. Takenaka, H. Oldiges, H. Konig, et al., 'Carcinogenicity of cadmium chloride aerosols in wistar rats', *Journal of the National Cancer Institute*, vol. 70, 1983, pp. 367–73.

43. L. Stayner, R. Smith, M. Thun, et al., 'A dose-response analysis and quantitative assessment of lung cancer risk and occupational cadmium exposure', *Annals of Epidemiology*, vol. 2, no. 3, 1992, pp. 177–94.

44. Ibid.

45. H. A. Schroeder, J. J. Balassa, et al., 'Chromium, Cadmium and Lead in Rats: Effects on Life Span, Tumours and Tissue Levels', *Journal of Nutrition*, vol. 86, 1965, p. 63.

46. G. Maroni, D. Lastowski–Perry, et al., 'Effects of Heavy Metals on *Drosophila* Larvae and a Metallothionein cDNA', *Environmental Health Perspectives*, vol. 65, 1986, p. 107.

47. B. L. Carson, et al., op. cit., p. 52.

48. M. Nordberg, 'General Aspects of Cadmium: Transport, Uptake and Metabolism by the Kidney', *Environmental Health Perspectives*, vol. 54, 1984, p. 13.

49. K. T. Suzuki, 'Studies of Cadmium Uptake and Metabolism by the Kidney', *Environmental Health Perspectives*, vol. 54, 1984, pp. 21–30.

50. Ibid.

51. M. Fleischer, et al., op. cit., pp. 257–8.

52. M. Choudhary, L. D. Bailey, C. A. Grant and D. Leisle, 'Effects of zinc on the concentration of Cd and Zn in plant tissue of two durum wheat lines', *Canadian Journal of Plant Science*, vol. 75, no. 2, 1995, pp. 445–8.

53. S. Onosaka, K. Tanaka and M. G. Cherian, 'Effects of Cadmium and Zinc on Tissue Levels of Metallothionein', *Environmental Health Perspectives*, vol. 54, 1984, pp. 67–72.

54. K. Kostial, 'Effect of Age and Diet on Renal Cadmium Retention in Rats', *Environmental Health Perspectives*, vol. 54, 1984, pp. 51–6.

55. M. Fleischer, et al., op. cit., p. 297.

56. K. Vreman, N. G. van der Veen, et al., 'Transfer of Cadmium, Lead, Mercury and Arsenic From Feed into Milk and Various Tissues of Dairy Cows: Chemical and Pathological Data', *FSTA*, vol. 19, no. 6, 1987, p. 20.

57. J. M. Scarlett–Kranz, B. S. Shane, et al., ' Survey of Nitrate, Cadmium and Selenium in Baby Foods—Health Considerations', *FSTA*, vol. 19, no. 4, 1987, p. 29.

CHAPTER 11

1. R. Carson, *Silent Spring*, Hamish Hamilton, London, 1963, p. 48.

2. Ibid., p. 201.

3. Ibid., p. 90.

4. Ibid., p. 91.

5. Ibid., pp. 84–105.

6. Ibid., p. 241.

7. National Research Council, *Alternative Agriculture*, National Academy Press, Washington DC, 1989, p. 122.

8. L. Wickenden, *Our Daily Poison*, Hillman Books, New York, 1961, pp. 17–34.

9. R. Carson, op. cit., pp. 178–99.

10. Ibid., p. 174.
11. See for example W. F. Durham and C. H. Williams, 'Mutagenic, Teratogenic, and Carcinogenic Properties of Pesticides', *Annual Review of Entomology*, vol. 17, 1972, pp. 123–48.
12. 'Deadly sprays worse than useless', *New Scientist*, 23 November 1996, p. 7.
13. National Research Council, op. cit., pp. 121–2. See also: J. Reisner, 'Nader's New Beef', *Food Business*, 5 June 1989, p. 30.
14. S. K. Hoar, A. Blair, F. F. Holmes, et al., 'Agricultural herbicide use and risk of lymphomes and soft-tissue sarcoma', *Journal of the American Medical Association*, vol. 256, no. 9, 1986, pp. 1141–7.
15. National Research Council, *Alternative Agriculture*, National Academy Press, Washington DC, 1989, pp. 121–2.
16. L. Roberts, 'Pesticides and Kids', *Science* (USA), vol. 243, 10 March 1989, pp. 1280–1.
17. 'EPA says ban aldicarb', *Food Business*, 17 April 1989, p. 30.
18. J. M. Harrington, 'Safety and chemicals in the environment: Effects and epidemiology', *Chemistry and Industry*, 3 November 1979, p. 727.
19. 'Garden Cancer', *New Scientist*, 11 March 1995, p. 13.
20. 'Soviet Union planned paddy fields for Aral Sea', *New Scientist*, 20 May 1989, p. 9.
21. 'Deformities found in North Sea fish embryos', *New Scientist*, 27 January 1990, p. 4.
22. C. Lenghaus, 'Good enough to eat?' *New Scientist*, 6 May 1989, p. 52.
23. State of the Environment Advisory Council, *Australia State of the Environment 1996*, CSIRO Publishing, Collingwood, Victoria, 1996, p. 8–11.
24. Ibid.
25. H. V. Kuhnlein, O. Receveur, D. C. G. Muir, et al., 'Arctic indigenous women consume greater than acceptable levels of organochlorines', *FSTA*, vol. 28, no. 3, 1996, p. 77.
26. B. Hileman, 'Chemically induced endocrine disruption still concerns Great Lakes commission, *Chemical and Engineering News*, 9 October 1995, p. 30.
27. R. H. Dunstan, M. Donohoe, W. Taylor, et al., 'A preliminary investigation of chlorinated hydrocarbons and chronic fatigue syndrome', *Medical Journal of Australia*, vol. 163, 1995, pp. 294–7.
28. E. Carlsen, A. Givercman and N. E. Skakkeback, 'Evidence for decreasing quality of semen during past 50 years', *British Medical Journal*, vol. 305, 1992, pp. 609–13.
29. N. Bauman, 'Panic over falling sperm counts may be premature', *New Scientist*, 11 May 1996, p. 10.
30. A. Abell, E. Ernst and J. P. Bonde, 'High sperm density among members of organic farmers' association', *The Lancet*, 11 June 1994, p. 1498.
31. B. Hileman, 'Chemically Induced endocrine disruption still concerns Great Lakes commission', op. cit.
32. State of the Environment Advisory Council, op. cit., p. 6–33.
33. Ibid, p. 4–14
34. Ibid, p. 6–33.
35. E. L. Gunderson, 'FDA Total diet study, July 1986–April 1991, Dietary Intakes of Pesticides, Selected Elements and other chemicals', *Journal of AOAC International*, vol. 78, no. 6, 1995, pp. 1353–63.

36. B. Hileman, 'Pesticides in food may be higher than FDA says', *Chemical and Engineering News*, 27 February 1995, pp. 8–9.
37. Ibid.
38. 'Deadly sprays worse than useless', op. cit.
39. Ibid.
40. Ibid.
41. Ibid.
42. A. Howard, *An Agricultural Testament*, Oxford University Press, London, 1943, p. 30.
43. National Research Council, *Alternative Agriculture*, National Academy Press, Washington, DC, 1989. See also: J. Hodges, *Natural Gardening and Farming in Australia*, Viking O'Neil, Ringwood, Victoria, 1991.
44. B. Mollison, *Permaculture, A designers manual*, Tagari Publications, Tyalgum, Australia, 1992.

CHAPTER 12

1. L. Hodges, *Environmental Pollution*, 2nd ed., Holt, Rinehart and Winston, New York, 1977, p. 189.
2. W. J. Llewellyn, *Journal of the American Medical Association*, vol. 146, no. 13, 1951, p. 1273.
3. Ibid.
4. L. Clark, *Get Well Naturally*, ARC Books, New York, 1971, p. 327.
5. R. A. Passwater, *Super-Nutrition for Healthy Hearts*, Jove Publications, New York, 1978, p. 155.
6. J. M. Price, *Coronaries, Cholesterol, Chlorine*, Pyramid Publications Ltd., Banhadlog Hall, Tylwich, Llanidloes, 1984, pp. 32, 33.
7. N. W. Revis, P. McCauley, R. Bull and G. Holdsworth, 'Relationship of drinking water disinfectants to plasma cholesterol and thyroid hormone levels in experimental studies', *Proceedings of the National Academy of Sciences*, USA., vol. 83, March 1986, p. 1485–9.
8. P. H. S. Zierler, L. Feingold, et al., 'Bladder cancer in Massachusetts related to chlorinated and chloraminated drinking water—a case-control study', *Archives of Environmental Health*, vol. 43, no. 2, 1988, pp. 195–200.
9. K. P. Cantor, R. Hoover, et al., 'Bladder cancer, drinking water source, and tap water consumption: a case-control study', *Journal of the National Cancer Institute*, vol. 79, no. 6, 1987, pp. 1269– 79.
10. M. A. Mc Geehin, J. S. Reif, J. C. Becher and E. J. Mangione, 'Case-control study of bladder cancer and water disinfection methods in Colorado', *American Journal of Epidemiology*, vol. 138, no. 7, 1993, pp. 492–501.
11. J. K. Dunnick and R. L. Melnick, 'Assessment of the carcinogenic potential of chlorinated water: experimental studies of chlorine, chloramine, and trihalomethanes', *Journal of the National Cancer Institute*, vol. 85, no. 10, 1993, pp. 817–22.
12. M. K. Smith, et al., 'Developmental Toxicity of Halogenated Aentonitriles: Drinking Water By-Products of Chlorine Disinfection', *Toxicology*, vol. 46, 1987, pp. 83–93.
13. P. Backlund, E. Wondergem, et al., 'Mutagenic activity and presence of the strong mutagen 3–chloro–4–(dichloromethyl)–5–hydroxy–2 (5H)–furanone (MX) in chlorinated raw and drinking waters in the Netherlands', *Science of the Total Environment*, vol. 84, 1989, pp. 273–82.

14. J. R. Meier, R. B. Knohl, et al., 'Studies on the potent bacterial mutagen, 3–chloro–4–(dichloromethyl)–5 hydroxy–2 (5H)–furanone: aqueous stability, XAD recovery and analytical determination in drinking water and in chlorinated humic acid solutions', *Mutation Research, Genetic Toxicology Testing*, vol. 189, no. 4, 1987, pp. 363–73.

15. B. C. Nicholson, 'Disinfection and Water Treatment: Formation of chemical by-products', *Chemistry in Australia*, vol. 61, no. 8, 1994, pp. 442–4.

16. R. D. Morris, A. M. Audet, I. F. Anelillo, et al., 'Chlorination, chlorination by-products, and cancer: a meta–analysis', *American Journal of Public Health*, vol. 82, no. 7, 1992, pp. 955–63.

17. L. S. Pilotto, 'Disinfection of drinking water, disinfection by-products and cancer: what about Australia?' *Australian Journal of Public Health*, vol. 19, no. 1, 1995, pp. 89–93.

18. N. W. Revis, et al., op. cit.

19. C. P. Weisel and W. J. Chen, 'Exposure to chlorination by-products from hot water uses', *Risk Analysis*, vol. 14, no. 1, 1994, pp. 101–6.

20. F. H. J. Rampen, P. J. Nelemans and A. L. M. Verbeek, 'Is water pollution a cause of cutaneous melanoma?' *Epidemiology*, vol. 3, no. 3, 1992, pp. 263–5.

21. P. J. Nelemans, F. H. Rampen, H. Groenendal, et al., 'Swimming and the risk of cutaneous melanoma', *Melanoma Research*, vol. 4, no. 5, 1994, pp. 281–6.

22. J. O.'Connell, 'Organochlorines in the spotlight', *Chemistry in Australia*, January/February, 1997, pp. 14–15.

23. Ibid.

24. M. Sim, C. Fairley and J. Mc Neil, 'Drinking water quality: new challenges for an old problem', *Occupational and Environmental Medicine,* vol. 53, 1994, pp. 649–51.

25. W. R. MacKenzie, N. J. Hoxie, M. E. Proctor, et al., 'A massive outbreak in Milwaukee of Cryptosporidium infection transmitted through the public water supply', *New England Journal of Medicine*, vol. 331, 1994, pp. 161–7.

26. A. Acra, M. Jurdi, H. Mu'allem, et al., 'Sunlight as disinfectant', *Lancet*, (i), 1989, p. 280.

27. T. M. Joyce, K. G. Mc Guigan, M. Elmore-Meegan and R. M. Conroy, 'Inactivation of fecal bacteria in drinking water by solar heating', *Applied and Environmental Microbiology*, vol. 62, no. 2, 1996 pp. 399–402.

28. J. Hecht, 'Sunlight gives toxic waste a tanning', *New Scientist*, 14 April 1990, p. 16.

29. V. Hasselbarth, 'Hygienic requirements for disinfection of drinking water and water for food factories by UV irradiation, *FSTA*, vol. 28, no. 12, 1996, p. 95.

30. State of the Environment Advisory Council, *Australia State of the Environment*, CSIRO Publishing, Collingwood, Victoria, 1996, pp. 5-10–11, 5-18–19.

CHAPTER 13

1. N. I. Sax, *Dangerous Properties of Industrial Materials*, 2nd ed., Reinhold Publishing Corp., New York, 1963, p. 1187.

2. L. Hodges, *Environmental Pollution*, 2nd ed., Holt, Rinehart and Winston, New York, 1977, p. 64.

3. H. J. Sanders, 'Tooth decay', *Chemical and Engineering News*, 25 February 1980, p. 36.

4. H. T. Dean, 'Studies on the Minimal Threshold of the Dental Sign of Chronic Endemic Fluorosis', *Public Health Reports*, 1934; vol. 50, pp. 1719–29. See also: H. T. Dean, 'Chronic endemic dental fluorosis (mottled enamel)', *Journal of the American Medical Association*, vol. 107, 1936, pp. 1269–72.

5. H. T. Dean, 'Dental fluorosis and its relation to dental caries', *Public Health Reports*, vol. 53, 1938, pp. 1443–52. See also: H. T. Dean, E. Elvove and R. F. Poston, 'Mottled enamel in South Dakota', *Public Health Reports*, vol. 54, 1939, pp. 212–28.

6. National Health and Medical Research Council, *The Effectivemness of Water Fluoridation*, Australian Government Publishing Service, Canberra, 1991, pp. 28–34, 147.

7. W. Varney, *Fluoride in Australia—A Case to Answer*, Hale and Iremonger, Sydney, 1986, p. 14.

8. J. J. Murray, A. J. Rugg-Gunn, *Fluoridesand Dental Caries*, 2nd ed., Wright, Bristol. UK, 1982.

9. E. Newbrun, 'Effectiveness of Water Fluoridation', *Journal of Public Health Dentistry*, vol. 49, no. 5, 1989, pp. 279–88.

10. National Helath and Medical Research Council, 'Fluorides in the control of dental caries', *Journal of Food and Nutrition*, vol. 42, no. 4, 1985, pp. 178–88.

11. M. Diesendorf, 'The mystery of declining tooth decay', *Nature*, vol. 322, 1986, pp. 125–9.

12. M. Diesendorf, 'Fluoridation: New Evidence of harm to young teeth and old bones', *International Clinical Nutrition Review*, vol. 12, no. 1, 1992, pp. 1–8.

13. National Health and Medical Research Council, 1991, op. cit., pp. 179–82.

14. Ibid., p. 31.

15. Ibid., p. 27.

16. Ibid., pp. 19–165.

17. M. Diesendorf, 'Have the benefits of water fluoridation been overestimated', *International Clinical Nutrition Review*, vol. 10, no. 2, 1990, pp. 292–303.

18. Committee on Food Protection, Food and Nutritional Board National Research Council, *Toxicants Occurring Naturally in Foods*, National Academy of Sciences, Washington DC, 1973, pp. 72–4.

19. Y. Z. Han, J. Q. Zhang, X. Y. Lui, et al., 'High fluoride content of food and endemic fluorosis', *Fluoride*, vol. 28, no. 4, 1995, pp. 201–2.

20. National Health and Medical Research Council, 1991, op. cit., p. 125.

21. State of the Environment Advisory Council, *Australia State of the Environment 1996*, CSIRO Publishing, Collinwood, Victoria, 1996, p. 5–30.

22. B. D. Gessner, M. Beller, J. P. Middaugh and G. M. Whitford, 'Accute fluoride poisoning form a public water system', *New England Journal of Medicine*, vol. 330, 1994, pp. 95–9.

23. Ibid.

24. J. Colquhoun, 'Flawed foundation: A re-examination of the scientifc basis for a dental benefit from fluoridation', *Community Health Studies*, vol. xiv, no. 3, 1990, pp. 288–96.

25. J. Colquhoun, 'Fluoridation: New Evidence of harm to young teeth and old bones', *International Clinical Nutrition Review*, vol. 12, no. 1, 1992, pp. 1–8.

26. J. Yiamouyiannis and D. Burk, 'Fluoridation and Cancer. Age Dependence of Cancer Mortality Related to Artificial Fluoridation', *Fluoride*, vol. 10, 1977, pp. 102–23.

27. Ibid, p. 105.

28. Ibid, p. 120.
29. G. Smith, 'Fluoridation—are the dangers resolved?', *New Scientist*, 5 May 1983, p. 286.
30. J. Ylamouyiannis, *Fluoride The Aging Eactor*, Health Action Press, Delaware, Ohio, 1983. See also: ibid.
31. M. Diesendorf, 'Fluoride: New Risk?', *Search*, vol. l6, no. 54, 1985, p. 129.
32. Ibid.
33. B. Hileman, 'Fluoride/Cancer: Equivocal link in rats endorsed', *Chemical and Engineering News*, 7 May 1990, p. 4.
34. J. A. Yiomouyiannis, 'Fluoridation and cancer—the biology and epidemiology of bone and oral cancer related to fluoridation', *Fluoride*, vol. 26, no. 2, 1993, pp. 83–96.
35. J. R. Lee, 'Fluoridation and bone cancer', *Fluoride*, vol. 26, no. 2, 1993, pp. 79–82,
36. J. A. Yiomouyiannis, 1993, op. cit.
37. J. R. Lee, 'Fluoridation and hip fracture. According to the National Research Council Report: Health effects of ingested fluoride', *Fluoride*, vol. 26, no. 4, 1993, pp. 274–7.
38. S. J. Jacobsen, et al., 'Regional variation in the incidence of hip fracture', *Journal of the American Medical Association*, vol. 264, 1990, pp. 500–2.
39. P. R. Rockwood, et al., 'Hip fractures: a future epidemic?' *Journal of Orthopedic Trauma*, vol. 4, 1990, pp. 388–93.
40. M. R. Sowers, et al., 'A prospective study of bone mineral content and fracture in communities with differential fluoride exposure', *American Journal of Epidemiology*, vol. 133, 1991, pp. 649–60.
41. A. W. Burgstahler and J. Colquhoun, 'Neurotoxicity of fluoride', *Fluoride*, vol. 29, 1996, pp. 57–8.
42. Ibid.
43. H. C. Hodge and F. A. Smith, 'Minerals: Fluorine and Dental Caries' in R. F. Gould (ed.) *Dietary Chemicals Vs Dental Caries*, American Chemical Society, Washington DC, 1970, p. 93.
44. E. J. Underwood, *Toxicants Occurring Naturally in Foods*, National Academy of Sciences, Washington DC, 1973, p. 72. See also: ibid. p. 94.
45. H. Sinclair, 'Fluoridation Falsified', *British Medical Journal*, 29 February 1964, p. 554.
46. H. C. Hodge and F. A. Smith, op. cit., p. 98.
47. D. Pepin and M. P. Sauvant, 'Fluoride and mineral water: influence of the mineralisation of water on the urinary excretion of fluoride in the healthy adult', *FSTA*, vol. 28, no. 12, 1996, p. 93.
48. E. J. Underwood, op. cit., p. 73.
49. J. Colquhoun, 1992, op. cit.
50. S. P. S. Teotia and M. Teotia, 'Endemic skeletal fluorosis: clinical and radiological variants (Review of 25 years of personal research), *Fluoride*, vol. 21, 1988, pp. 39–44.
51. G. Smith, op. cit., p. 287.
52. G. N. Jenkins, 'Recent advances in work on fluorides and teeth', *British Medical Bulletin*, vol. 31, no. 2, 1975, pp. 142–5.
53. H. C. Hodge and F. A. Smith, op. cit., p. 99.
54. E. J. Underwood, op. cit., p. 72. See also: ibid., pp. 96–7.
55. H. Sinclair, op. cit.

56. A. Singh, S. S. Jolly and B. C. Bansal, 'Skeletal fluorosis and its neurological complications', *Lancet*, (I), 1961, pp. 197–200.
57. H. Sinclair, op. cit.
58. J. A. Yiamouyiannis, 1983, op. cit.
59. S. G. Srikantia, 'Endemic fluorosis in Andhra Pradesh', *Fluoride*, vol. 18, 1985, pp. 66–7.
60. P. Park, 'Soy cheese for a sunny smile with fewer fillings', *New Scientist*, 6 April 1991, p. 13.
61. A. K. Adatia, 'Dental caries and periodontal disease', in D. P. Burkitt and H. C. Troqwell (eds), *Refined Carbohydrate Foods and Disease, Some Implications of Dietary Fibre*, Academy Press, London, 1975, p. 263.
62. Ibid., pp. 266, 268.
63. Ibid., pp. 269–70.
64. Ibid., p. 270.
65. F. Pearce, 'Tooth decay linked to persistent lead pollution', *New Scientist*, 18 January, 1997, p. 5.
66. Ibid.
67. W. Varbey, op. cit.
68. H. Bloom, I. C. Lewis and R. Kishimoto, 'A study of lead concentrations in blood of children (and some adults) of southern Tasmania', *Medical Journal of Australia*, vol. 2, 1974, pp. 275–9.
69. M. Diesendorf, 1986, op. cit.
70. Y. S. Ong, B. Williams and R. Holt, 'The effect of domestic water filters on water fluoride content', *British Dental Journal*, vol. 181, no. 2, 1996, pp. 59–63.

CHAPTER 14

1. A. Bernatzky, 'Ecological Principles in Town Planning: The Impact of Vegetation on the Quality of Life in the City' in R. Krieps (ed), *Environment and Health: A Holistic Approach*, Gower Publishing Co., Aldershot, England, 1989, p. 136.
2. Hippocrates, 'Airs, Waters, Places' in G. E. R. Lloyd (ed), *Hippocratic Writings*, Pelican Classic Edition, Penguin, Harmondsworth, England, 1978, pp. 148–69.
3. T. J. Dondero, R. C. Rendtorrf, et al., 'An Outbreak of Legionnaire's Disease Associated With a Contaminated Air Conditioning Cooling Tower', *New England Journal of Medicine*, vol. 302, 1980, pp. 365–70.
4. D. W. Fraser, T. R. Tsai, et al., 'Legionnaire's Disease. Description of an Epidemic of Pneumonia', *NewEngland Journal of Medicine*, vol. 297, 1977, pp. 1189–97.
5. M. J. Finnegan, 'Air Conditioning and Disease', *The Practitioner*, vol. 231, 8 April 1987, pp. 482–5.
6. Ibid.
7. Ibid.
8. P. Carter, 'Tobacco Smoke, Ventilation and Indoor Air Quality', *Australian Refrigeration, Air Conditioning and Heating*, April 1983, pp. 13–16.
9. J. L. Repace and A. H. Lowrey, 'Indoor Air Pollution, Tobacco Smoke, and Public Health', *Science*, vol. 208, 1980, p. 464.
10. P. Carter, op. cit., p. 16.
11. See for example: F. Soyka and A. Edmonds, *The Ion Effect*, Bantam Books, New York, 1977.

12. Editorial, 'Airs, Waters, Places', *The Lancet*, 7 November 1987, pp. 1062–3.
13. M. J. Finnegan, 1987, op. cit., p. 482.
14. Editorial, 'Airs, Waters, Places', op. cit.
15. M. J. Finnegan and C. A. C. Pickering, 'Review. Building Related Illness', *Clinical Allergy*, vol. 16, 1986, pp. 389–405.
16. Editorial, 'Airs, Waters, Places', op. cit.
17. M. Glaskin, 'Kill the bugs to cure the building', *New Scientist*, 11 March 1995, p. 22.
18. Ibid .
19. A. T. Purcell and R. H. Thorne, 'The Thermal Environment and Level of Arousal: Field Studies to Ascertain the Effects of Air–conditioned Environments on Office Workers', *Architectural Science Review*, vol. 30, 1987, pp. 91–116.
20. H. Steven and A. Lindley, 'The Race to Heal the Ozone Hole', *New Scientist*, 16 June 1990, pp. 30–3.
21. R. A. Plumb, 'Atmospheric Ozone—Physics and Chemistry', *Transactions of the Menzies Foundation*, vol. 15, 1989, p. 3–13.
22. See for example related article on refrigeration: D. Olivier, 'A Cool Solution to Global Warming', *New Scientist*, 12 May 1990, pp. 20–3.
23. A. Bernatzky, op. cit., p. 140.
24. Ibid., p. 137.
25. Ibid.
26. Ibid., p. 144.
27. T. G. Randolph, 'Domiciliary chemical air pollution in the etiology of ecologic mental illness', *International Journal of Social Psychology*, vol. 16, 1970, pp. 223–65.
28. S. A. Rogers, 'Diagnosing the Tight Building Syndrome', *Environmental Health Perspectives*, vol. 76, 1987, p. 195.
29. Ibid.
30. Ibid.
31. N. Nelson, R. J. Levine, et al., 'Contribution of Formaldehyde to Respiratory Cancer', *Environmental Health Perspectives*, vol. 70, 1986, pp. 23–35. See also: R. B. Conolly and M. E. Andersen, 'An approach to mechanism-based cancer risk assessment for formaldehyde', *Environmental Health Perspectives*, vol. 101, suppl. 6, 1993, pp. 169–76.
32. 'Services improve occupational health and safety', *The Australian Government Analytical Laboratories Newsletter*, no. 4, Summer 1996, pp. 1, 3.
33. J. Scheirs and S. W. Bigger, 'Volatile organic emissions from polymeric materials', *Chemistry in Australia*, June 1996, pp. 261–4.
34. Ibid.
35. Ibid.
36. S. Brown, 'Volatile organic compounds in indoor air: sources and control', *Chemistry in Australia*, January/February 1997, pp. 10–13.
37. Ibid.
38. Ibid.
39. Ibid.
40. National Health and Medical Research Council, *National indoor air quality goal for total volatile organic compounds*, NHMEC, Canberra, 1992.
41. Purkis and J. Wilson, 'Cleaning up the cooling towers', *New Scientist*, 16 September 1989, p. 34.

42. Cited by R. Dubos, *Mirage of Health*, Harper and Row, New York, 1959, p. 148.

CHAPTER 15

1. C. F. Garland, F. C. Garland and E. D. Gorham, 'Could sunscreens increase melanoma risk', *American Journal of Public Health*, vol. 82, no. 2, 1992, pp. 614–15. See also: D. J. Leffell and D. E. Brash, 'Sunlight and skin cancer', *Scientific American*, July 1996, pp. 38–43.
2. M. S. Coates and J. M. Ford, 'Skin cancer in New South Wales' *Transactions of the Menzies Foundation*, vol. 15, 1989, pp. 149–52.
3. I. T. Ring, S. M. Walker, D. Firman, et al., 'Melanoma in Queensland', *Transactions of the Menzies Foundation*, vol. 15, 1989, pp. 157–63.
4. State of the Environment Advisory Council, *Australia State of the Environment 1996*, CSIRO Publishing, Collingwood, Victoria, 1996, pp. 3–30.
5. G. Giles, T. Dwyer, M. Coates, et al., 'Trends in Skin Cancer in Australia, an Overview of the Available Data', *Transactions of the Menzies Foundation*, vol. 15, 1989, pp. 143–7. See also: B. Cox and B. Coombs, 'Trends in Melanoma of the Skin in New Zealand', ibid., pp. 137–40.
6. J. A. H. Lee and D. Strickland, 'Malignant Melanoma: Social Status and Outdoor Work', *British Journal of Cancer*, vol. 41, 1980, pp. 757–63.
7. K. J. Goodman, M. L. Bible, S. London and T. M. Mack, 'Proportional Melanoma incidence and occupation among white males in Los Angeles County', *Cancer Causes and Control*, vol. 6, no. 5, 1995, pp. 451–9.
8. R. J. Reiter, 'Melatonin suppression by static and extremely low frequency electromagnetic fields: Relationship to the reported increased incidence of cancer', *Reviews on Environmental Health*, vol. 10, no. 3–4, 1994, pp. 171–86.
9. A. L. Rouco and F. Halberg, 'The pineal gland and cancer', *Anticancer Research*, vol. 16, no. 4A, 1996, pp. 2033–9.
10. R. Olin, D. Vagero and A. Ahlbom, 'Mortality experience of electrical engineers', *British Journal of Industrial Medicine*, vol. 42, 1985, pp. 211–13.
11. J. A. G. Holt, 'The cause of cancer: biochemical defects in the cancer cell demonstrated by the effects of electromagnetic radiation, glucose and oxygen', *Medical Hypotheses*, vol. 5, 1979, pp. 109–43. See also: J. A. G. Holt, 'Some characteristics of the glutathione cycle revealed by ionising and non-ionising electromagnetic radiation', *Medical Hypotheses*, vol. 45, 1995, pp. 345–68.
12. J. A. G. Holt, 'Changing epidemiology of malignant melanoma in Queensland', *Medical Journal of Australia*, vol. 1, no. 12, 1980, pp. 619–20.
13. P. N. Karnauchow, 'Melanoma and sun exposure', *The Lancet*, 30 September 1995, p. 915. See also: J. M. Elwood, 'The Epidemiology of Melanoma: Its relationship to Ultraviolet radiation and ozone depletion, in 'The Ozone Layer and Health', *Transactions of the Menzies Foundation*, vol. 15, 1989, pp. 95–106; and Editorial, 'The Aetiology of Melanoma', *The Lancet*, 31 January 1981, pp. 253–5.
14. R. MacKie, J. A. A. Hunter, T. C. Aitchison, et al., 'Cutaneous malignant melanoma, Scotland, 1979–89', *The Lancet*, vol. 339, 1992, pp. 971–5.
15. M. Balter, 'Europe: as many cancers as cuisines', *Science*, vol. 254, 1991, pp. 114–15.
16. J. F. Ashton and R. S. Laura, 'Let there be light', *Search*, vol. 21, no. 8, 1990, pp. 268–9.

17. R. J. Wurtman, 'Effects of light on the human body', *Scientific American*, July 1975, pp. 69–77.

18. R. J. Wurtman, 'The Action of light on Man and mammals', in T. B. Fitzpatrick (ed.), *Sunlight and Man*, University of Tokyo Press, Tokyo, 1974, pp. 237–40. See also: R. L. Edelson and J. M. Fink, 'The Immunologic Function of Skin', *Scientific American*, June 1985, pp. 34–41.

19. W. E. Stumpf, 'The endocrinology of sunlight and darkness. Complementary roles of vitamin D and pineal hormones', *Naturwissenschaften*, vol. 75, no. 5, 1988, pp. 247–51.

20. C. S. Baker, 'Photosensitivity', *Medical Journal of Australia*, vol. 165, 15 July, 1996, pp. 96–101.

21. S. W. Young, 'A List of Photosensitizing Agents of Interest to the Dermatologist', *Bulletin of the Association of Military Dermatologists*, vol. 13, March 1964, pp. 33–5. See also: J. H. Epstein, 'Phototoxicity and Photoallergy: Clinical Syndromes', in *Sunlight and Man*, op. cit., pp. 459–69.

22. M. J. Ashwood-Smith, 'Possible Cancer Hazard Associated With 5–Methoxypsoralen in Suntan Preparations', *British Medical Journal*, 3 November 1979, p. 1144.

23. F. Daniels, 'Physiological and pathological extracutaneous effects of light on man and mammals not mediated by pineal or other neuroendocrine mechanisms' in T. B. Fitzpatrick (ed) *Sunlight and Man*, University of Tokyo Press, Tokyo, 1974, p. 253.

24. S. I. Lamberg, 'A New Photosensitizer', *Journal of the American Medical Association*, vol. 201, no. 10, 1967, pp. 1214.

25. C. S. Baker, op. cit., p. 100.

26. T. B. Fitzpatrick, M. A. Pathak, L. C. Harber, et al., 'An introduction to the problem of normal and abnormal responses of man's skin to solar radiation', in T. B. Fitzpatrick (ed.), *Sunlight and Man*, University of Tokyo Press, Tokyo, 1974, pp. 3–14.

27. J. A. H. Lee and B. E. Storer, 'Excess of Malignant Melanomas in Women in the British Isles', *The Lancet*, 20/27 December 1980, pp. 1337–9; See also: V. Beral, S. Ramcharan and R. Faris, 'Malignant Melanoma and Oral Contraceptive Use Among Women in California', *British Journal of Cancer*, vol. 36, 1977, pp. 804–9.

28. P. J. Nelemans, F. H. Rampen, H. Groenendal, et al., 'Swimming and the risk of cutaneous melanoma', *Melanoma Research*, vol. 4, no. 5, 1994, pp. 281–6.

29. C. A. Baumann and H. P. Rusch, 'Effect of Diet on Tumors Induced by Ultraviolet Light', *American Journal of Cancer*, vol. 35, 1939, pp. 213, 220. See also: V. E. Reeve, M. Matheson, G. E. Greenoak, et al., 'Effect of dietary lipid on UV light carcinogenesis in the hairless mouse', *Photochemisty and Photobiology*, vol. 48, no. 5, 1988, pp. 689–96.

30. B. S. Mackie, A. R. Johnson, L. E. Mackie, et al., 'Dietary Polyunsaturated Fats and Malignant Melanoma', *Medical Journal of Australia*, 23 February 1980, pp. 159–63; M. R. Chedekel and L. Zeise, 'Sunlight, Melanogenesis and Radicals in the Skin', *Lipids*, vol. 23, no. 6, 1988, p. 590.

31. K. L. Erickson, 'Dietary fat influences on murine melanoma growth and lymphocyte-mediated cytotoxicity', *Journal of the National Cancer Institute*, vol. 72, 1984, pp. 115–20.

32. R. A. Buist, 'Malignant Melanoma and dietary polyunsaturated fats', *International Clinical Nutrition Review*, vol. 8, no. 2, 1988, pp. 53–4.

33. R. B. Cope. M. Bosnic, C. Boehm-Wilcox, et al., 'Dietary butter protects against ulraviolet radiation-induced suppression of contact hypersensitivity in Skh: HR–1 hairless mice', *Journal of Nutrition*, vol. 126, 1996, pp. 681–92.

34. V. E. Reeve, R. B. Cope, M. Bosnic, et al., 'Protection from ultraviolet radiation pathology by topical and systemic agents', *Environmental Dermatology*, vol. 3, suppl. 1, 1996, pp. 20–5. See also: V. E. Reeve, M. Bosnic and C. Boehm-Wilcox, 'Dependence of photocarcinogenesis and photoimmunosuppression in the hiarless mouse on dietary polyunsaturated fat', *Cancer Letters*, 1997.

35. B. A. Martin, R. R. Wilson and R. W. Rinne, 'Temperature effects upon the expression of a high oleic trait in soybean', *Journal of the American Oil Chemists Society*, vol. 63, no. 3, 1986, pp. 346–52. See also: A. G. Green, 'Effect of temperature during seed maturation on the oil composition of low-linoleic genotypes of flax', *Crop Science*, vol. 26, September–October 1986, pp. 961–5.

36. L. Thomas and M. Rowney, 'Australian milkfat survey—fatty acid composition', *Australian Journal of Dairy Technology*, vol. 51, October 1996, pp. 112–17.

37. B. W. Simpson, C. M. McLeod and D. L. George, 'Selection for high linoleic acid content in sunflower (*Helianthus annus* L.)', *Australian Journal of Experimental Agriculture*, vol. 29, 1989, pp. 233–9.

38. US Council For Coconut Research/Information, A pamphlet on coconut fats (no. title), Washington DC, 1990.

39. J. A. Driskell, 'Vitamin B6', in *Handbook of Vitamins*, L. J. Machlin (ed.), Marcel Dekker, New York, 1984, p. 385.

40. H. M. Sinclair, 'Essential Fatty Acids and Their Relation to Pyridoxine', in 'Lipid Metabolism', *Biochemical Society Symposia*, no. 9, 1952, p. 92.

41. V. E. Reeve, M. Bosnic and C. Boehm-Wilcox, 'Pyridoxine supplementation protects mice from suppression of contact hypersensitivity induced by 2–acetyl–4–tetrahydroxybutyl–imidazole (THI), ultraviolet B radiation (280–320nm) or cia–urocanic acid', *American Journal of Clinical Nutrition*, vol. 61, 1995, pp. 571–6.

42. C. Bain, A. Green, V. Sisking, et al., 'Diet and melanoma: an exploratory case-control study', *Annals of Epidemiology*, vol. 3, 1993, pp. 235–8.

43. *National Dietary Survey of Adults—No2, Nutrient Intakes*, Australian Government Publishing Service, Canberra, 1983.

44. K. I. Baghurst, I. Dreosti, J. Syrette, et al., *Nutrition Research*, vol. 11, 1991, pp. 23–32.

45. G. Block and B. Abrams, *Annals of the New York Academy of Sciences*, vol. 678, 1993, pp. 244–54.

46. C. Bain, A. Green, V. Sisking, et al., op. cit.

47. M. Bosnic, N. Nishimura and V. Reeve, 'Zinc, Metallothionein (MT) and UV radiation (UVR)', *Australian Society for Experimental Pathology Conference Abstracts*, University of New South Wales, Sydney, October 1996.

48. H. L. Gensler, K. E. Gerrish, T. Williams, et al., 'Prevention of photo-carcinogenesis and UV–induced immunosuppression in mice by topical tannic acid', *Nutrtion and Cancer*, vol. 22, 1994, pp. 121–30.

49. V. E. Reeve, M. Bosnic, E. Rozininova and C. Boehm-Wilcox, 'A garlic extract protects from ultra violet B (280–320nm) radiation-induced suppression of contact hypersensitivity', *Photochemistry and Photobiology*, vol. 58, no. 6, 1993, pp. 813–17.

50. A. C. Giese, op. cit., p. 111.
51. C. Bain, A. Green, V. Sisking, et al., op. cit.
52. Editorial, 'The Aetiology of Melanoma', *The Lancet*, 31 January 1981, pp. 254–5.
53. J. F. Ashton and R. S. Laura, 'Environmental factors and the etiology of melanoma', *Cancer Causes and Control*, vol. 4, no. 1, 1993, pp. 59–62.
54. C. H. Gallagher, V. E. Reeve, G. E. Greenoak and P. J. Cranfield, 'The influence of sunscreens on UV-induced skin cancer', *Transactions of the Menzies Foundation*, vol. 15, 1989, pp. 201–10.
55. P. Autier, J-F. Dore, E. Schifflers, et al., 'Melanoma and use of sunscreens: a case-control study in Germany, Belgium and France', *International Journal of Cancer*, vol. 61, 1995, pp. 749–55. See also: T. Patel, 'We're browned off says suntan firm', *New Scientist*, 20 May 1995, p. 7.
56. Ibid. See also: K. Colston, M. J. Colston and D. Feldman, '1, 25–Dihydroxyvitamin D3 and Malignant Melanoma: The Presence of Receptors and Inhibition of Cell Growth in Culture', *Endocrinology*, vol. 108, no. 3, 1981, pp. 1083–6; C. Garland, R. B. Shekelle, et al., 'Dietary Vitamin D and Calcium and Risk of Colorectal Cancer: A 19-Year Prospective Study in Men', *The Lancet*, 9 February 1985, pp. 307–9.
57. O. Klepp and K. Magnus, 'Some environmental and bodily characteristics of melanoma patients. A case-control study', *International Journal of Cancer*, vol. 23, 1979, pp. 482–6.
58. S. Graham, J. Marshall, B. Haughey, et al., 'An enquiry into the epidemiology of melanoma', *American Journal of Epidemiology*, vol. 122, 1985, pp. 606–19.
59. H. Beitner, S. E. Norell, U. Ringborg, et al., 'Malignant melanoma: aetiological importance of individual pigmentation and sun exposure', *British Journal of Dermatology*, vol. 122, 1990, pp. 43–51.
60. J. Westerdahl, H. Olsson, A. Masback, et al., 'Is the use of sunscreens a risk factor for malignant melanoma? *Melanoma Research*, vol. 5, 1995, pp. 59–65.
61. Ibid., p. 63.
62. M. Peak, 'But does it last?', *New Scientist*, 26 March 1994, p. 51.
63. P. Wolf, C. K. Donawho and M. L. Kripke, 'Effect of sunscreens on UV radiation-induced enhancement of melanoma growth in mice', *Journal of the National Cancer Institute*, vol. 86, no. 2, 1994, pp. 99–105.
64. C. F. Garland, F. C. Garland and E. D. Gorham, 'Could suncreens increase melanoma risk?, *American Journal of Public Health*, vol. 82, no. 4, 1992, pp. 614–15.
65. Ibid.
66. Ibid.
67. V. E. Reeve, M. Bosnic and C. Kerr, 'Do suscreens fail to protect from skin carcinogenesis? A study in the hairless mouse', *NSW State Cancer Council Report*, Sydney, 1995.
68. Ibid.
69. L. Y. Matsuoko, L. Ide, J. Wortsman, et al., 'Sunscreens suppress cutaneous vitamin D3 synthesis', *Journal of Clinical Endocrinology and Metabolism*, vol. 64, 1987, pp. 1165–8. See also: L. Y. Matuoka, R. Freaney, A. Meade, et al., 'Chronic sunscreen use decreases circulating concentrations of 1, 25–hydroxyvitamin D', *Archives of Dermatology*, vol. 124, 1988, pp. 1802–4.

70. R. J. Frampton, S. A. Omond, J. A. Eisman, 'Inhibition of human cancer cell growth by 1, 25–dihydroxyvitamin D₃ metabolites', *Cancer Research*, vol. 43, 1983, pp. 4443–7. See also: K. Colson, M. J. Colston and D. Feldman, op. cit.
71. W. E. Stumf, 'The endocrinology of sunlight and darkness', *Naturwissenschaften*, vol. 75, 1988, pp. 247–51.
72. F. Daniels and B. E. Johnson, 'Normal, Physiologic, and Pathologic Effects of Solar Radiation on the Skin', in *Sunlight and Man*, op. cit., p. 127.
73. M. Luckiesh, *Artificial Sunlight*, D. Van Nostrand Co., New York, 1930, p. 33.
74. A. C. Giese, *Living with Our Sun's Ultraviolet Rays*, Plenum Press, New York, 1976, p. 20.
75. R. J. Wurtman and J. J. Wurtman, 'Carbohydrates and Depression', *Scientific American*, January 1989, p. 54.
76. F. Hollwich, *The Influence of Ocular Light Perception on Metabolism in Man and in Animal*, Springer–Verlag, New York, 1979, p. 89.
77. P. C. Hughes, 'The Use of Light and Colour in Health' in *Health for the Whole Person*, A. C. Hastings, et al. (eds), Westview Press, Boulder, Colorado, 1980, p. 287.
78. A. C. Giese, op. cit., p. 20.
79. Ibid., p. 28. See also: ibid., p. 2.
80. R. J. Wurtman, 'Effects of Light on the Human Body', *Scientific American*, July 1975, p. 70. See also: ibid., p. 28.
81. R. H. Warring, *All About Home Lighting*, Model and Allied Publications, Argus Books Ltd., Watford, England, 1977, p. 13.
82. R. J. Wurtman and J. J. Wurtman, op. cit., p. 54.
83. J. Arendt, *Melatonin and the Mammalian Pineal*, Chapman and Hall, London, 1996.
84. A. L. Ronco and F. Halberg, 'The Pineal Gland and Cancer', *Anticancer Research*, vol. 16, 1996, pp. 2033–40.
85. R. J. Wurtman, 'The Action of Light on Man and Mammals: Normal Physiologic and Pathologic Extracutaneous Effects' in *Sunlight and Man*, T. B. Fitzpatrick, et al. (eds), University of Tokyo Press, Tokyo, 1974, p. 234.
86. F. Hollwich, op. cit., p. 90.
87. 'Testing Facility for Sunscreens', *Laboratory News*, Sydney, April 1989, p. 13.
88. D. Downing, *Day Light Robbery*, Arrow Books Ltd., London, 1988, p. 102. See also: C. R. Roy, 'Ultraviolet Radiation: Sources, Biological Interaction and Personal Protection', in *Non-Ionising Radiations*, M. H. Repacholi (ed.), International Radiation Protection Association, Yallambie, Victoria, 1988, p. 64.
89. J. M. Elwood, A. J. Swerdlow and B. Cox, 'Trends in Incidence and Mortality from Cutaneous Melanoma in England and Wales' in 'The Ozone Layer and Health', *Transactions of the Menzies Foundation*, vol. 15, 1989, p. 134.
90. R. R. Bresler, 'Cutaneous Burns Due to Fluorescent Light', *Journal of the American Medical Association*, vol. 140, no. 17, 27 August 1949, pp. 1334–6.
91. V. Beral, S. Evans, H. Shaw and G. Milton, 'Malignant Melanoma and Exposure to Fluorescent Lighting at Work', *The Lancet*, 7 August 1982, pp. 290–3. See also: A. R. Kennedy, M. A. Ritter and J. B. Little, 'Fluorescent Light Induces Malignant Transformation in Mouse Embryo Cell *187* Cultures', *Science*, vol. 207, 1980, pp. 1209–11.
92. 'Ultraviolet Radiation', *IARC Monographs on the Evaluation of the Carcinogenic Risk of Chemicals to Humans*, vol. 40, WHO, 1986, p. 400.

93. M. F. Goldsmith, 'Medical News and Perspectives', *Journal of the American Medical Association*, vol. 257, no. 7, 20 February 1987, p. 893.
94. M. F. Goldsmith, 'Medical News and Perspectives', *Journal of the American Medical Association*, vol. 258, no. 10, 11 September 1987, p. 37.
95. F. Hollwich, op. cit., p 93. See also: D. Downing, op. cit., p. 107.
96. A. Knutsson, T. Akerstedt, et al., 'Increased Risk of Ischaemic Heart Disease in Shift Workers', *The Lancet*, 12 July 1986, pp. 89–92.
97. P. C. Hughes, op. cit., p. 292.
98. J. N. Ott, *Light, Radiation and You*, Devin-Adair Co., Old Greenwich, Connecticut, 1982, pp. 100–4.
99. J. A. Johnson, 'The Etiology of Hyperactivity', *Exceptional Children*, vol. 47, no. 5, February 1981, p. 352. See also: J. N. Ott, op. cit., pp. 130–3.
100. B. R. East, 'Mean annual hours of sunshine and incidence of dental caries', *American Journal of Public Health*, vol. 29, 1939, pp. 777. See also: E. C. McBeath and T. F. Zuker, 'The role of vitamin D in the control of Dental Caries in children', *Journal of Nutrition*, vol. 15, 1938, p. 547.
101. J. Liberman, *Light, Medicine of the Future*, Bear and Company, Santa Fe, New Mexico, 1991, pp. 58–9.
102. P. Park, 'Say cheese for a sunny smile with fewer fillings', *New Scientist*, 6 April 1991, p. 13.
103. P. C. Hughes, op. cit., pp. 292–3.
104. J. Liberman, op. cit., pp. 47–8.
105. A. G. Schauss, 'The physiological effect of colour on the suppression of human aggression: research on Baker-Miller Pink', *International Journal of Biosocial Research*, vol. 72, 1085, pp. 55–64. See also: D. Downing, op. cit., pp. 142–3; J. N. Ott, op. cit., pp. 154–5.
106. G. Legwold, 'Colour-boosted energy: How lights affect muscle action', *American Health*, May, 1988.
107. N. Henbest, 'Save our Skies', *New Scientist*, 11 February 1989, p. 44.
108. M. Luckiesh, op. cit., p. 16.
109. F. Daniels, 'Physiological and Pathological Extracutaneous Effects of Light on Man and Mammals, Not Mediated by Pineal or Other Neuroendocrine Mechanisms' in *Sunlight and Man*, op. cit., p. 251.
110. Ibid., p. 251–2.
111. Ibid., p. 253.
112. D. Downing, op. cit., pp. 117–8. See also: F. Daniels, op. cit., p. 254.
113. D. Downing, op. cit., p. 4.
114. B. E. Miller and A. W. Norman, 'Vitamin D' in L. J. Machlin (ed.), *Handbook of Vitamins*, Marcel Dekker Inc., New York, 1984, pp. 48–50.
115. Ibid., p. 80.
116. R. J. Wurtman, 1975, op. cit., p. 73. See also: F. Daniels, op. cit., pp. 248–50.
117. R. J. Wurtman, 1975, op. cit., pp. 74–5.
118. F. Daniels, op. cit., p. 254.
119. J. N. Ott, op. cit., pp. 94–116.
120. Ibid. pp. 98–9.
121. Ibid.pp. 100–103.
122. R. Pool, 'Illuminating Jet lag', *Science*, vol. 244, 1989, pp. 1256–7. See also: J. Miller, 'Bright lights can reset the human clock', *New Scientist*, 15 July 1989, p. 15.

123. R. J. Reiter, 1994, op. cit.
124. Ibid.
125. F. C. Carland, C. F. Garland, E. D. Gorham and J. F. Young, 'Geographic variation in breast cancer mortality in the United States: A hypothesis Involving Exposure to Solar Radiation', *Preventive Medicine*, vol. 19, 1990, pp. 614–22.
126. H. D. Foster, *Health, Disease and the Environment*, Belhaven Press, London, 1992.
127. F. Hollwich, op. cit., p. 1.
128. Ibid., p. 2.
129. F. Hollwich, op. cit., p. 10. See also: W. C. Quevedo, et al., 'Light and Skin Colour' in *Sunlight and Man*, op. cit., pp. 182–4.
130. F. Hollwich, op. cit. See also: R. J. Wurtman, 1974, op. cit., pp. 241–6, and R. J. Wurtman, op. cit., 1975.
131. J. F. Ashton and R. S. Laura, 'Sunglasses and Melanoma, is there a connection?', *Nautral Health*, December 1995/January 1996, pp. 10–11.
132. R. L. Edelson and J. M. Fink, 'The Immunologic Function of Skin', *Scientific American*, June 1985, p. 39.
133. F. Daniels, op. cit., p. 255. See also: D. Downing, op. cit., pp. 92–3.
134. F. Daniels and B. E. Johnson, op. cit., p. 127.
135. M. A. Pathak, T. B. Fitzpatrick, 'The Role of Natural Photoprotective Agents in Human Skin', in *Sunlight and Man*, op. cit., pp. 727, 736.
136. M. A. Pathak, 'Activation of the Melanocyte System by Ultraviolet Radiation and Cell Transformation', *Annals of the New York Academy of Sciences*, vol. 435, 1985, p. 32.
137. A. C. Giese, op. cit., p. 38.
138. R. J. Wurtman, 1974, op. cit., p. 235.
139. D. Downing, op. cit., pp. 80–1.
140. R. J. Wurtman, 1975, op. cit., p. 72.
141. R. J. Wurtman, 1975, op. cit., p. 74.
142. D. Downing, op. cit., p. 51. See also: P. Park, 'Say cheese for a sunny smile with fewer fillings', *New Scientist*, 6 April 1991, p. 13.
143. D. Downing, op. cit., p. 13. See also: R. Altschul and I. H. Herman, 'Ultraviolet Irradiation and Cholesterol metabolism', *Circulation*, vol. 8, 1953, p. 438.
144. W. N. Tapp and B. H. Natelson, 'Life Extension in Heart Disease: An Animal Model', *The Lancet*, February 1986, pp. 238–40.
145. R. Scragg, R. Jackson, I. M. Holdaway, et al., 'Myocardial infarction is inversely associated with plasma 25–hydroxyvitamin D3 levels: A community based study', *International Journal of Epidemiology*, vol. 19, no. 3, 1990, pp. 559–63.
146. F. Hollwich, op. cit., pp. 60–3
147. D. Downing, op. cit., p. 106
148. B. E. Miller and A. W. Norman, op. cit., pp. 88.150.
149. R. J. Wurtman and J. J. Wurtman, op. cit., p. 54.
150. C. A. Czeisler, R. E. Kronauer, J. S. Allan, et al., 'Bright Light Induction of Strong (Type O) Resetting of the Human Circadian Pacemaker', *Science*, vol. 244, 1989, pp. 1328–32.
151. R. Pool, op. cit. See also: J. Miller, op. cit.

152. G. C. Brainard, et al., 'Effect of Light Wavelength on the Suppression of Nocturnal Plasma Melatonin in Normal volunteers', see no. 70, *Annals of the New York Academy of Sciences*, vol. 435, 1985, pp. 376–8.
153. F. Hollwich, op. cit., p. vii.
154. W. E. Stumpf, op. cit., p. 247.
155. A. C. Giese, op. cit., p. 54.

CHAPTER 16

1. S. McDonagh, *To Care For the Earth*, Geoffrey Chapman, London, 1989, pp. 149–51.
2. G. Atherley and W. Noble, 'Occupational Deafness: The Continuing Challenge of Early German and Scottish Research', *American Journal of Industrial Medicine*, vol. 8, 1985, pp. 101–17.
3. S. Cohen, D. S. Krantz, et al., 'Cardiovascular and Behavioural Effects of Community Noise', *American Scientist*, vol. 69, Sept–Oct 1981, p. 528.
4. A. Hede, D. Meagher and D. Watkins, 'National noise survey—1986', *Acoustics Australia*, vol. 15, no. 2, 1986, pp. 39–42.
5. A. Brown, 'Exposure of the Australlian Population to Road Traffic Noise', *Report prepared for ANZEC*, Australian Government Printing Service, Canberra, 1993.
6. Environment Protection Agency NSW, *NSW State of the Environment*, EPA, Sydney, 1993.
7. M. Bond, 'Plagued by noise', *New Scientist*, 16 November 1996, pp. 14–15.
8. L. Kavaler, *Noise: the New Menace*, The John Day Co., New York, p. 37.
9. T. Emmet, 'Putting Strategies Into Effect', *Australian Safety News*, July 1990, p 41.
10. '10 000 Comp Cases for Deafness Every Year', *Australian Safety News*, October 1989, p, 56. See Also: Australian Hearing Services, 'Noises: It's murder on your ears', *The Sun Herald* (Sydney), 7 April 1996, p. 35.
11. L. Ponte, 'How Noise Can Harm You', *Reader's Digest*, August 1989.
12. 'Personal Stereos May Damage Your Job Prospects', *New Scientist*, 13 January 1990, p. 5.
13. T. Patel, 'Falling on deaf ears', *New Scientist*, 29 June 1996, pp. 12–13.
14. T. Patel, 'France takes steps to turn down the volume', *New Scientist*, 23 March, 1996, p. 5.
15. T. Patel, 'Live rock is hardest on the ears', *New Scientist*, 27 January 1996, p. 5.
16. 'Personal Stereos May Damage Your Job Prospects', op. cit.
17. L. Kavaler, op. cit., p. 49.
18. B. L. Plakke, 'Noise Levels of Electronic Arcade Games: A Potential Hearing Hazard to Children', *Ear Hear*, vol. 4, no. 4, July–Aug. 1983, pp. 202–3.
19. R. E. Meyer, T. E. Aldrich and C. E. Easterly, 'Effects of Noise and Electromagnetic Fields on Reproductive Outcomes', *Environmental Health Perspectives*, vol. 81, 1989, p. 193.
20. G. Bugliarello, A. Alexandre, et al., *The Impact of Noise Pollution*, Pergamon Press, New York, 1976, pp. 67–8. See also: L. Kavaler, op. cit., p. 57.
21. Ibid.
22. L. Kavaler, op. cit., p. 57.
23. S. Cohen, et al., op. cit., p. 529.
24. M. Bond, op. cit., p. 15. See also: ibid., p. 530.
25. L. Kavaler, op. cit., p. 57.

26. S. Cohen, et al., op. cit., p. 529.
27. Ibid.
28. Ibid., p. 531.
29. M. Bond, op. cit., p. 14.
30. V. Regecova and E. Kellerova, 'Effects of urban noise pollution on blood pressure and heart rate in pre-school children', *Journal of Hypertension*, vol. 13, no. 4, 1995, pp. 405–12.
31. S. Cohen, et al., op cit., p. 532.
32. Ibid., p. 532.
33. Ibid.
34. M. Bond, op. cit., p 14.
35. Ibid.
36. R. E. Meyer, et al., op. cit., p. 194.
37. Ibid.
38. Ibid., p. 195.
39. Ibid.
40. Ibid.
41. M. Bond, op. cit., p 15.
42. R. E. Meyer, et al., op. cit., p. 194.
43. M. Bond, op. cit., p 15.
44. E. F. Brown and W. R. Hendee, 'Adolescents and Their Music', *Journal of the American Medical Association*, vol. 262, no. 12, 1989, p. 1659.
45. E. F. Brown, et al., op. cit., p. 1661.
46. Ibid.
47. D. L. Peterson and K. S. Post, 'Influence of Rock Videos on Attitudes of Violence Against Women', *Psychological Reports*, vol. 64, 1989, pp. 319–22.
48. K. Allen and J. Blascovich, 'Effects of music on cardiovascular reactivity among surgeons', *Journal of the American Medical Association*, vol. 272, 1994, pp. 882–4.
49. Ibid.
50. A. Bernatzky, 'Ecological Principles in Town Planning: The Impact of Vegetation on the Quality of Life in the City', in R. Krieps (ed.), *Environment and Health: A Holistic Approach*, Avebury, Aldershot, England, 1989, p. 139.
51. J. Geary, 'Saving the sounds of silence', *New Scientist*, 13 April 1996, p. 45.
52. M. Bond, op. cit., p 15.

CHAPTER 17

1. F. Pearce, 'Big freeze digs a deeper hole in ozone layer', *New Scientist*, 13 November 1996, p. 7.
2. P. Goodrich, 'Darwinian Fever', *New Scientist*, 13 November 1993, p. 53.
3. L. J. Henderson, *The Fitness of the Environment*, Peter Smith, Gloucester, Mass., 1970. (Original edition published 1913.)
4. L. J. Henderson, *The Order of Nature*, Harvard University Press, Cambridge, Mass., 1917.
5. Ibid., p. 185.
6. Ibid., p. 192.
7. J. D. Barrow and F. J. Tipler, *The Anthropic Cosmological Principle*, Clarendon Press, Oxford, 1986, pp. 250–5.
8. B. J. Carr and M. J. Rees, 'The anthropic principle and the structure of the physical world', *Nature*, vol. 278, 1979, p. 612.

9. J. D. Barrow and J. Silk, *The Left Hand of Creation*, Heinemann, London, 1983, p. 227.
10. R. Hannah, 'Chaos in Australia', *Graduate Review*, University of New South Wales, Sydney, Autumn 1990, p. 23.
11. J. Gleik, *Chaos*, Cardinal, London, 1988, p. 316–17.
12. W. T. Keeton, J. L. Gould and C. G. Gould, *Biological Science*, Fourth Edition, W. W. Norton and Company, New York, 1986, p. 940.
13. R. Moy, 'The chaotic rhythms of life', *New Scientist*, 18 November 1989, pp. 21–5.
14. I. Stewart, 'Portraits of choas', *New Scientist*, 4 November 1989, pp. 22–7.
15. P. Davies, *The Cosmic Blueprint*, Unwin Paperbacks, London, 1989, p. 53.
16. J. P. Crutchfield, J. D. Farmer, N. H. Pachard and R. S. Shaw, 'Chaos', *Scientific American*, December 1989, pp. 38–49.
17. P. Poole, 'Microbes on the move', *New Scientist*, 3 March 1990. pp. 22–4.
18. J. Polkinghorne, *Science and Creation*, Shambhala, Boston, 1989, p. 45.
19. O. Sattaur, 'Termites change the face of Africa', *New Scientist*, 29 September 1990, p. 9.
20. S. Hughes, 'Antelope activate the acacia's alarm system', *New Scientist*, 29 September 1990, p. 9.
21. P. Davies, op. cit., pp. 6–7.
22. J. Monod, *Chance and Necessity.*, Alfred A. Knopf, New York, 1971, p. 9.
23. Ibid., pp. 21–22.
24. A. Smith, *Sugar Gliders, Wattles and Rural Eucalypt Dieback*, University of New England, Armidale, NSW, 1992.
25. E. L. Wynder, S. D. Stellman andE. A. Zang, 'High fibre intake. Indicator of a healthy lifestyle', *Journal of the American Medical Associatoin*, vol. 275, 1996, pp. 486–7. See also: C. L. Williams, M. Bollella and E. L. Wynder, 'A new recommendation for dietary fibre in childhood', *Pediatrics*, vol. 96, 1995, pp. 985–8.
26. G. Vines, 'Secret power of pets', *New Scientist*, 9 October 1993, pp. 30–34.
27. C. Birch, *On Purpose*, New South Wales University Press, Kensington, NSW, 1990, pp. 128–33, 174.
28. Ibid., p. 144.
29. N. Herbert, *Quantum Reality*, Rider and Co., London, 1985, p. 250.

GLOSSARY and ACRONYMS

AC	alternating current; an electric current, the flow of which alternates in direction
AM	amplitude modulation, as in radio waves
analog	refers to radio communication transmissions where the signals are not digitised
bioavailability	the extent to which a substance is available for use in a living system
biodiversity	biological diversity in an environment as indicated by number of different species of plants and animals
CFC	chlorofluorocarbon; a compound made up of chlorine, fluorine and carbon, and sometimes hydrogen
chronobiology	the study of biological rhythms
circadian rhythms	biological activity or function occurring in cycles
CRT	Cathode ray tube, as used, for example, in TVs
CSIRO	Commonwealth Scientific and Industrial Research Organisation
DC	direct current; an electric current that flows in one direction only and does not have any appreciable pulsations in its magnitude
decibel (dB)	a unit of measure for sound intensity
dielectric constant	a measure of a substance's ability to absorb electrical energy
digital	refers to radio communication signals transmitted in binary code
ELF	extremely low frequency
EMF	electromagnetic field
fly ash	fine solid particles of ash and soot emitted from smoke stacks when coal and other fossil fuels are burned
FM	frequency modulation as in radio waves
free radical	a chemically reactive group of atoms
gamma ray	form of ionising radiation
ganglia	masses of nerve tissue

gauss (G)	a unit of measure for magnetic field strength
gray (Gy)	a unit of measure of ionising radiation absorbed
hertz (Hz)	a unit of frequency; one cycle per second
infrared radiation	heat radiation
IPCC	Intergovernmental Panel of Climate Change
IRPA	International Radiation Protection Association
kelvin (K)	unit of temperature measurement beginning at absolute zero
kilovolt (kV)	1000 volts
kV/m	kilovolts per metre; a unit of measure for electric field strength
kilowatt-hour (KW-hour)	a unit of energy equal to the work performed in one hour by a power of 1000 watts
lux (lx)	a unit of illumination
magnetotelluric field	a weak fluctuating magnetic field produced near the surface of the Earth by lightening, etc.
melanoma	a pigmented malignant tumour of the skin
meta-analysis	a study incorporating the results of several similar studies
microwave	electromagnetic waves with wavelengths less than 1 metre
milligauss (mG)	1000th of a gauss
millisievert	a unit of measurement of ionising radiation absorbed
mV/v	millivolts per metre
nanometre	1 billionth of a metre
NHMRC	National Health and Medical Research Council (Australia)
OECD	Organisation for Economic Co-operation and Development
oncostatic	stopping or holding the growth of tumours
permaculture	the practice of co-operation with nature involving the design of environments which have the diversity, stability and resilience of natural ecosystems
picogauss (pG)	1 millionth of a millionth of a gauss (i.e. 10^{-12} gauss); also 1 billionth of a mG

phon	a unit of objective loudness of sound
pineal gland	a small gland behind the third ventricle of the brain
pulsed electric field	an electric field of pulsing intensity
PVC polyvinylchloride	a plastic containing chlorine
rad	a unit of absorbed dose of ionised radiation equal to 1 hundredth of a gray
radio wave	an electromagnetic wave of radio frequency
radiounclide	a radioactive element
RF	radio frequency
sonic	relating to sound or sound waves
static electric field	a non-fluctuating electric field
tera-bequerel (Tbq)	a unit of radioactivity
tesla	a unit of magnetic field strength
UHF	ultra high frequency
ultrasonic	mechanical vibrations or sound waves with frequencies higher than those that can be heard by ear
USFDA	United States Food and Drug Administration
uT	microtesla; 1 millionth of a tesla; a unit of magnetic field strength
UV-A	ultraviolet A; a longer wavelength form of ultra violet radiation
UV-B	ultraviolet B; a shorter wavelength form of ultraviolet radiation
V/m	volts per metre; a unit of electric field strength
VDU	visual display unit
VHF	very high frequency
voltage	the value of an electric force, expressed in volts
watt	a unit of electric power
x-radiation	radiation composed of x-rays
x-rays	a form of highly penetrating ionising radiation

FURTHER READING

J.F. Ashton and R.S. Laura,*The Life Enhancement Handbook,* Simon and Schuster, Sydney, 1997.

R. Augros and G. Stanciu,*The New Biology: Discovering the Wisdom in Nature,* Shambhala, Boston, 1988.

D. Barlett and J. Steele, *America: What Went Wrong?,* Andrews and McNeel, Kansas City, Mo, 1992.

J. Beniger, *The Control Revolution,* Harvard University Press, Cambridge, 1986.

C. Birch, *On Purpose,* UNSW Press, Sydney, 1990.

C.L. Brod, *Techno Stress: The Human Cost of the Computer Revolution,* Addison-Wesley, Reading, MA, 1984.

L.R. Brown, *State of the World 1997,* Earthscan Publications, London, 1997.

R. Callahan,*Education and the Cult of Efficiency,* University of Chicago Press, Chicago, 1964.

J. Corn (ed.), *Imagining Tomorrow,* Harvard University Press, Cambridge, 1986.

G. Cross,*Time and Money: The Making of Consumer Culture,* Routledge, New York, 1993.

J. Ellul,*The Technological Society,* Random House, New York, 1984.

G. Fjermedal,*The Tomorrow Makers,* Macmillan, New York, 1986.

H.D. Foster, *Health, Disease and the Environment,* Belhaven Press, London, 1992.

L. Garrett, *The Coming Plague: Newly Emerging Diseases in a World Out of Balance,* Penguin Books, New York, 1994.

P. Hawken, *The Ecology of Commerce,* Phoenix, London, 1990.

R.S. Laura and J. Ashton, *101 Vital Tips for a Healthy Lifestyle,* Harper Collins, Sydney, 1993.

R.S. Laura and S. Heaney, *Philosophical Foundations of Health Education,* Routledge, New York, 1992.

J.L. Liberman, *Light Medicine of the Future,* Bear and Company, Santa Fe, NM, 1991.

J. Lovelock, *The Ages of Gaia,* Oxford University Press, 1989.

B. Mollison, *Permaculture: A Designer's Manual,* Tagari Publications, Tyalgum, NSW, 1992.

National Research Council, *Alternative Agriculture,* National Academy Press, Washington, DC, 1989.

S. Roach, *Making Technology Work,* Morgan Stanley, New York, 1993.

J. Rifkin, *Declaration of a Heretic,* Routledge and Kegan Paul, Boston, 1985.

C. Sagan and R. Turco, *A Path Where No Man Thought,* Random Century, New York, 1991.

R. Sheldrake, *The Rebirth of Nature: The Greening of Science and God,* Random/Century, London, 1990.

G. Simons, *Robots: The Quest for Living Machines,* Basil Blackwell, New York, 1992.

State of the Environment Advisory Council, *Australia State of the Environment 1996,* CSIRO, Collingwood, Vic., 1996.

D. Suzuki, *Inventing the Future,* Allen and Unwin, Sydney, 1990.

P. Taylor, *Respect for Nature,* Princeton University Press, Princeton, NJ, 1986.

E.O. Wilson (ed.), *Biodiversity,* National Academy Press, Washington, DC, 1988.

J. Young, *Sustaining the Earth: The Past, Present and Future of the Green Revolution,* UNSW Press, Sydney, 1991.

INDEX